STUDENT SOLUTIONS MANUAL

to accompany

CALCULUS

SINGLE VARIABLE   **FOURTH EDITION**

Deborah Hughes-Hallett
University of Arizona
Andrew M. Gleason
Harvard University
William G. McCallum
University of Arizona

et al.

Prepared by:

Rick Cangelosi
Scott Clark
Cristi Guevara
Elliot Marks
Igor Padure
Seung-Hye Song
Aaron Wootton

WILEY

John Wiley & Sons, Inc.

©Pete Turner/The Image Bank/Getty Images

To order books or for customer service call 1-800-CALL-WILEY (225-5945).

ISBN 0-471-65997-5

Printed in the United States of America

10 9 8 7 6 5 4 3 2 1

Printed and bound by Courier Kendallville, Inc.

CONTENTS

CHAPTER ONE

Solutions for Section 1.1

Exercises

1. $f(35)$ means the value of P corresponding to $t = 35$. Since t represents the number of years since 1950, we see that $f(35)$ means the population of the city in 1985. So, in 1985, the city's population was 12 million.

5. The slope is $(3 - 2)/(2 - 0) = 1/2$. So the equation of the line is $y = (1/2)x + 2$.

9. Rewriting the equation of the line as

$$y = \frac{12}{6}x - \frac{4}{6}$$
$$y = 2x - \frac{2}{3},$$

we see that the line has slope 2 and vertical intercept $-2/3$.

13. The line $y + 4x = 7$ has slope -4. Therefore the parallel line has slope -4 and equation $y - 5 = -4(x - 1)$ or $y = -4x + 9$. The perpendicular line has slope $\frac{-1}{(-4)} = \frac{1}{4}$ and equation $y - 5 = \frac{1}{4}(x - 1)$ or $y = 0.25x + 4.75$.

17. Since x goes from 1 to 5 and y goes from 1 to 6, the domain is $1 \le x \le 5$ and the range is $1 \le y \le 6$.

21. Factoring gives
$$g(x) = \frac{(2 - x)(2 + x)}{x(x + 1)}.$$

The values of x which make $g(x)$ undefined are $x = 0$ and $x = -1$, when the denominator is 0. So the domain is all $x \ne 0, -1$. Solving $g(x) = 0$ means one of the numerator's factors is 0, so $x = \pm 2$.

Problems

25. (a) The flat tire corresponds to a part of the graph where the velocity is zero. This could be (II) or (III). Since the velocity in (II) is higher for the later part of the interval, corresponding to speeding up, the answer is (II).
(b) This is (I), as the graph shows a positive velocity followed by a zero velocity.
(c) In (IV), the velocity is positive, zero (while the package is being dropped off) and then negative (the drive home).
(d) Graph (III) could represent a drive to the country, stopping to have lunch (velocity zero), and continuing on at a slower speed to look at the scenery. Other stories are possible.

29. (a) We find the slope m and intercept b in the linear equation $C = b + mw$. To find the slope m, we use

$$m = \frac{\Delta C}{\Delta w} = \frac{48 - 32}{180 - 100} = 0.2.$$

We substitute to find b:

$$C = b + mw$$
$$32 = b + (0.2)(100)$$
$$b = 12.$$

The linear formula is $C = 12 + 0.2w$.
(b) The slope is 0.2 dollars per kilogram. Each additional kilogram of waste costs 20 cents.
(c) The intercept is 12 dollars. The flat monthly fee to subscribe to the waste collection service is $12. This is the amount charged even if there is no waste.

33. We are looking for a linear function $y = f(x)$ that, given a time x in years, gives a value y in dollars for the value of the refrigerator. We know that when $x = 0$, that is, when the refrigerator is new, $y = 950$, and when $x = 7$, the refrigerator is worthless, so $y = 0$. Thus $(0, 950)$ and $(7, 0)$ are on the line that we are looking for. The slope is then given by

$$m = \frac{950}{-7}$$

It is negative, indicating that the value decreases as time passes. Having found the slope, we can take the point $(7, 0)$ and use the point-slope formula:

$$y - y_1 = m(x - x_1).$$

So,

$$y - 0 = -\frac{950}{7}(x - 7)$$
$$y = -\frac{950}{7}x + 950.$$

37. Given $l - l_0 = al_0(t - t_0)$ with l_0, t_0 and a all constant,

(a) We have $l = al_0(t - t_0) + l_0 = al_0t - al_0t_0 + l_0$, which is a linear function of t with slope al_0 and y-intercept at $(0, -al_0t_0 + l_0)$.

(b) If $l_0 = 100$, $t_0 = 60°$F and $a = 10^{-5}$, then

$$l = 10^{-5}(100)t - 10^{-5}(100)(60) + 100 = 10^{-3}t + 99.94$$
$$= 0.001t + 99.94$$

(c) If the slope is positive, (as in (b)), then as the temperature rises, the length of the metal increases: it expands. If the slope were negative, then the metal would contract as the temperature rises.

Solutions for Section 1.2

Exercises

1. Initial quantity $= 5$; growth rate $= 0.07 = 7\%$.

5. (a) The function is linear with initial population of 1000 and slope of 50, so $P = 1000 + 50t$.
(b) This function is exponential with initial population of 1000 and growth rate of 5%, so $P = 1000(1.05)^t$.

9. This graph is neither concave up or down.

Problems

13. (a) We have

$$\text{Reduced size } = (0.80) \cdot \text{Original size}$$

or

$$\text{Original size} = \frac{1}{(0.80)} \text{ Reduced size } = (1.25) \text{ Reduced size,}$$

so the copy must be enlarged by a factor of 1.25, which means it is enlarged to 125% of the reduced size.

(b) If a page is copied n times, then

$$\text{New size } = (0.80)^n \cdot \text{Original.}$$

We want to solve for n so that

$$(0.80)^n = 0.15.$$

By trial and error, we find $(0.80)^8 = 0.168$ and $(0.80)^9 = 0.134$. So the page needs to be copied 9 times.

17. We look for an equation of the form $y = y_0 a^x$ since the graph looks exponential. The points $(0, 3)$ and $(2, 12)$ are on the graph, so

$$3 = y_0 a^0 = y_0$$

and

$$12 = y_0 \cdot a^2 = 3 \cdot a^2, \quad \text{giving} \quad a = \pm 2.$$

Since $a > 0$, our equation is $y = 3(2^x)$.

21. (a) Let $Q = Q_0 a^t$. Then $Q_0 a^5 = 75.94$ and $Q_0 a^7 = 170.86$. So

$$\frac{Q_0 a^7}{Q_0 a^5} = \frac{170.86}{75.94} = 2.25 = a^2.$$

So $a = 1.5$.

(b) Since $a = 1.5$, the growth rate is $r = 0.5 = 50\%$.

25. $P = P_0 (e^{0.2})^t = P_0 (1.2214)^t$. Exponential growth because $0.2 > 0$ or $1.2214 > 1$.

29. (a) We compound the daily inflation rate 30 times to get the desired monthly rate r:

$$\left(1 + \frac{r}{100}\right)^1 = \left(1 + \frac{1.3}{100}\right)^{30}.$$

Solving for r, we get $r = 47.3$, so the inflation rate for June was 47.3%.

(b) We compound the daily inflation rate 366 times to get a yearly rate R for 1988, a leap year:

$$\left(1 + \frac{R}{100}\right)^1 = \left(1 + \frac{1.3}{100}\right)^{366}.$$

Solving for R, we get $R = 111.994$, so the yearly rate was 11,199.4% during 1988. We could have obtained approximately the same result by compounding the monthly rate 12 times. Computing the annual rate from the monthly gives a lower result, because 12 months of 30 days each is only 360 days.

33. Because the population is growing exponentially, the time it takes to double is the same, regardless of the population levels we are considering. For example, the population is 20,000 at time 3.7, and 40,000 at time 6.0. This represents a doubling of the population in a span of $6.0 - 3.7 = 2.3$ years.

How long does it take the population to double a second time, from 40,000 to 80,000? Looking at the graph once again, we see that the population reaches 80,000 at time $t = 8.3$. This second doubling has taken $8.3 - 6.0 = 2.3$ years, the same amount of time as the first doubling.

Further comparison of any two populations on this graph that differ by a factor of two will show that the time that separates them is 2.3 years. Similarly, during any 2.3 year period, the population will double. Thus, the doubling time is 2.3 years.

Suppose $P = P_0 a^t$ doubles from time t to time $t + d$. We now have $P_0 a^{t+d} = 2 P_0 a^t$, so $P_0 a^t a^d = 2 P_0 a^t$. Thus, canceling P_0 and a^t, d must be the number such that $a^d = 2$, no matter what t is.

37. We see that $\frac{1.09}{1.06} \approx 1.03$, and therefore $h(s) = c(1.03)^s$; c must be 1. Similarly $\frac{2.42}{2.20} = 1.1$, and so $f(s) = a(1.1)^s$; $a = 2$. Lastly, $\frac{3.65}{3.47} \approx 1.05$, so $g(s) = b(1.05)^s$; $b \approx 3$.

Solutions for Section 1.3

Exercises

1. (a) $f(g(1)) = f(1 + 1) = f(2) = 2^2 = 4$
(b) $g(f(1)) = g(1^2) = g(1) = 1 + 1 = 2$
(c) $f(g(x)) = f(x + 1) = (x + 1)^2$
(d) $g(f(x)) = g(x^2) = x^2 + 1$
(e) $f(t)g(t) = t^2(t + 1)$

5. This graph is the graph of $m(t)$ shifted upward by two units. See Figure 1.1.

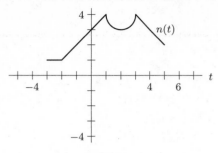

Figure 1.1

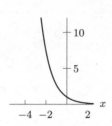

Figure 1.2

9. For $f(-x)$, the graph is reflected in the y-axis. See Figure 1.2.

13. For $f(5x)$, the values increase 5 times as fast as we move right. See Figure 1.3.

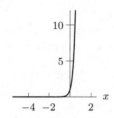

Figure 1.3

17. $m(z+1) - m(z) = (z+1)^2 - z^2 = 2z + 1$.

21. (a) $f(25)$ is q corresponding to $p = 25$, or, in other words, the number of items sold when the price is 25.
(b) $f^{-1}(30)$ is p corresponding to $q = 30$, or the price at which 30 units will be sold.

25. The function is not invertible since there are many horizontal lines which hit the function twice.

29. This looks like a shift of the graph $y = x^3$. The graph is shifted to the right 2 units and down 1 unit, so a possible formula is $y = (x-2)^3 - 1$.

33. Since

$$f(-x) = (-x)^3 + 1 = -x^3 + 1,$$

we see $f(-x) \neq f(x)$ and $f(-x) \neq -f(x)$, so f is neither even nor odd

37. Since

$$f(-x) = e^{-x} + x,$$

we see $f(-x) \neq f(x)$ and $f(-x) \neq -f(x)$, so f is neither even nor odd

Problems

41. Invertible. Since at $4°$C, the mass of 1 liter of water is 1 kilogram, the mass of x liters is x kilograms. So $f(x) = x$ and therefore, $f^{-1}(x) = x$.

45. $f(f(1)) \approx f(-0.4) \approx -0.9$.

49. $f(x) = x^3, \quad g(x) = x + 1$.

53. (a) For each 2.2 pounds of weight the object has, it has 1 kilogram of mass, so the conversion formula is

$$k = f(p) = \frac{1}{2.2}p.$$

(b) The inverse function is

$$p = 2.2k,$$

and it gives the weight of an object in pounds as a function of its mass in kilograms.

Solutions for Section 1.4

Exercises

1. Using the identity $e^{\ln x} = x$, we have $e^{\ln(1/2)} = \frac{1}{2}$.

5. Using the rules for ln, we have

$$\ln\left(\frac{1}{e}\right) + \ln AB = \ln 1 - \ln e + \ln A + \ln B$$
$$= 0 - 1 + \ln A + \ln B$$
$$= -1 + \ln A + \ln B.$$

9. Isolating the exponential term

$$20 = 50(1.04)^x$$
$$\frac{20}{50} = (1.04)^x.$$

Taking logs of both sides

$$\log \frac{2}{5} = \log(1.04)^x$$
$$\log \frac{2}{5} = x \log(1.04)$$
$$x = \frac{\log(2/5)}{\log(1.04)} = -23.4.$$

13.
$$\ln(2e^{3x}) = \ln(4e^{5x})$$
$$\ln 2 + \ln(e^{3x}) = \ln 4 + \ln(e^{5x})$$
$$0.693 + 3x = 1.386 + 5x$$
$$x = -0.347$$

17. Using the rules for ln, we get

$$\ln 9^x = \ln 2e^{x^2}$$
$$x \ln 9 = \ln 2 + x^2$$
$$x^2 - x \ln 9 + \ln 2 = 0.$$

We can use the quadratic formula to get $x = \dfrac{\ln 9 \pm \sqrt{(\ln 9)^2 - 4 \ln 2}}{2}$, so $x \approx 0.382$, $x \approx 1.815$.

21. Collecting similar terms yields

$$\left(\frac{a}{b}\right)^t = \frac{Q_0}{P_0}.$$

Hence

$$t = \frac{\log\left(\frac{Q_0}{P_0}\right)}{\log\left(\frac{a}{b}\right)}.$$

25. We want $1.7^t = e^{kt}$ so $1.7 = e^k$ and $k = \ln 1.7 = 0.5306$. Thus $P = 10e^{0.5306t}$.

29. Since f is increasing, f has an inverse. To find the inverse of $f(t) = 50e^{0.1t}$, we replace t with $f^{-1}(t)$, and, since $f(f^{-1}(t)) = t$, we have

$$t = 50e^{0.1f^{-1}(t)}.$$

We then solve for $f^{-1}(t)$:

$$t = 50e^{0.1f^{-1}(t)}$$
$$\frac{t}{50} = e^{0.1f^{-1}(t)}$$
$$\ln\left(\frac{t}{50}\right) = 0.1f^{-1}(t)$$
$$f^{-1}(t) = \frac{1}{0.1} \ln\left(\frac{t}{50}\right) = 10 \ln\left(\frac{t}{50}\right).$$

Problems

33. Since the factor by which the prices have increased after time t is given by $(1.05)^t$, the time after which the prices have doubled solves

$$2 = (1.05)^t$$
$$\log 2 = \log(1.05^t) = t\log(1.05)$$
$$t = \frac{\log 2}{\log 1.05} \approx 14.21 \text{ years.}$$

37. Let B represent the sales (in millions of dollars) at Borders bookstores t years since 1991. Since $B = 78$ when $t = 0$ and we want the continuous growth rate, we write $B = 78e^{kt}$. We use the information from 1994, that $B = 412$ when $t = 3$, to find k:

$$412 = 78e^{k \cdot 3}$$
$$5.282 = e^{3k}$$
$$\ln(5.282) = 3k$$
$$k = 0.555.$$

We have $B = 78e^{0.555t}$, which represents a continuous growth rate of 55.5% per year.

41. (a) Since the initial amount of caffeine is 100 mg and the exponential decay rate is -0.17, we have $A = 100e^{-0.17t}$.
    (b) See Figure 1.4. We estimate the half-life by estimating t when the caffeine is reduced by half (so $A = 50$); this occurs at approximately $t = 4$ hours.

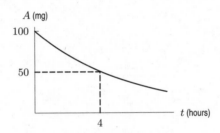

Figure 1.4

(c) We want to find the value of t when $A = 50$:

$$50 = 100e^{-0.17t}$$
$$0.5 = e^{-0.17t}$$
$$\ln 0.5 = -0.17t$$
$$t = 4.077.$$

The half-life of caffeine is about 4.077 hours. This agrees with what we saw in Figure 1.4.

45. Since the amount of strontium-90 remaining halves every 29 years, we can solve for the decay constant;

$$0.5P_0 = P_0 e^{-29k}$$
$$k = \frac{\ln(1/2)}{-29}.$$

Knowing this, we can look for the time t in which $P = 0.10P_0$, or

$$0.10P_0 = P_0 e^{\ln(0.5)t/29}$$
$$t = \frac{29\ln(0.10)}{\ln(0.5)} = 96.34 \text{ years.}$$

Solutions for Section 1.5

Exercises

1. See Figure 1.5.

$$\sin\left(\frac{3\pi}{2}\right) = -1 \quad \text{is negative.}$$

$$\cos\left(\frac{3\pi}{2}\right) = 0$$

$$\tan\left(\frac{3\pi}{2}\right) \quad \text{is undefined.}$$

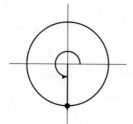

Figure 1.5

5. See Figure 1.6.

$$\sin\left(\frac{\pi}{6}\right) \quad \text{is positive.}$$

$$\cos\left(\frac{\pi}{6}\right) \quad \text{is positive.}$$

$$\tan\left(\frac{\pi}{6}\right) \quad \text{is positive.}$$

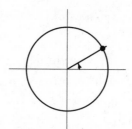

Figure 1.6

9. -1 radian $\cdot \dfrac{180°}{\pi \text{ radians}} = -\left(\dfrac{180°}{\pi}\right) \approx -60°$. See Figure 1.7.

$$\sin(-1) \quad \text{is negative}$$

$$\cos(-1) \quad \text{is positive}$$

$$\tan(-1) \quad \text{is negative.}$$

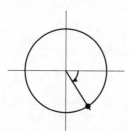

Figure 1.7

13. (a) We determine the amplitude of y by looking at the coefficient of the cosine term. Here, the coefficient is 1, so the amplitude of y is 1. Note that the constant term does not affect the amplitude.

(b) We know that the cosine function $\cos x$ repeats itself at $x = 2\pi$, so the function $\cos(3x)$ must repeat itself when $3x = 2\pi$, or at $x = 2\pi/3$. So the period of y is $2\pi/3$. Here as well the constant term has no effect.

(c) The graph of y is shown in the figure below.

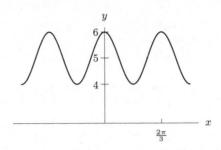

17. The period is $2\pi/\pi = 2$, since when t increases from 0 to 2, the value of πt increases from 0 to 2π. The amplitude is 0.1, since the function oscillates between 1.9 and 2.1.

21. This graph is an inverted sine curve with amplitude 4 and period π, so it is given by $f(x) = -4\sin(2x)$.

25. The graph is an inverted sine curve with amplitude 1 and period 2π, shifted up by 2, so it is given by $f(x) = 2 - \sin x$.

29. We first divide by 5 and then use inverse sine:

$$\frac{2}{5} = \sin(3x)$$
$$\sin^{-1}(2/5) = 3x$$
$$x = \frac{\sin^{-1}(2/5)}{3} \approx 0.1372.$$

There are infinitely many other possible solutions since the sine is periodic.

33. We first isolate $\sin(5x)$ and then use inverse sine:

$$8 = 4\sin(5x)$$
$$2 = \sin(5x).$$

But this equation has no solution since $-1 \le \sin(5x) \le 1$.

Problems

37. Suppose P is at the point $(3\pi/2, -1)$ and Q is at the point $(5\pi/2, 1)$. Then

$$\text{Slope} = \frac{1 - (-1)}{5\pi/2 - 3\pi/2} = \frac{2}{\pi}.$$

If P had been picked to the right of Q, the slope would have been $-2/\pi$.

41. The US voltage has a maximum value of 156 volts and has a period of $1/60$ of a second, so it executes 60 cycles a second.

The European voltage has a higher maximum of 339 volts, and a slightly longer period of $1/50$ seconds, so it oscillates at 50 cycles per second.

45. (a) Reading the graph of θ against t shows that $\theta \approx 5.2$ when $t = 1.5$. Since the coordinates of P are $x = 5\cos\theta$, $y = 5\sin\theta$, when $t = 1.5$ the coordinates are

$$(x, y) \approx (5\cos 5.2, 5\sin 5.2) = (2.3, -4.4).$$

(b) As t increases from 0 to 5, the angle θ increases from 0 to about 6.3 and then decreases to 0 again. Since $6.3 \approx 2\pi$, this means that P starts on the x-axis at the point $(5, 0)$, moves counterclockwise the whole way around the circle (at which time $\theta \approx 2\pi$), and then moves back clockwise to its starting point.

Solutions for Section 1.6

Exercises

1. Exponential growth dominates power growth as $x \to \infty$, so $10 \cdot 2^x$ is larger.

5. **(I)** **(a)** Minimum degree is 3 because graph turns around twice.
   (b) Leading coefficient is negative because $y \to -\infty$ as $x \to \infty$.
 (II) **(a)** Minimum degree is 4 because graph turns around three times.
   (b) Leading coefficient is positive because $y \to \infty$ as $x \to \infty$.
(III) **(a)** Minimum degree is 4 because graph turns around three times.
   (b) Leading coefficient is negative because $y \to -\infty$ as $x \to \infty$.
 (IV) **(a)** Minimum degree is 5 because graph turns around four times.
   (b) Leading coefficient is negative because $y \to -\infty$ as $x \to \infty$.
  (V) **(a)** Minimum degree is 5 because graph turns around four times.
   (b) Leading coefficient is positive because $y \to \infty$ as $x \to \infty$.

9. $f(x) = k(x+2)(x-1)(x-3)(x-5) = k(x^4 - 7x^3 + 5x^2 + 31x - 30)$, where $k > 0$. ($k \approx \frac{1}{15}$ if the horizontal and vertical scales are equal; otherwise one can't tell how large k is.)

Problems

13. **(a)** Because our cubic has a root at 2 and a double root at -2, it has the form

$$y = k(x+2)(x+2)(x-2).$$

Since $y = 4$ when $x = 0$,

$$4 = k(2)(2)(-2) = -8k,$$
$$k = -\frac{1}{2}.$$

Thus our equation is

$$y = -\frac{1}{2}(x+2)^2(x-2).$$

17. Let $D(v)$ be the stopping distance required by an Alpha Romeo as a function of its velocity. The assumption that stopping distance is proportional to the square of velocity is equivalent to the equation

$$D(v) = kv^2$$

where k is a constant of proportionality. To determine the value of k, we use the fact that $D(70) = 177$.

$$D(70) = k(70)^2 = 177.$$

Thus,

$$k = \frac{177}{70^2} \approx 0.0361.$$

It follows that

$$D(35) = \left(\frac{177}{70^2}\right)(35)^2 = \frac{177}{4} = 44.25 \text{ ft}$$

and

$$D(140) = \left(\frac{177}{70^2}\right)(140)^2 = 708 \text{ ft}.$$

Thus, at half the speed it requires one fourth the distance, whereas at twice the speed it requires four times the distance, as we would expect from the equation. (We could in fact have figured it out that way, without solving for k explicitly.)

21. The pomegranate is at ground level when $f(t) = -16t^2 + 64t = -16t(t-4) = 0$, so when $t = 0$ or $t = 4$. At time $t = 0$ it is thrown, so it must hit the ground at $t = 4$ seconds. The symmetry of its path with respect to time may convince you that it reaches its maximum height after 2 seconds. Alternatively, we can think of the graph of $f(t) = -16t^2 + 64t = -16(t-2)^2 + 64$, which is a downward parabola with vertex (i.e., highest point) at $(2, 64)$. The maximum height is $f(2) = 64$ feet.

25. The function is a cubic polynomial with positive leading coefficient. Since the figure given in the text shows that the function turns around once, we know that the function has the shape shown in Figure 1.8. The function is below the x-axis for $x = 5$ in the given graph, and we know that it goes to $+\infty$ as $x \to +\infty$ because the leading coefficient is positive. Therefore, there are exactly three zeros. Two zeros are shown, and occur at approximately $x = -1$ and $x = 3$. The third zero must be to the right of $x = 10$ and so occurs for some $x > 10$.

Figure 1.8

29. (a) III
   (b) IV
   (c) I
   (d) II

Solutions for Section 1.7

Exercises

1. Yes, because $2x + x^{2/3}$ is defined for all x.

5. Yes, because $2x - 5$ is positive for $3 \leq x \leq 4$.

9. No, because $e^x - 1 = 0$ at $x = 0$.

13. We have that $f(0) = -1 < 0$ and $f(1) = 1 - \cos 1 > 0$ and that f is continuous. Thus, by the Intermediate Value Theorem applied to $k = 0$, there is a number c in $[0, 1]$ such that $f(c) = k = 0$.

Problems

17. For any value of k, the function is continuous at every point except $x = 2$. We choose k to make the function continuous at $x = 2$.

   Since $3x^2$ takes the value $3(2^2) = 12$ at $x = 2$, we choose k so that kx goes through the point $(2, 12)$. Thus $k = 6$.

21.

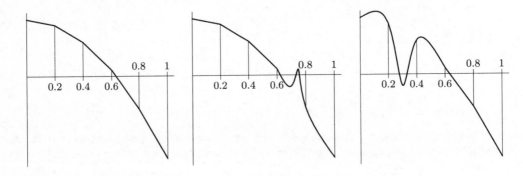

Solutions for Section 1.8

Exercises

1. (a) As x approaches -2 from either side, the values of $f(x)$ get closer and closer to 3, so the limit appears to be about 3.

   (b) As x approaches 0 from either side, the values of $f(x)$ get closer and closer to 7. (Recall that to find a limit, we are interested in what happens to the function near x but not at x.) The limit appears to be about 7.

   (c) As x approaches 2 from either side, the values of $f(x)$ get closer and closer to 3 on one side of $x = 2$ and get closer and closer to 2 on the other side of $x = 2$. Thus the limit does not exist.

   (d) As x approaches 4 from either side, the values of $f(x)$ get closer and closer to 8. (Again, recall that we don't care what happens right at $x = 4$.) The limit appears to be about 8.

5. For $-90° \leq \theta \leq 90°$, $0 \leq y \leq 0.02$, the graph of $y = \dfrac{\sin \theta}{\theta}$ is shown in Figure 1.9. Therefore, by tracing along the curve, we see that in degrees, $\lim\limits_{\theta \to 0} \dfrac{\sin \theta}{\theta} = 0.01745\ldots$.

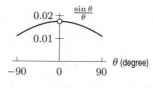

Figure 1.9

9. From Table 1.1, it appears the limit is 0. This is confirmed by Figure 1.10. An appropriate window is $-0.005 < x < 0.005$, $-0.01 < y < 0.01$.

Table 1.1

x	$f(x)$
0.1	0.1987
0.01	0.0200
0.001	0.0020
0.0001	0.0002

x	$f(x)$
-0.0001	-0.0002
-0.001	-0.0020
-0.01	-0.0200
-0.1	-0.1987

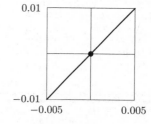

Figure 1.10

13. From Table 1.2, it appears the limit is 1. This is confirmed by Figure 1.11. An appropriate window is $-0.0198 < x < 0.0198$, $0.99 < y < 1.01$.

Table 1.2

x	$f(x)$
0.1	1.0517
0.01	1.0050
0.001	1.0005
0.0001	1.0001

x	$f(x)$
-0.0001	1.0000
-0.001	0.9995
-0.01	0.9950
-0.1	0.9516

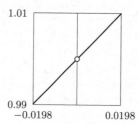

Figure 1.11

17. From Table 1.3, it appears the limit is 0. Figure 1.12 confirms this. An appropriate window is $1.55 < x < 1.59$, $-0.01 < y < 0.01$.

Table 1.3

x	$f(x)$
1.6708	−0.0500
1.5808	−0.0050
1.5718	−0.0005
1.5709	−0.0001
1.5707	0.0001
1.5698	0.0005
1.5608	0.0050
1.4708	0.0500

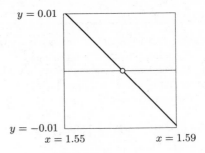

Figure 1.12

21. $f(x) = \begin{cases} x^2 - 2 & 0 < x < 3 \\ 2 & x = 3 \\ 2x + 1 & 3 < x \end{cases}$

Figure 1.13 confirms that $\lim_{x \to 3^-} f(x) = \lim_{x \to 3^-} (x^2 - 2) = 7$ and that $\lim_{x \to 3+} f(x) = \lim_{x \to 3+} (2x + 1) = 7$, so $\lim_{x \to 3} f(x) = 7$. Note, however, that $f(x)$ is not continuous at $x = 3$ since $f(3) = 2$.

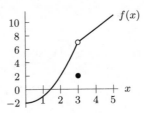

Figure 1.13

Problems

25. The only change is that, instead of considering all x near c, we only consider x near to and less than c. Thus the phrase "$|x - c| < \delta$" must be replaced by "$c - \delta < x < c$." Thus, we define

$$\lim_{x \to c^-} f(x) = L$$

to mean that for any $\epsilon > 0$ (as small as we want), there is a $\delta > 0$ (sufficiently small) such that if $c - \delta < x < c$, then $|f(x) - L| < \epsilon$.

29. The limit appears to be 1; a graph and table of values is shown below.

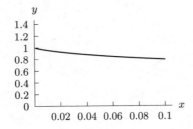

x	x^x
0.1	0.7943
0.01	0.9550
0.001	0.9931
0.0001	0.9990
0.00001	0.9999

33. Divide numerator and denominator by x^3, giving

$$f(x) = \frac{2x^3 - 16x^2}{4x^2 + 3x^3} = \frac{2 - 16/x}{4/x + 3},$$

so

$$\lim_{x \to \infty} f(x) = \lim_{x \to \infty} \frac{2 - 16/x}{4/x + 3} = \frac{\lim_{x \to \infty}(2 - 16/x)}{\lim_{x \to \infty}(4/x + 3)} = \frac{2}{3}.$$

37. Because the denominator equals 0 when $x = 4$, so must the numerator. This means $k^2 = 16$ and the choices for k are 4 or -4.

41. For the numerator, $\displaystyle\lim_{x \to -\infty}\left(e^{2x} - 5\right) = -5$. If $k > 0$, $\displaystyle\lim_{x \to -\infty}\left(e^{kx} + 3\right) = 3$, so the quotient has a limit of $-5/3$. If $k = 0$, $\displaystyle\lim_{x \to -\infty}\left(e^{kx} + 3\right) = 4$, so the quotient has limit of $-5/4$. If $k < 0$, the limit of the quotient is given by $\displaystyle\lim_{x \to -\infty}\left(e^{2x} - 5\right)/(e^{kx} + 3) = 0$.

45. By tracing on a calculator or solving equations, we find the following values of δ:
For $\epsilon = 0.2$, $\delta \le 0.1$.
For $\epsilon = 0.1$, $\delta \le 0.05$.
For $\epsilon = 0.02$, $\delta \le 0.01$.
For $\epsilon = 0.01$, $\delta \le 0.005$.
For $\epsilon = 0.002$, $\delta \le 0.001$.
For $\epsilon = 0.001$, $\delta \le 0.0005$.

49. (a) If $b = 0$, then the property says $\lim_{x \to c} 0 = 0$, which is easy to see is true.

(b) If $|f(x) - L| < \frac{\epsilon}{|b|}$, then multiplying by $|b|$ gives

$$|b||f(x) - L| < \epsilon.$$

Since

$$|b||f(x) - L| = |b(f(x) - L)| = |bf(x) - bL|,$$

we have

$$|bf(x) - bL| < \epsilon.$$

(c) Suppose that $\lim_{x \to c} f(x) = L$. We want to show that $\lim_{x \to c} bf(x) = bL$. If we are to have

$$|bf(x) - bL| < \epsilon,$$

then we will need

$$|f(x) - L| < \frac{\epsilon}{|b|}.$$

We choose δ small enough that

$$|x - c| < \delta \quad \text{implies} \quad |f(x) - L| < \frac{\epsilon}{|b|}.$$

By part (b), this ensures that

$$|bf(x) - bL| < \epsilon,$$

as we wanted.

53. Since $f(x) = x$ is continuous, Theorem 1.3 on page 54 shows that products of the form $f(x) \cdot f(x) = x^2$ and $f(x) \cdot x^2 = x^3$, etc., are continuous. By a similar argument, x^n is continuous for any $n > 0$.

Solutions for Chapter 1 Review

Exercises

1. (a) The domain of f is the set of values of x for which the function is defined. Since the function is defined by the graph and the graph goes from $x = 0$ to $x = 7$, the domain of f is $[0, 7]$.

(b) The range of f is the set of values of y attainable over the domain. Looking at the graph, we can see that y gets as high as 5 and as low as -2, so the range is $[-2, 5]$.

(c) Only at $x = 5$ does $f(x) = 0$. So 5 is the only zero of $f(x)$.

(d) Looking at the graph, we can see that $f(x)$ is decreasing on $(1, 7)$.

(e) The graph indicates that $f(x)$ is concave up at $x = 6$.

(f) The value $f(4)$ is the y-value that corresponds to $x = 4$. From the graph, we can see that $f(4)$ is approximately 1.

(g) This function is not invertible, since it fails the horizontal-line test. A horizontal line at $y = 3$ would cut the graph of $f(x)$ in two places, instead of the required one.

5. To solve for x, we first divide both sides by 5 and then take the natural logarithm of both sides.

$$\frac{7}{5} = e^{0.2x}$$
$$\ln(7/5) = 0.2x$$
$$x = \frac{\ln(7/5)}{0.2} \approx 1.68.$$

9. Starting with the general exponential equation $y = Ae^{kx}$, we first find that for $(0, 1)$ to be on the graph, we must have $A = 1$. Then to make $(3, 4)$ lie on the graph, we require

$$4 = e^{3k}$$
$$\ln 4 = 3k$$
$$k = \frac{\ln 4}{3} \approx 0.4621.$$

Thus the equation is

$$y = e^{0.4621x}.$$

Alternatively, we can use the form $y = a^x$, in which case we find $y = (1.5874)^x$.

13. $z = 1 - \cos\theta$

17. This looks like a fourth degree polynomial with roots at -5 and -1 and a double root at 3. The leading coefficient is negative, and so a possible formula is

$$y = -(x + 5)(x + 1)(x - 3)^2.$$

21. This graph has period 5, amplitude 1 and no vertical shift or horizontal shift from $\sin x$, so it is given by

$$f(x) = \sin\left(\frac{2\pi}{5}x\right).$$

25. The graph of $g(x)$ is shown in Figure 1.14. It has a break at $x = 0$, so $g(x)$ is not continuous on $[-1, 1]$.

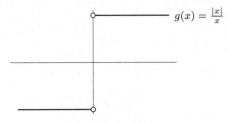

Figure 1.14

29. From Table 1.4, it appears the limit is 0. This is confirmed by Figure 1.15. An appropriate window is $-0.0029 < x < 0.0029$, $-0.01 < y < 0.01$.

Table 1.4

x	$f(x)$
0.1	0.3365
0.01	0.0337
0.001	0.0034
0.0001	0.0004

x	$f(x)$
-0.0001	-0.0004
-0.001	-0.0034
-0.01	-0.0337
-0.1	-0.3365

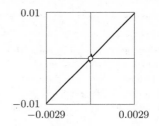

Figure 1.15

33. $f(x) = \begin{cases} e^x & -1 < x < 0 \\ 1 & x = 0 \\ \cos x & 0 < x < 1 \end{cases}$

Figure 1.16 confirms that $\lim\limits_{x \to 0^-} f(x) = \lim\limits_{x \to 0^-} e^x = e^0 = 1$, and that $\lim\limits_{x \to 0^+} f(x) = \lim\limits_{x \to 0^+} \cos x = \cos 0 = 1$, so $\lim\limits_{x \to 0} f(x) = 1$.

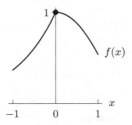

Figure 1.16

Problems

37. (a) We find the slope m and intercept b in the linear equation $S = b + mt$. To find the slope m, we use

$$m = \frac{\Delta S}{\Delta t} = \frac{66 - 113}{50 - 0} = -0.94.$$

When $t = 0$, we have $S = 113$, so the intercept b is 113. The linear formula is

$$S = 113 - 0.94t.$$

(b) We use the formula $S = 113 - 0.94t$. When $S = 20$, we have $20 = 113 - 0.94t$ and so $t = 98.9$. If this linear model were correct, the average male sperm count would drop below the fertility level during the year 2038.

41. Since we are told that the rate of decay is *continuous*, we use the function $Q(t) = Q_0 e^{rt}$ to model the decay, where $Q(t)$ is the amount of strontium-90 which remains at time t, and Q_0 is the original amount. Then

$$Q(t) = Q_0 e^{-0.0247t}.$$

So after 100 years,

$$Q(100) = Q_0 e^{-0.0247 \cdot 100}$$

and

$$\frac{Q(100)}{Q_0} = e^{-2.47} \approx 0.0846$$

so about 8.46% of the strontium-90 remains.

45. (a) Let the height of the can be h. Then

$$V = \pi r^2 h.$$

The surface area consists of the area of the ends (each is πr^2) and the curved sides (area $2\pi rh$), so

$$S = 2\pi r^2 + 2\pi rh.$$

Solving for h from the formula for V, we have

$$h = \frac{V}{\pi r^2}.$$

Substituting into the formula for S, we get

$$S = 2\pi r^2 + 2\pi r \cdot \frac{V}{\pi r^2} = 2\pi r^2 + \frac{2V}{r}.$$

(b) For large r, the $2V/r$ term becomes negligible, meaning $S \approx 2\pi r^2$, and thus $S \to \infty$ as $r \to \infty$.

(c) The graph is in Figure 1.17.

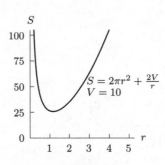

Figure 1.17

49. (a) Yes, f is invertible, since f is increasing everywhere.

(b) $f^{-1}(400)$ is the year in which 400 million motor vehicles were registered in the world. From the picture, we see that $f^{-1}(400)$ is around 1979.

(c) Since the graph of f^{-1} is the reflection of the graph of f over the line $y = x$, we get Figure 1.18.

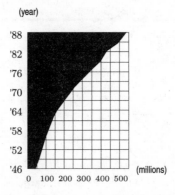

Figure 1.18: Graph of f^{-1}

53. (a) $r(p) = kp(A - p)$, where $k > 0$ is a constant.

(b) $p = A/2$.

CAS Challenge Problems

57. (a) A CAS gives $f(x) = (x - a)(x + a)(x + b)(x - c)$.

(b) The graph of $f(x)$ crosses the x-axis at $x = a$, $x = -a$, $x = -b$, $x = c$; it crosses the y-axis at a^2bc. Since the coefficient of x^4 (namely 1) is positive, the graph of f looks like that shown in Figure 1.19.

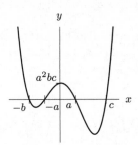

Figure 1.19: Graph of
$$f(x) =$$
$$(x-a)(x+a)(x+b)(x-c)$$

61. (a) A CAS or division gives
$$f(x) = \frac{x^3 - 30}{x - 3} = x^2 + 3x + 9 - \frac{3}{x - 3},$$
so $p(x) = x^2 + 3x + 9$, and $r(x) = -3$, and $q(x) = x - 3$.

(b) The vertical asymptote is $x = 3$. Near $x = 3$, the values of $p(x)$ are much smaller than the values of $r(x)/q(x)$. Thus
$$f(x) \approx \frac{-3}{x - 3} \qquad \text{for } x \text{ near } 3.$$

(c) For large x, the values of $p(x)$ are much larger than the value of $r(x)/q(x)$. Thus
$$f(x) \approx x^2 + 3x + 9 \qquad \text{as } x \to \infty, x \to -\infty.$$

(d) Figure 1.20 shows $f(x)$ and $y = -3/(x - 3)$ for x near 3. Figure 1.21 shows $f(x)$ and $y = x^2 + 3x + 9$ for $-20 \le x \le 20$. Note that in each case the graphs of f and the approximating function are close.

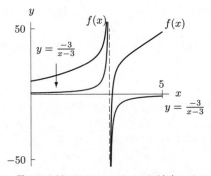

Figure 1.20: Close-up view of $f(x)$ and
$y = -3/(x - 3)$

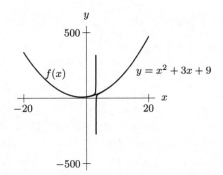

Figure 1.21: Far-away view of $f(x)$ and
$y = x^2 + 3x + 9$

CHECK YOUR UNDERSTANDING

1. False. A line can be put through any two points in the plane. However, if the line is vertical, it is not the graph of a function.

5. True. The highest degree term in a polynomial determines how the polynomial behaves when x is very large in the positive or negative direction. When n is odd, x^n is positive when x is large and positive but negative when x is large and negative. Thus if a polynomial $p(x)$ has odd degree, it will be positive for some values of x and negative for other values of x. Since every polynomial is continuous, the Intermediate Value Theorem then guarantees that $p(x) = 0$ for some value of x.

9. False. Suppose $y = 5^x$. Then increasing x by 1 increases y by a factor of 5. However increasing x by 2 increases y by a factor of 25, not 10, since

$$y = 5^{x+2} = 5^x \cdot 5^2 = 25 \cdot 5^x.$$

(Other examples are possible.)

13. True. The period is $2\pi/(200\pi) = 1/100$ seconds. Thus, the function executes 100 cycles in 1 second.

17. False. A counterexample is given by $f(x) = \sin x$, which has period 2π, and $g(x) = x^2$. The graph of $f(g(x)) = \sin(x^2)$ in Figure 1.22 is not periodic with period 2π.

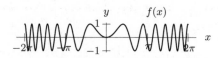

Figure 1.22

21. True. If f is increasing then its reflection about the line $y = x$ is also increasing. An example is shown in Figure 1.23. The statement is true.

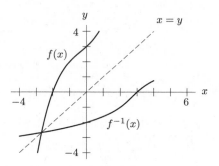

Figure 1.23

25. True. If $b > 1$, then $ab^x \to 0$ as $x \to -\infty$. If $0 < b < 1$, then $ab^x \to 0$ as $x \to \infty$. In either case, the function $y = a + ab^x$ has $y = a$ as the horizontal asymptote.

29. False. A counterexample is given by $f(x) = x^2$ and $g(x) = x + 1$. The function $f(g(x)) = (x + 1)^2$ is not even because $f(g(1)) = 4$ and $f(g(-1)) = 0 \neq 4$.

33. Let $f(x) = \dfrac{1}{x + 7\pi}$. Other answers are possible.

37. Let $f(x) = x$ and $g(x) = -2x$. Then $f(x) + g(x) = -x$, which is decreasing. Note f is increasing since it has positive slope, and g is decreasing since it has negative slope.

41. False. For example, $f(x) = x/(x^2 + 1)$ has no vertical asymptote since the denominator is never 0.

45. False. For example, if $y = 4x + 1$ (so $m = 4$) and $x = 1$, then $y = 5$. Increasing x by 2 units gives 3, so $y = 4(3) + 1 = 13$. Thus, y has increased by 8 units, not $4 + 2 = 6$. (Other examples are possible.)

49. True. The constant function $f(x) = 0$ is the only function that is both even and odd. This follows, since if f is both even and odd, then, for all x, $f(-x) = f(x)$ (if f is even) and $f(-x) = -f(x)$ (if f is odd). Thus, for all x, $f(x) = -f(x)$ i.e. $f(x) = 0$, for all x. So $f(x) = 0$ is both even and odd and is the only such function.

53. True, by Properties 2 and 3 of limits in Theorem 1.2.

$$\lim_{x \to 3} g(x) = \lim_{x \to 3} (f(x) + g(x) + (-1)f(x)) = \lim_{x \to 3} (f(x) + g(x)) + (-1) \lim_{x \to 3} f(x) = 12 + (-1)7 = 5.$$

57. False. For some functions we need to pick smaller values of δ. For example, if $f(x) = x^{1/3} + 2$ and $c = 0$ and $L = 2$, then $f(x)$ is within 10^{-3} of 2 if $|x^{1/3}| < 10^{-3}$. This only happens if x is within $(10^{-3})^3 = 10^{-9}$ of 0. If $x = 10^{-3}$ then $x^{1/3} = (10^{-3})^{1/3} = 10^{-1}$, which is too large.

61. False. The definition of the limit says that if x is within δ of c, then $f(x)$ is within ϵ of L, not the other way round.

CHAPTER TWO

Solutions for Section 2.1

Exercises

1. For t between 2 and 5, we have

$$\text{Average velocity} = \frac{\Delta s}{\Delta t} = \frac{400 - 135}{5 - 2} = \frac{265}{3} \text{ km/hr.}$$

The average velocity on this part of the trip was $265/3$ km/hr.

5.

Slope	-3	-1	0	$1/2$	1	2
Point	F	C	E	A	B	D

9. Using $h = 0.1, 0.01, 0.001$, we see

$$\frac{7^{0.1} - 1}{0.1} = 2.148$$

$$\frac{7^{0.01} - 1}{0.01} = 1.965$$

$$\frac{7^{0.001} - 1}{0.001} = 1.948$$

$$\frac{7^{0.0001} - 1}{0.0001} = 1.946.$$

This suggests that $\lim\limits_{h \to 0} \dfrac{7^h - 1}{h} \approx 1.9$.

Problems

13. See Figure 2.1.

Figure 2.1

17.

$$\begin{pmatrix} \text{Average velocity} \\ 0 < t < 0.2 \end{pmatrix} = \frac{s(0.2) - s(0)}{0.2 - 0} = \frac{0.5}{0.2} = 2.5 \text{ ft/sec.}$$

$$\begin{pmatrix} \text{Average velocity} \\ 0.2 < t < 0.4 \end{pmatrix} = \frac{s(0.4) - s(0.2)}{0.4 - 0.2} = \frac{1.3}{0.2} = 6.5 \text{ ft/sec.}$$

A reasonable estimate of the velocity at $t = 0.2$ is the average: $\frac{1}{2}(6.5 + 2.5) = 4.5$ ft/sec.

21. $\lim\limits_{h \to 0} \dfrac{(3 + h)^2 - (3 - h)^2}{2h} = \lim\limits_{h \to 0} \dfrac{9 + 6h + h^2 - 9 + 6h - h^2}{2h} = \lim\limits_{h \to 0} \dfrac{12h}{2h} = \lim\limits_{h \to 0} 6 = 6.$

Solutions for Section 2.2

Exercises

1. The derivative, $f'(2)$, is the rate of change of x^3 at $x = 2$. Notice that each time x changes by 0.001 in the table, the value of x^3 changes by 0.012. Therefore, we estimate

$$f'(2) = \begin{array}{c} \text{Rate of change} \\ \text{of } f \text{ at } x = 2 \end{array} \approx \frac{0.012}{0.001} = 12.$$

The function values in the table look exactly linear because they have been rounded. For example, the exact value of x^3 when $x = 2.001$ is 8.012006001, not 8.012. Thus, the table can tell us only that the derivative is approximately 12. Example 5 on page 85 shows how to compute the derivative of $f(x)$ exactly.

5.

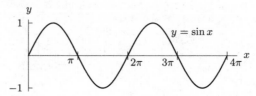

Since $\sin x$ is decreasing for values near $x = 3\pi$, its derivative at $x = 3\pi$ is negative.

9. One possible choice of points is shown below.

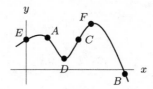

Problems

13. (a) Since the point $B = (2, 5)$ is on the graph of g, we have $g(2) = 5$.

(b) The slope of the tangent line touching the graph at $x = 2$ is given by

$$\text{Slope} = \frac{\text{Rise}}{\text{Run}} = \frac{5 - 5.02}{2 - 1.95} = \frac{-0.02}{0.05} = -0.4.$$

Thus, $g'(2) = -0.4$.

17. (a) $f(4)/4$ is the slope of the line connecting $(0,0)$ to $(4, f(4))$. (See Figure 2.2.)

(b) It is clear from Figure 2.2 that $f(3)/3 > f(4)/4$.

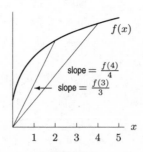

Figure 2.2

21. We want $f'(2)$. The exact answer is

$$f'(2) = \lim_{h \to 0} \frac{f(2+h) - f(2)}{h} = \lim_{h \to 0} \frac{(2+h)^{2+h} - 4}{h},$$

but we can approximate this. If $h = 0.001$, then

$$\frac{(2.001)^{2.001} - 4}{0.001} \approx 6.779$$

and if $h = 0.0001$ then

$$\frac{(2.0001)^{2.0001} - 4}{0.0001} \approx 6.773,$$

so $f'(2) \approx 6.77$.

25. (a) From Figure 2.3, it appears that the slopes of the tangent lines to the two graphs are the same at each x. For $x = 0$, the slopes of the tangents to the graphs of $f(x)$ and $g(x)$ at 0 are

$$\begin{aligned} f'(0) &= \lim_{h \to 0} \frac{f(0+h) - f(0)}{h} \\ &= \lim_{h \to 0} \frac{f(h) - 0}{h} \\ &= \lim_{h \to 0} \frac{\frac{1}{2}h^2}{h} \\ &= \lim_{h \to 0} \frac{1}{2}h \\ &= 0, \end{aligned}$$

$$\begin{aligned} g'(0) &= \lim_{h \to 0} \frac{g(0+h) - g(0)}{h} \\ &= \lim_{h \to 0} \frac{g(h) - g(0)}{h} \\ &= \lim_{h \to 0} \frac{\frac{1}{2}h^2 + 3 - 3}{h} \\ &= \lim_{h \to 0} \frac{\frac{1}{2}h^2}{h} \\ &= \lim_{h \to 0} \frac{1}{2}h \\ &= 0. \end{aligned}$$

For $x = 2$, the slopes of the tangents to the graphs of $f(x)$ and $g(x)$ are

$$\begin{aligned} f'(2) &= \lim_{h \to 0} \frac{f(2+h) - f(2)}{h} \\ &= \lim_{h \to 0} \frac{\frac{1}{2}(2+h)^2 - \frac{1}{2}(2)^2}{h} \\ &= \lim_{h \to 0} \frac{\frac{1}{2}(4 + 4h + h^2) - 2}{h} \\ &= \lim_{h \to 0} \frac{2 + 2h + \frac{1}{2}h^2 - 2}{h} \\ &= \lim_{h \to 0} \frac{2h + \frac{1}{2}h^2}{h} \\ &= \lim_{h \to 0} \left(2 + \frac{1}{2}h\right) \\ &= 2, \end{aligned}$$

$$\begin{aligned} g'(2) &= \lim_{h \to 0} \frac{g(2+h) - g(2)}{h} \\ &= \lim_{h \to 0} \frac{\frac{1}{2}(2+h)^2 + 3 - (\frac{1}{2}(2)^2 + 3)}{h} \\ &= \lim_{h \to 0} \frac{\frac{1}{2}(2+h)^2 - \frac{1}{2}(2)^2}{h} \\ &= \lim_{h \to 0} \frac{\frac{1}{2}(4 + 4h + h^2) - 2}{h} \\ &= \lim_{h \to 0} \frac{2 + 2h + \frac{1}{2}(h^2) - 2}{h} \\ &= \lim_{h \to 0} \frac{2h + \frac{1}{2}(h^2)}{h} \\ &= \lim_{h \to 0} \left(2 + \frac{1}{2}h\right) \\ &= 2. \end{aligned}$$

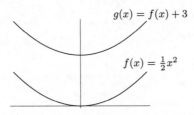

$g(x) = f(x) + 3$

$f(x) = \frac{1}{2}x^2$

Figure 2.3

For $x = x_0$, the slopes of the tangents to the graphs of $f(x)$ and $g(x)$ are

$$
\begin{aligned}
f'(x_0) &= \lim_{h \to 0} \frac{f(x_0 + h) - f(x_0)}{h} \\
&= \lim_{h \to 0} \frac{\frac{1}{2}(x_0 + h)^2 - \frac{1}{2}x_0^2}{h} \\
&= \lim_{h \to 0} \frac{\frac{1}{2}(x_0^2 + 2x_0 h + h^2) - \frac{1}{2}x_0^2}{h} \\
&= \lim_{h \to 0} \frac{x_0 h + \frac{1}{2}h^2}{h} \\
&= \lim_{h \to 0} \left(x_0 + \frac{1}{2}h\right) \\
&= x_0,
\end{aligned}
\qquad
\begin{aligned}
g'(x_0) &= \lim_{h \to 0} \frac{g(x_0 + h) - g(x_0)}{h} \\
&= \lim_{h \to 0} \frac{\frac{1}{2}(x_0 + h)^2 + 3 - (\frac{1}{2}(x_0)^2 + 3)}{h} \\
&= \lim_{h \to 0} \frac{\frac{1}{2}(x_0 + h)^2 - \frac{1}{2}(x_0)^2}{h} \\
&= \lim_{h \to 0} \frac{\frac{1}{2}(x_0^2 + 2x_0 h + h^2) - \frac{1}{2}x_0^2}{h} \\
&= \lim_{h \to 0} \frac{x_0 h + \frac{1}{2}h^2}{h} \\
&= \lim_{h \to 0} \left(x_0 + \frac{1}{2}h\right) \\
&= x_0.
\end{aligned}
$$

(b)

$$
\begin{aligned}
g'(x) &= \lim_{h \to 0} \frac{g(x + h) - g(x)}{h} \\
&= \lim_{h \to 0} \frac{f(x + h) + C - (f(x) + C)}{h} \\
&= \lim_{h \to 0} \frac{f(x + h) - f(x)}{h} \\
&= f'(x).
\end{aligned}
$$

29. $\lim\limits_{h \to 0} \dfrac{1}{h}\left(\dfrac{1}{1 + h} - 1\right) = \lim\limits_{h \to 0} \dfrac{1 - (1 + h)}{(1 + h)h} = \lim\limits_{h \to 0} \dfrac{-1}{1 + h} = -1$

33. Using the definition of the derivative, we have

$$
\begin{aligned}
f'(10) &= \lim_{h \to 0} \frac{f(10 + h) - f(10)}{h} \\
&= \lim_{h \to 0} \frac{5(10 + h)^2 - 5(10)^2}{h} \\
&= \lim_{h \to 0} \frac{500 + 100h + 5h^2 - 500}{h} \\
&= \lim_{h \to 0} \frac{100h + 5h^2}{h} \\
&= \lim_{h \to 0} \frac{h(100 + 5h)}{h} \\
&= \lim_{h \to 0} 100 + 5h \\
&= 100.
\end{aligned}
$$

37.

$$
\begin{aligned}
g'(2) &= \lim_{h \to 0} \frac{g(2 + h) - g(2)}{h} = \lim_{h \to 0} \frac{\frac{1}{2+h} - \frac{1}{2}}{h} \\
&= \lim_{h \to 0} \frac{2 - (2 + h)}{h(2 + h)2} = \lim_{h \to 0} \frac{-h}{h(2 + h)2} \\
&= \lim_{h \to 0} \frac{-1}{(2 + h)2} = -\frac{1}{4}
\end{aligned}
$$

41. We know that the slope of the tangent line to $f(x) = x$ when $x = 20$ is 1. When $x = 20$, $f(x) = 20$ so $(20, 20)$ is on the tangent line. Thus the equation of the tangent line is $y = 1(x - 20) + 20 = x$.

Solutions for Section 2.3

Exercises

1. The graph is that of the line $y = -2x + 2$. The slope, and hence the derivative, is -2. See Figure 2.4.

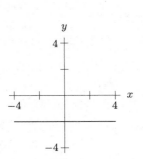

Figure 2.4

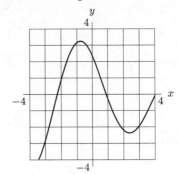

Figure 2.5

5. See Figure 2.5.

9. See Figure 2.6.

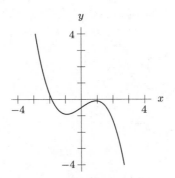

Figure 2.6

13. Since $1/x = x^{-1}$, using the power rule gives

$$k'(x) = (-1)x^{-2} = -\frac{1}{x^2}.$$

Using the definition of the derivative, we have

$$k'(x) = \lim_{h \to 0} \frac{k(x+h) - k(x)}{h} = \lim_{h \to 0} \frac{\frac{1}{x+h} - \frac{1}{x}}{h} = \lim_{h \to 0} \frac{x - (x+h)}{h(x+h)x}$$

$$= \lim_{h \to 0} \frac{-h}{h(x+h)x} = \lim_{h \to 0} \frac{-1}{(x+h)x} = -\frac{1}{x^2}.$$

17.

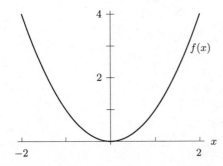

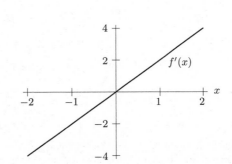

Problems

21. We know that $f'(x) \approx \dfrac{f(x+h) - f(x)}{h}$. For this problem, we'll take the average of the values obtained for $h = 1$ and $h = -1$; that's the average of $f(x+1) - f(x)$ and $f(x) - f(x-1)$ which equals $\dfrac{f(x+1) - f(x-1)}{2}$. Thus,

$f'(0) \approx f(1) - f(0) = 13 - 18 = -5.$
$f'(1) \approx [f(2) - f(0)]/2 = [10 - 18]/2 = -4.$
$f'(2) \approx [f(3) - f(1)]/2 = [9 - 13]/2 = -2.$
$f'(3) \approx [f(4) - f(2)]/2 = [9 - 10]/2 = -0.5.$
$f'(4) \approx [f(5) - f(3)]/2 = [11 - 9]/2 = 1.$
$f'(5) \approx [f(6) - f(4)]/2 = [15 - 9]/2 = 3.$
$f'(6) \approx [f(7) - f(5)]/2 = [21 - 11]/2 = 5.$
$f'(7) \approx [f(8) - f(6)]/2 = [30 - 15]/2 = 7.5.$
$f'(8) \approx f(8) - f(7) = 30 - 21 = 9.$

The rate of change of $f(x)$ is positive for $4 \le x \le 8$, negative for $0 \le x \le 3$. The rate of change is greatest at about $x = 8$.

25. See Figure 2.7.

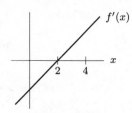

Figure 2.7

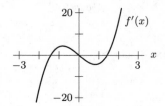

Figure 2.8

29. See Figure 2.8.

33. (a) x_3 (b) x_4 (c) x_5 (d) x_3

37. (a) $t = 3$
  (b) $t = 9$
  (c) $t = 14$
  (d)

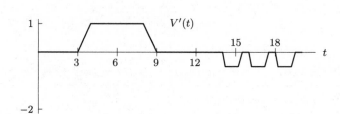

41. Figure 2.9 shows a possible graph – yours may be different.

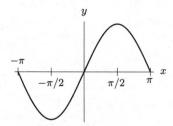

Figure 2.9

Solutions for Section 2.4

Exercises

1. (a) As the cup of coffee cools, the temperature decreases, so $f'(t)$ is negative.
   (b) Since $f'(t) = dH/dt$, the units are degrees Celsius per minute. The quantity $f'(20)$ represents the rate at which the coffee is cooling, in degrees per minute, 20 minutes after the cup is put on the counter.

5. Since B is measured in dollars and t is measured in years, dB/dt is measured in dollars per year. We can interpret dB as the extra money added to your balance in dt years. Therefore dB/dt represents how fast your balance is growing, in units of dollars/year.

9. The units of $f'(x)$ are feet/mile. The derivative, $f'(x)$, represents the rate of change of elevation with distance from the source, so if the river is flowing downhill everywhere, the elevation is always decreasing and $f'(x)$ is always negative. (In fact, there may be some stretches where the elevation is more or less constant, so $f'(x) = 0$.)

Problems

13. Since $f(t) = 1.15(1.014)^t$, we have
$$f(6) = 1.15(1.014)^6 = 1.25.$$
To estimate $f'(6)$, we use a small interval around 6:
$$f'(6) \approx \frac{f(6.001) - f(6)}{6.001 - 6} = \frac{1.15(1.014)^{6.001} - 1.15(1.014)^6}{0.001} = 0.0174.$$
We see that $f(6) = 1.25$ billion people and $f'(6) = 0.0174$ billion people per year. This model tells us that the population of China was about 1,250,000,000 people in 1999 and was growing at a rate of about 17,400,000 people per year at that time.

17. Units of $g'(55)$ are mpg/mph. The statement $g'(55) = -0.54$ means that at 55 miles per hour the fuel efficiency (in miles per gallon, or mpg) of the car decreases at a rate of approximately one half mpg as the velocity increases by one mph.

21. (a) The units of compliance are units of volume per units of pressure, or liters per centimeter of water.
   (b) The increase in volume for a 5 cm reduction in pressure is largest between 10 and 15 cm. Thus, the compliance appears maximum between 10 and 15 cm of pressure reduction. The derivative is given by the slope, so
$$\text{Compliance} \approx \frac{0.70 - 0.49}{15 - 10} = 0.042 \text{ liters per centimeter.}$$
   (c) When the lung is nearly full, it cannot expand much more to accommodate more air.

Solutions for Section 2.5

Exercises

1. (a) Since the graph is below the x-axis at $x = 2$, the value of $f(2)$ is negative.
   (b) Since $f(x)$ is decreasing at $x = 2$, the value of $f'(2)$ is negative.
   (c) Since $f(x)$ is concave up at $x = 2$, the value of $f''(2)$ is positive.

5. The graph must be everywhere decreasing and concave up on some intervals and concave down on other intervals. One possibility is shown in Figure 2.10.

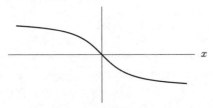

Figure 2.10

9. $f'(x) < 0$
$f''(x) = 0$

13. The velocity is the derivative of the distance, that is, $v(t) = s'(t)$. Therefore, we have

$$v(t) = \lim_{h \to 0} \frac{s(t+h) - s(t)}{h}$$

$$= \lim_{h \to 0} \frac{(5(t+h)^2 + 3) - (5t^2 + 3)}{h}$$

$$= \lim_{h \to 0} \frac{10th + 5h^2}{h}$$

$$= \lim_{h \to 0} \frac{h(10t + 5h)}{h} = \lim_{h \to 0} (10t + 5h) = 10t$$

The acceleration is the derivative of velocity, so $a(t) = v'(t)$:

$$a(t) = \lim_{h \to 0} \frac{10(t+h) - 10t}{h}$$

$$= \lim_{h \to 0} \frac{10h}{h} = 10.$$

Problems

17. (a)

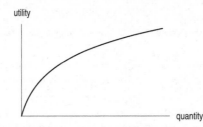

(b) As a function of quantity, utility is increasing but at a decreasing rate; the graph is increasing but concave down. So the derivative of utility is positive, but the second derivative of utility is negative.

21. Since f' is everywhere positive, f is everywhere increasing. Hence the greatest value of f is at x_6 and the least value of f is at x_1. Directly from the graph, we see that f' is greatest at x_3 and least at x_2. Since f'' gives the slope of the graph of f', f'' is greatest where f' is rising most rapidly, namely at x_6, and f'' is least where f' is falling most rapidly, namely at x_1.

Solutions for Section 2.6

Exercises

1. (a) Function f is not continuous at $x = 1$.
(b) Function f appears not differentiable at $x = 1, 2, 3$.

5. Yes.

Problems

9. We can see from Figure 2.11 that the graph of f oscillates infinitely often between the curves $y = x^2$ and $y = -x^2$ near the origin. Thus the slope of the line from $(0, 0)$ to $(h, f(h))$ oscillates between h (when $f(h) = h^2$ and $\frac{f(h) - 0}{h - 0} = h$) and $-h$ (when $f(h) = -h^2$ and $\frac{f(h) - 0}{h - 0} = -h$) as h tends to zero. So, the limit of the slope as h tends to zero is 0, which is the derivative of f at the origin. Another way to see this is to observe that

$$\lim_{h \to 0} \frac{f(h) - f(0)}{h} = \lim_{h \to 0} \left(\frac{h^2 \sin(\frac{1}{h})}{h} \right)$$

$$= \lim_{h \to 0} h \sin\left(\frac{1}{h}\right)$$

$$= 0,$$

since $\lim_{h \to 0} h = 0$ and $-1 \le \sin(\frac{1}{h}) \le 1$ for any h. Thus f is differentiable at $x = 0$, and $f'(0) = 0$.

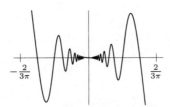

Figure 2.11

13. (a) Notice that B is a linear function of r for $r \le r_0$ and a reciprocal for $r > r_0$. The constant B_0 is the value of B at $r = r_0$ and the maximum value of B. See Figure 2.12.

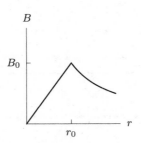

Figure 2.12

(b) B is continuous at $r = r_0$ because there is no break in the graph there. Using the formula for B, we have

$$\lim_{r \to r_0^-} B = \frac{r_0}{r_0} B_0 = B_0 \quad \text{and} \quad \lim_{r \to r_0^+} B = \frac{r_0}{r_0} B_0 = B_0.$$

(c) The function B is not differentiable at $r = r_0$ because the graph has a corner there. The slope is positive for $r < r_0$ and the slope is negative for $r > r_0$.

17. (a) The graph of

$$f(x) = \begin{cases} 0 & \text{if } x < 0. \\ x^2 & \text{if } x \ge 0. \end{cases}$$

is shown to the right. The graph is continuous and has no vertical segments or corners, so $f(x)$ is differentiable everywhere.

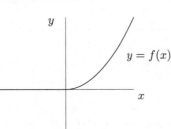

By Example 4 on page 84,

$$f'(x) = \begin{cases} 0 & \text{if } x < 0 \\ 2x & \text{if } x \geq 0 \end{cases}$$

So its graph is shown to the right.

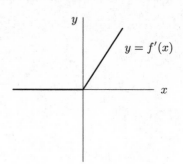

(b) The graph of the derivative has a corner at $x = 0$ so $f'(x)$ is not differentiable at $x = 0$. The graph of

$$f''(x) = \begin{cases} 0 & \text{if } x < 0 \\ 2 & \text{if } x > 0 \end{cases}$$

looks like:

The second derivative is not defined at $x = 0$. So it is certainly neither differentiable nor continuous at $x = 0$.

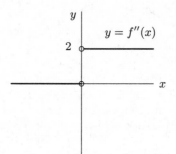

Solutions for Chapter 2 Review

Exercises

1. See Figure 2.13.

Figure 2.13

5. See Figure 2.14.

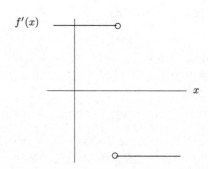

Figure 2.14

9. $\lim\limits_{h \to 0} \dfrac{(a+h)^2 - a^2}{h} = \lim\limits_{h \to 0} \dfrac{a^2 + 2ah + h^2 - a^2}{h} = \lim\limits_{h \to 0}(2a + h) = 2a$

13. We combine terms in the numerator and multiply top and bottom by $\sqrt{a} + \sqrt{a+h}$.

$$\frac{1}{\sqrt{a+h}} - \frac{1}{\sqrt{a}} = \frac{\sqrt{a} - \sqrt{a+h}}{\sqrt{a+h}\sqrt{a}} = \frac{(\sqrt{a} - \sqrt{a+h})(\sqrt{a} + \sqrt{a+h})}{\sqrt{a+h}\sqrt{a}(\sqrt{a} + \sqrt{a+h})}$$

$$= \frac{a - (a+h)}{\sqrt{a+h}\sqrt{a}(\sqrt{a} + \sqrt{a+h})}$$

Therefore $\lim\limits_{h \to 0} \dfrac{1}{h}\left(\dfrac{1}{\sqrt{a+h}} - \dfrac{1}{\sqrt{a}}\right) = \lim\limits_{h \to 0} \dfrac{-1}{\sqrt{a+h}\sqrt{a}(\sqrt{a} + \sqrt{a+h})} = \dfrac{-1}{2(\sqrt{a})^3}$

Problems

17. Using the approximation $\Delta y \approx f'(x)\Delta x$ with $\Delta x = 2$, we have $\Delta y \approx f'(20) \cdot 2 = 6 \cdot 2$, so

$$f(22) \approx f(20) + f'(20) \cdot 2 = 345 + 6 \cdot 2 = 357.$$

21. A possible graph of $y = f(x)$ is shown in Figure 2.15.

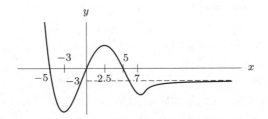

Figure 2.15

25. (a) $f'(0.6) \approx \dfrac{f(0.8) - f(0.6)}{0.8 - 0.6} = \dfrac{4.0 - 3.9}{0.2} = 0.5.$ $f'(0.5) \approx \dfrac{f(0.6) - f(0.4)}{0.6 - 0.4} = \dfrac{0.4}{0.2} = 2.$

 (b) Using the values of f' from part (a), we get $f''(0.6) \approx \dfrac{f'(0.6) - f'(0.5)}{0.6 - 0.5} = \dfrac{0.5 - 2}{0.1} = \dfrac{-1.5}{0.1} = -15.$

 (c) The maximum value of f is probably near $x = 0.8$. The minimum value of f is probably near $x = 0.3$.

29. (a) The population varies periodically with a period of 12 months (i.e. one year).

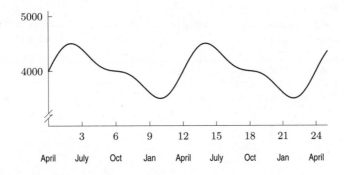

 (b) The herd is largest about June 1st when there are about 4500 deer.
 (c) The herd is smallest about February 1st when there are about 3500 deer.
 (d) The herd grows the fastest about April 1st. The herd shrinks the fastest about July 15 and again about December 15.
 (e) It grows the fastest about April 1st when the rate of growth is about 400 deer/month, i.e about 13 new fawns per day.

33. (a)

Table 2.1

x	$\frac{\sinh(x+0.001)-\sinh(x)}{0.001}$	$\frac{\sinh(x+0.0001)-\sinh(x)}{0.0001}$	so $f'(0) \approx$	$\cosh(x)$
0	1.00000	1.00000	1.00000	1.00000
0.3	1.04549	1.04535	1.04535	1.04534
0.7	1.25555	1.25521	1.25521	1.25517
1	1.54367	1.54314	1.54314	1.54308

(b) It seems that they are approximately the same, i.e. the derivative of $\sinh(x) = \cosh(x)$ for $x = 0, 0.3, 0.7$, and 1.

CAS Challenge Problems

37. (a) The CAS gives the same derivative, $1/x$, in all three cases.

(b) From the properties of logarithms, $g(x) = \ln(2x) = \ln 2 + \ln x = f(x) + \ln 2$. So the graph of g is the same shape as the graph of f, only shifted up by $\ln 2$. So the graphs have the same slope everywhere, and therefore the two functions have the same derivative. By the same reasoning, $h(x) = f(x) + \ln 3$, so h and f have the same derivative as well.

CHECK YOUR UNDERSTANDING

1. False. For example, the car could slow down or even stop at one minute after 2 pm, and then speed back up to 60 mph at one minute before 3 pm. In this case the car would travel only a few miles during the hour, much less than 50 miles.

5. True. By definition, Average velocity = Distance traveled/Time.

9. False. If $f'(x)$ is increasing then $f(x)$ is concave up. However, $f(x)$ may be either increasing or decreasing. For example, the exponential decay function $f(x) = e^{-x}$ is decreasing but $f'(x)$ is increasing because the graph of f is concave up.

13. False. The function $f(x)$ may be discontinuous at $x = 0$, for instance $f(x) = \begin{cases} 0 \text{ if } x \leq 0 \\ 1 \text{ if } x > 0 \end{cases}$. The graph of f may have a vertical tangent line at $x = 0$, for instance $f(x) = x^{1/3}$.

17. True. Instantaneous acceleration is a derivative, and all derivatives are limits of difference quotients. More precisely, instantaneous acceleration $a(t)$ is the derivative of the velocity $v(t)$, so

$$a(t) = \lim_{h \to 0} \frac{v(t+h) - v(t)}{h}.$$

21. True. If a function is differentiable at a point, then it is continuous at that point. For example, $f(x) = x^2$ is both differentiable and continuous on any interval. However, *one* example does not establish the truth of this statement; it merely illustrates the statement.

25. (a) This is not a counterexample, since it does not satisfy the conditions of the statement, and therefore does not have the potential to contradict the statement.

(b) This contradicts the statement, because it satisfies its conditions but not its conclusion. Hence it is a counterexample. Notice that this counterexample could not actually exist, since the statement is true.

(c) This is an example illustrating the statement; it is not a counterexample.

(d) This is not a counterexample, for the same reason as in part (a).

CHAPTER THREE

Solutions for Section 3.1

Exercises

1. The derivative, $f'(x)$, is defined as

$$f'(x) = \lim_{h \to 0} \frac{f(x+h) - f(x)}{h}.$$

If $f(x) = 7$, then

$$f'(x) = \lim_{h \to 0} \frac{7 - 7}{h} = \lim_{h \to 0} \frac{0}{h} = 0.$$

5. $y' = -12x^{-13}$.

9. $y' = \frac{3}{4}x^{-1/4}$.

13. Since $f(z) = -\frac{1}{z^{6.1}} = -z^{-6.1}$, we have $f'(z) = -(-6.1)z^{-7.1} = 6.1z^{-7.1}$.

17. Since $h(\theta) = \frac{1}{\sqrt[3]{\theta}} = \theta^{-1/3}$, we have $h'(\theta) = -\frac{1}{3}\theta^{-4/3}$.

21. $f'(t) = 6t - 4$.

25. $h'(w) = 6w^{-4} + \frac{3}{2}w^{-1/2}$

29. Since $y = \sqrt{x}(x+1) = x^{1/2}x + x^{1/2} \cdot 1 = x^{3/2} + x^{1/2}$, we have $\frac{dy}{dx} = \frac{3}{2}x^{1/2} + \frac{1}{2}x^{-1/2}$.

33. $y = x + \frac{1}{x}$, so $y' = 1 - \frac{1}{x^2}$.

37. $j'(x) = \frac{3x^2}{a} + \frac{2ax}{b} - c$

41. Since $4/3$, π, and b are all constants, we have

$$\frac{dV}{dr} = \frac{4}{3}\pi(2r)b = \frac{8}{3}\pi rb.$$

Problems

45. So far, we can only take the derivative of powers of x and the sums of constant multiples of powers of x. Since we cannot write $\sqrt{x+3}$ in this form, we cannot yet take its derivative.

49. We cannot write $\frac{1}{3x^2+4}$ as the sum of powers of x multiplied by constants.

53. Differentiating gives

$$f'(x) = 6x^2 - 4x \quad \text{so} \quad f'(1) = 6 - 4 = 2.$$

Thus the equation of the tangent line is $(y - 1) = 2(x - 1)$ or $y = 2x - 1$.

57. Decreasing means $f'(x) < 0$:

$$f'(x) = 4x^3 - 12x^2 = 4x^2(x - 3),$$

so $f'(x) < 0$ when $x < 3$ and $x \neq 0$. Concave up means $f''(x) > 0$:

$$f''(x) = 12x^2 - 24x = 12x(x - 2)$$

so $f''(x) > 0$ when

$$12x(x - 2) > 0$$
$$x < 0 \quad \text{or} \quad x > 2.$$

So, both conditions hold for $x < 0$ or $2 < x < 3$.

61. Since $f(t) = 700 - 3t^2$, we have $f(5) = 700 - 3(25) = 625$ cm. Since $f'(t) = -6t$, we have $f'(5) = -30$ cm/year. In the year 2000, the sand dune was 625 cm high and it was eroding at a rate of 30 centimeters per year.

65. (a) $T = 2\pi\sqrt{\dfrac{l}{g}} = \dfrac{2\pi}{\sqrt{g}}\left(l^{\frac{1}{2}}\right)$, so $\dfrac{dT}{dl} = \dfrac{2\pi}{\sqrt{g}}\left(\dfrac{1}{2}l^{-\frac{1}{2}}\right) = \dfrac{\pi}{\sqrt{gl}}$.

(b) Since $\dfrac{dT}{dl}$ is positive, the period T increases as the length l increases.

69. Since $f(x) = ax^n$, $f'(x) = anx^{n-1}$. We know that $f'(2) = (an)2^{n-1} = 3$, and $f'(4) = (an)4^{n-1} = 24$. Therefore,

$$\frac{f'(4)}{f'(2)} = \frac{24}{3}$$

$$\frac{(an)4^{n-1}}{(an)2^{n-1}} = \left(\frac{4}{2}\right)^{n-1} = 8$$

$$2^{n-1} = 8, \text{ and thus } n = 4.$$

Substituting $n = 4$ into the expression for $f'(2)$, we get $3 = a(4)(8)$, or $a = 3/32$.

Solutions for Section 3.2

Exercises

1. $f'(x) = 2e^x + 2x$.

5. $y' = 10x + (\ln 2)2^x$.

9. Since $y = 2^x + \dfrac{2}{x^3} = 2^x + 2x^{-3}$, we have $\dfrac{dy}{dx} = (\ln 2)2^x - 6x^{-4}$.

13. $f'(t) = (\ln(\ln 3))(\ln 3)^t$.

17. $f'(x) = 3x^2 + 3^x \ln 3$

21. Since e and k are constants, e^k is constant, so we have $f'(x) = (\ln k)k^x$.

25. $y'(x) = a^x \ln a + ax^{a-1}$.

29. We can take the derivative of the sum $x^2 + 2^x$, but not the product.

33. The exponent is x^2, and we haven't learned what to do about that yet.

Problems

37.

$$\frac{dP}{dt} = 35{,}000 \cdot (\ln 0.98)(0.98^t).$$

At $t = 23$, this is $35{,}000(\ln 0.98)(0.98^{23}) \approx -444.3$ people/year. (Note: the negative sign indicates that the population is decreasing.)

41. (a) $f(x) = 1 - e^x$ crosses the x-axis where $0 = 1 - e^x$, which happens when $e^x = 1$, so $x = 0$. Since $f'(x) = -e^x$, $f'(0) = -e^0 = -1$.

(b) $y = -x$

(c) The negative of the reciprocal of -1 is 1, so the equation of the normal line is $y = x$.

45. The equation $2^x = 2x$ has solutions $x = 1$ and $x = 2$. (Check this by substituting these values into the equation). The graph below suggests that these are the only solutions, but how can we be sure?

Let's look at the slope of the curve $f(x) = 2^x$, which is $f'(x) = (\ln 2)2^x \approx (0.693)2^x$, and the slope of the line $g(x) = 2x$ which is 2. At $x = 1$, the slope of $f(x)$ is less than 2; at $x = 2$, the slope of $f(x)$ is more than 2. Since slope of $f(x)$ is always increasing, there can be no other point of intersection. (If there were another point of intersection, the graph f would have to "turn around".)

Here's another way of seeing this. Suppose $g(x)$ represents the position of a car going a steady 2 mph, while $f(x)$ represents a car which starts ahead of g (because the graph of f is above g) and is initially going slower than g. The car f is first overtaken by g. All the while, however, f is speeding up until eventually it overtakes g again. Notice that the two cars will only meet twice (corresponding to the two intersections of the curve): once when g overtakes f and once when f overtakes g.

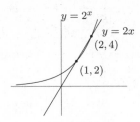

Solutions for Section 3.3

Exercises

1. By the product rule, $f'(x) = 2x(x^3 + 5) + x^2(3x^2) = 2x^4 + 3x^4 + 10x = 5x^4 + 10x$. Alternatively, $f'(x) = (x^5 + 5x^2)' = 5x^4 + 10x$. The two answers should, and do, match.

5. $y' = \frac{1}{2\sqrt{x}}2^x + \sqrt{x}(\ln 2)2^x$.

9. $f'(y) = (\ln 4)4^y(2 - y^2) + 4^y(-2y) = 4^y((\ln 4)(2 - y^2) - 2y)$.

13. $\dfrac{dy}{dx} = \dfrac{1 \cdot 2^t - (t+1)(\ln 2)2^t}{(2^t)^2} = \dfrac{2^t(1 - (t+1)\ln 2)}{(2^t)^2} = \dfrac{1 - (t+1)\ln 2}{2^t}$

17. $\dfrac{dz}{dt} = \dfrac{3(5t+2) - (3t+1)5}{(5t+2)^2} = \dfrac{15t+6 - 15t - 5}{(5t+2)^2} = \dfrac{1}{(5t+2)^2}$.

21. $w = y^2 - 6y + 7.\quad w' = 2y - 6, y \neq 0$.

25. $h'(r) = \dfrac{d}{dr}\left(\dfrac{r^2}{2r+1}\right) = \dfrac{(2r)(2r+1) - 2r^2}{(2r+1)^2} = \dfrac{2r(r+1)}{(2r+1)^2}$.

29.

$$f'(x) = \frac{(2 + 3x + 4x^2)(1) - (1+x)(3+8x)}{(2 + 3x + 4x^2)^2}$$

$$= \frac{2 + 3x + 4x^2 - 3 - 11x - 8x^2}{(2 + 3x + 4x^2)^2}$$

$$= \frac{-4x^2 - 8x - 1}{(2 + 3x + 4x^2)^2}.$$

Problems

33. Using the quotient rule, we know that $j'(x) = (g'(x) \cdot f(x) - g(x) \cdot f'(x))/(f(x))^2$. We use slope to compute the derivatives. Since $f(x)$ is linear on the interval $0 < x < 2$, we compute the slope of the line to see that $f'(x) = 2$ on this interval. Similarly, we compute the slope on the interval $2 < x < 4$ to see that $f'(x) = -2$ on the interval $2 < x < 4$. Since $f(x)$ has a corner at $x = 2$, we know that $f'(2)$ does not exist.

Similarly, $g(x)$ is linear on the interval shown, and we see that the slope of $g(x)$ on this interval is -1 so we have $g'(x) = -1$ on this interval.

(a) We have

$$j'(1) = \frac{g'(1) \cdot f(1) - g(1) \cdot f'(1)}{(f(1))^2} = \frac{(-1)2 - 3 \cdot 2}{2^2} = \frac{-2 - 6}{4} = \frac{-8}{4} = -2.$$

(b) We have $j'(2) = (g'(2) \cdot f(2) - g(2) \cdot f'(2))/(f(2)^2)$. Since $f(x)$ has a corner at $x = 2$, we know that $f'(2)$ does not exist. Therefore, $j'(2)$ does not exist.

(c) We have

$$j'(3) = \frac{g'(3) \cdot f(3) - g(3) \cdot f'(3)}{(f(3))^2} = \frac{(-1)2 - 1(-2)}{2^2} = \frac{-2 + 2}{4} = 0.$$

37. From the graphs, we estimate $f(2) \approx 0.3$, $f'(2) \approx 1.1$, $g(2) \approx 1.6$, and $g'(2) \approx -0.5$. By the quotient rule,

$$k'(2) = \frac{f'(2) \cdot g(2) - f(2) \cdot g'(2)}{(g(2))^2} \approx \frac{1.1(1.6) - 0.3(-0.5)}{(1.6)^2} = 0.75.$$

41.

$$f(t) = \frac{1}{e^t}$$

$$f'(t) = \frac{e^t \cdot 0 - e^t \cdot 1}{(e^t)^2}$$

$$= \frac{-1}{e^t} = -e^{-t}.$$

45. Using the quotient rule, we have

$$g'(x) = \frac{0 - 1(2x)}{(x^2 + 1)^2} = \frac{-2x}{(x^2 + 1)^2}$$

$$g''(x) = \frac{-2(x^2 + 1)^2 + 2x(4x^3 + 4x)}{(x^2 + 1)^4}$$

$$= \frac{-2(x^2 + 1)^2 + 8x^2(x^2 + 1)}{(x^2 + 1)^4}$$

$$= \frac{-2(x^2 + 1) + 8x^2}{(x^2 + 1)^3}$$

$$= \frac{2(3x^2 - 1)}{(x^2 + 1)^3}.$$

Since $(x^2 + 1)^3 > 0$ for all x, we have $g''(x) < 0$ if $(3x^2 - 1) < 0$, or when

$$3x^2 < 1$$

$$-\frac{1}{\sqrt{3}} < x < \frac{1}{\sqrt{3}}.$$

49. Since

$$x^{1/2} \cdot x^{1/2} = x,$$

we differentiate to obtain

$$\frac{d}{dx}(x^{1/2}) \cdot x^{1/2} + x^{1/2} \cdot \frac{d}{dx}(x^{1/2}) = 1.$$

Now solve for $d(x^{1/2})/dx$:

$$2x^{1/2} \frac{d}{dx}(x^{1/2}) = 1$$

$$\frac{d}{dx}(x^{1/2}) = \frac{1}{2x^{1/2}}.$$

53. (a) $f(140) = 15,000$ says that 15,000 skateboards are sold when the cost is \$140 per board.

$f'(140) = -100$ means that if the price is increased from \$140, roughly speaking, every dollar of increase will decrease the total sales by 100 boards.

(b) $\dfrac{dR}{dp} = \dfrac{d}{dp}(p \cdot q) = \dfrac{d}{dp}(p \cdot f(p)) = f(p) + pf'(p).$

So,

$$\frac{dR}{dp}\bigg|_{p=140} = f(140) + 140f'(140)$$

$$= 15,000 + 140(-100) = 1000.$$

(c) From (b) we see that $\dfrac{dR}{dp}\bigg|_{p=140} = 1000 > 0$. This means that the revenue will increase by about \$1000 if the price is raised by \$1.

57. Assume for $g(x) \neq f(x)$, $g'(x) = g(x)$ and $g(0) = 1$. Then for

$$h(x) = \frac{g(x)}{e^x}$$

$$h'(x) = \frac{g'(x)e^x - g(x)e^x}{(e^x)^2} = \frac{e^x(g'(x) - g(x))}{(e^x)^2} = \frac{g'(x) - g(x)}{e^x}.$$

But, since $g(x) = g'(x)$, $h'(x) = 0$, so $h(x)$ is constant. Thus, the ratio of $g(x)$ to e^x is constant. Since $\frac{g(0)}{e^0} = \frac{1}{1} = 1$, $\frac{g(x)}{e^x}$ must equal 1 for all x. Thus $g(x) = e^x = f(x)$ for all x, so f and g are the same function.

61. (a) Since $x = a$ is a double zero of a polynomial $P(x)$, we can write $P(x) = (x - a)^2 Q(x)$, so $P(a) = 0$. Using the product rule, we have

$$P'(x) = 2(x - a)Q(x) + (x - a)^2 Q'(x).$$

Substituting in $x = a$, we see $P'(a) = 0$ also.

(b) Since $P(a) = 0$, we know $x = a$ is a zero of P, so that $x - a$ is a factor of P and we can write

$$P(x) = (x - a)Q(x),$$

where Q is some polynomial. Differentiating this expression for P using the product rule, we get

$$P'(x) = Q(x) + (x - a)Q'(x).$$

Since we are told that $P'(a) = 0$, we have

$$P'(a) = Q(a) + (a - a)Q'(a) = 0$$

and so $Q(a) = 0$. Therefore $x = a$ is a zero of Q, so again we can write

$$Q(x) = (x - a)R(x),$$

where R is some other polynomial. As a result,

$$P(x) = (x - a)Q(x) = (x - a)^2 R(x),$$

so that $x = a$ is a double zero of P.

Solutions for Section 3.4 ━━━━

Exercises

1. $f'(x) = 99(x + 1)^{98} \cdot 1 = 99(x + 1)^{98}$.

5. $f'(x) = \frac{1}{2}(1 - x^2)^{-\frac{1}{2}}(-2x) = \dfrac{-x}{\sqrt{1 - x^2}}$.

9. We can write $w(r) = (r^4 + 1)^{1/2}$, so

$$w'(r) = \frac{1}{2}(r^4 + 1)^{-1/2}(4r^3) = \frac{2r^3}{\sqrt{r^4 + 1}}.$$

13. $g(x) = \pi e^{\pi x}$.

17. $f'(t) = 1 \cdot e^{5-2t} + te^{5-2t}(-2) = e^{5-2t}(1 - 2t)$.

21. $y' = \frac{3}{2}e^{\frac{3}{2}w}$.

25. $y' = 1 \cdot e^{-t^2} + te^{-t^2}(-2t)$

29. $w' = \frac{3}{2}\sqrt{x^2 \cdot 5^x}[2x(5^x) + (\ln 5)(x^2)(5^x)] = \frac{3}{2}x^2\sqrt{5^{3x}}(2 + x\ln 5)$.

33. $y' = 2\left(\dfrac{x^2 + 2}{3}\right)\left(\dfrac{2x}{3}\right) = \frac{4}{9}x\left(x^2 + 2\right)$

37. $h'(z) = \dfrac{-8b^4 z}{(a + z^2)^5}$

41. $f'(x) = 6(e^{5x})(5) + (e^{-x^2})(-2x) = 30e^{5x} - 2xe^{-x^2}$.

45. We write $y = (e^{-3t^2} + 5)^{1/2}$, so

$$\frac{dy}{dt} = \frac{1}{2}(e^{-3t^2} + 5)^{-1/2} \cdot \frac{d}{dt}(e^{-3t^2} + 5) = \frac{1}{2}(e^{-3t^2} + 5)^{-1/2} \cdot e^{-3t^2} \cdot \frac{d}{dt}(-3t^2)$$

$$= \frac{1}{2}(e^{-3t^2} + 5)^{-1/2} \cdot e^{-3t^2} \cdot (-6t) = -\frac{3te^{-3t^2}}{\sqrt{e^{-3t^2} + 5}}.$$

49. Since a and b are constants, we have $f'(x) = 3(ax^2 + b)^2(2ax) = 6ax(ax^2 + b)^2$.

Problems

53. Using the chain rule, we know that $h'(x) = f'(g(x)) \cdot g'(x)$. We use slope to compute the derivatives. Since $f(x)$ is linear on the interval $0 < x < 2$, we compute the slope of the line to see that $f'(x) = 2$ on this interval. Similarly, we compute the slope on the interval $2 < x < 4$ to see that $f'(x) = -2$ on the interval $2 < x < 4$. Since $f(x)$ has a corner at $x = 2$, we know that $f'(2)$ does not exist.

Similarly, $g(x)$ is linear on the interval shown, and we see that the slope of $g(x)$ on this interval is -1 so we have $g'(x) = -1$ on this interval.

(a) We have $h'(1) = f'(g(1)) \cdot g'(1) = (f'(3))(-1) = (-2)(-1) = 2$.

(b) We have $h'(2) = f'(g(2)) \cdot g'(2) = (f'(2))(-1)$. Since $f(x)$ has a corner at $x = 2$, we know that $f'(2)$ does not exist. Therefore, $h'(2)$ does not exist.

(c) We have $h'(3) = f'(g(3)) \cdot g'(3) = (f'(1))(-1) = 2(-1) = -2$.

57. The chain rule gives

$$\frac{d}{dx}f(g(x))\bigg|_{x=30} = f'(g(30))g'(30) = f'(55)g'(30) = (1)\left(\frac{1}{2}\right) = \frac{1}{2}.$$

61. We have $f(2) = (2 - 1)^3 = 1$, so $(2, 1)$ is a point on the tangent line. Since $f'(x) = 3(x - 1)^2$, the slope of the tangent line is

$$m = f'(2) = 3(2 - 1)^2 = 3.$$

The equation of the line is

$$y - 1 = 3(x - 2) \quad \text{or} \quad y = 3x - 5.$$

65.

$$f'(x) = [10(2x + 1)^9(2)][(3x - 1)^7] + [(2x + 1)^{10}][7(3x - 1)^6(3)]$$
$$= (2x + 1)^9(3x - 1)^6[20(3x - 1) + 21(2x + 1)]$$
$$= [(2x + 1)^9(3x - 1)^6](102x + 1)$$
$$f''(x) = [9(2x + 1)^8(2)(3x - 1)^6 + (2x + 1)^9(6)(3x - 1)^5(3)](102x + 1)$$
$$ + (2x + 1)^9(3x - 1)^6(102).$$

69. Yes. To see why, simply plug $x = \sqrt[3]{2t + 5}$ into the expression $3x^2\dfrac{dx}{dt}$ and evaluate it. To do this, first we calculate $\dfrac{dx}{dt}$. By the chain rule,

$$\frac{dx}{dt} = \frac{d}{dt}(2t + 5)^{\frac{1}{3}} = \frac{2}{3}(2t + 5)^{-\frac{2}{3}} = \frac{2}{3}[(2t + 5)^{\frac{1}{3}}]^{-2}.$$

But since $x = (2t + 5)^{\frac{1}{3}}$, we have (by substitution)

$$\frac{dx}{dt} = \frac{2}{3}x^{-2}.$$

It follows that $3x^2\dfrac{dx}{dt} = 3x^2\left(\dfrac{2}{3}x^{-2}\right) = 2$.

73. We have $h(0) = f(g(0)) = f(d) = d$. From the chain rule, $h'(0) = f'(g(0))g'(0)$. From the graph of g, we see that $g'(0) = 0$, so $h'(0) = f'(g(0)) \cdot 0 = 0$.

77. On the interval $-d < x < -b$, we see that the value of $g(x)$ increases from $-d$ to 0. On the interval $-d < x < 0$, the value of $f(x)$ decreases from d to $-d$. Thus, the value of $h(x) = f(g(x))$ decreases on the interval $-d < x < -b$ from

$$h(-d) = f(g(-d)) = f(-d) = d \quad \text{to} \quad h(-b) = f(g(-b)) = f(0) - d.$$

Confirming this using derivatives and the chain rule, we see

$$h'(x) = f'(g(x)) \cdot g'(x),$$

and since $g'(x)$ is negative on $-d < x < -b$ and $f'(g(x))$ is positive on this interval, the value of $h(x)$ is decreasing.

81. (a) For $t < 0, I = \dfrac{dQ}{dt} = 0$.

For $t > 0, I = \dfrac{dQ}{dt} = -\dfrac{Q_0}{RC} e^{-t/RC}$.

(b) For $t > 0, t \to 0$ (that is, as $t \to 0^+$),

$$I = -\frac{Q_0}{RC} e^{-t/RC} \to -\frac{Q_0}{RC}.$$

Since $I = 0$ just to the left of $t = 0$ and $I = -Q_0/RC$ just to the right of $t = 0$, it is not possible to define I at $t = 0$.

(c) Q is not differentiable at $t = 0$ because there is no tangent line at $t = 0$.

85. The problem tells us that

$$\frac{d}{dt} G(a - bt) = H(a - bt).$$

Since $\dfrac{d}{dt}(a - bt) = -b$, the chain rule tells us that

$$-bG'(a - bt) = H(a - bt),$$

so

$$G'(a - bt) = \left(-\frac{1}{b}\right) H(a - bt).$$

Replacing $a - bt$ by t, we have

$$G'(t) = \left(-\frac{1}{b}\right) H(t)$$

Solutions for Section 3.5

Exercises

1.

Table 3.1

x	$\cos x$	Difference Quotient	$-\sin x$
0	1.0	-0.0005	0.0
0.1	0.995	-0.10033	-0.099833
0.2	0.98007	-0.19916	-0.19867
0.3	0.95534	-0.296	-0.29552
0.4	0.92106	-0.38988	-0.38942
0.5	0.87758	-0.47986	-0.47943
0.6	0.82534	-0.56506	-0.56464

5. $f'(x) = \cos(3x) \cdot 3 = 3\cos(3x)$.

9. $f'(x) = (2x)(\cos x) + x^2(-\sin x) = 2x\cos x - x^2 \sin x$.

13. $z' = e^{\cos \theta} - \theta(\sin \theta)e^{\cos \theta}$.

17.
$$f(x) = (1 - \cos x)^{\frac{1}{2}}$$
$$f'(x) = \frac{1}{2}(1 - \cos x)^{-\frac{1}{2}}(-(-\sin x))$$
$$= \frac{\sin x}{2\sqrt{1 - \cos x}}.$$

21. $f'(x) = 2 \cdot [\sin(3x)] + 2x[\cos(3x)] \cdot 3 = 2\sin(3x) + 6x\cos(3x)$

25. $y' = 5\sin^4 \theta \cos \theta.$

29. $h'(t) = 1 \cdot (\cos t) + t(-\sin t) + \frac{1}{\cos^2 t} = \cos t - t\sin t + \frac{1}{\cos^2 t}.$

33. $y' = -2\cos w \sin w - \sin(w^2)(2w) = -2(\cos w \sin w + w\sin(w^2))$

37. Using the power and quotient rules gives

$$f'(x) = \frac{1}{2}\left(\frac{1 - \sin x}{1 - \cos x}\right)^{-1/2}\left[\frac{-\cos x(1 - \cos x) - (1 - \sin x)\sin x}{(1 - \cos x)^2}\right]$$

$$= \frac{1}{2}\sqrt{\frac{1 - \cos x}{1 - \sin x}}\left[\frac{-\cos x(1 - \cos x) - (1 - \sin x)\sin x}{(1 - \cos x)^2}\right]$$

$$= \frac{1}{2}\sqrt{\frac{1 - \cos x}{1 - \sin x}}\left[\frac{1 - \cos x - \sin x}{(1 - \cos x)^2}\right].$$

Problems

41. The pattern in the table below allows us to generalize and say that the $(4n)^{\text{th}}$ derivative of $\cos x$ is $\cos x$, i.e.,

$$\frac{d^4 y}{dx^4} = \frac{d^8 y}{dx^8} = \cdots = \frac{d^{4n} y}{dx^{4n}} = \cos x.$$

Thus we can say that $d^{48}y/dx^{48} = \cos x$. From there we differentiate twice more to obtain $d^{50}y/dx^{50} = -\cos x$.

n	1	2	3	4	$\cdots$	48	49	50
n^{th} derivative	$-\sin x$	$-\cos x$	$\sin x$	$\cos x$		$\cos x$	$-\sin x$	$-\cos x$

45. (a) $v(t) = \dfrac{dy}{dt} = \dfrac{d}{dt}(15 + \sin(2\pi t)) = 2\pi \cos(2\pi t).$

(b)

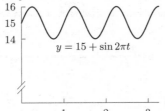
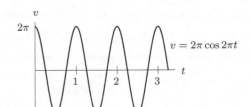

49. If the graphs of $y = \sin x$ and $y = ke^{-x}$ are tangent, then the y-values and the derivatives, $\dfrac{dy}{dx} = \cos x$ and $\dfrac{dy}{dx} = -ke^{-x}$, are equal at that point, so

$$\sin x = ke^{-x} \qquad \text{and} \qquad \cos x = -ke^{-x}.$$

Thus $\sin x = -\cos x$ so $\tan x = -1$. The smallest x-value is $x = 3\pi/4$, which leads to the smallest k value

$$k = \frac{\sin(3\pi/4)}{e^{-3\pi/4}} = 7.46.$$

When $x = \dfrac{3\pi}{4}$, we have $y = \sin\left(\dfrac{3\pi}{4}\right) = \dfrac{1}{\sqrt{2}}$ so the point is $\left(\dfrac{3\pi}{4}, \dfrac{1}{\sqrt{2}}\right).$

Solutions for Section 3.6

Exercises

1. $f'(t) = \dfrac{2t}{t^2 + 1}$.

5. $f'(x) = \dfrac{1}{1 - e^{-x}} \cdot (-e^{-x})(-1) = \dfrac{e^{-x}}{1 - e^{-x}}$.

9. $j'(x) = \dfrac{ae^{ax}}{(e^{ax} + b)}$

13. $f'(w) = \dfrac{1}{\cos(w - 1)}[-\sin(w - 1)] = -\tan(w - 1)$.

   [This could be done easily using the answer from Problem 6 and the chain rule.]

17. $g(\alpha) = \alpha$, so $g'(\alpha) = 1$.

21. $h'(w) = \arcsin w + \dfrac{w}{\sqrt{1 - w^2}}$.

25. $f'(x) = -\sin(\arctan 3x) \left(\dfrac{1}{1 + (3x)^2} \right) (3) = \dfrac{-3\sin(\arctan 3x)}{1 + 9x^2}$.

29. $\dfrac{dy}{dx} = 2(\ln x + \ln 2) + 2x\left(\dfrac{1}{x}\right) - 2 = 2(\ln x + \ln 2) = 2\ln(2x)$

33. Since $\ln \left[\left(\dfrac{1 - \cos t}{1 + \cos t} \right)^4 \right] = 4\ln \left[\left(\dfrac{1 - \cos t}{1 + \cos t} \right) \right]$ we have

$$a'(t) = 4 \left(\dfrac{1 + \cos t}{1 - \cos t} \right) \left[\dfrac{\sin t(1 + \cos t) + \sin t(1 - \cos t)}{(1 + \cos t)^2} \right]$$

$$= \left[\dfrac{1 + \cos t}{1 - \cos t} \right] \left[\dfrac{8 \sin t}{(1 + \cos t)^2} \right]$$

$$= \dfrac{8 \sin t}{1 - \cos^2 t}$$

$$= \dfrac{8}{\sin t}.$$

Problems

37. From the graphs, we estimate $g(2) \approx 1.6$, $g'(2) \approx -0.5$, and $f'(1.6) \approx 0.8$. Thus, by the chain rule,

$$h'(2) = f'(g(2)) \cdot g'(2) \approx f'(1.6) \cdot g'(2) \approx 0.8(-0.5) = -0.4.$$

41. Let

$$g(x) = \log x.$$

Then

$$10^{g(x)} = x.$$

Differentiating,

$$(\ln 10)[10^{g(x)}]g'(x) = 1$$

$$g'(x) = \dfrac{1}{(\ln 10)[10^{g(x)}]}$$

$$g'(x) = \dfrac{1}{(\ln 10)x}.$$

45. (a)

$$f'(x) = \frac{1}{1+x^2} + \frac{1}{1+\frac{1}{x^2}} \cdot \left(-\frac{1}{x^2}\right)$$

$$= \frac{1}{1+x^2} + \left(-\frac{1}{x^2+1}\right)$$

$$= \frac{1}{1+x^2} - \frac{1}{1+x^2}$$

$$= 0$$

(b) f is a constant function. Checking at a few values of x,

Table 3.2

x	$\arctan x$	$\arctan x^{-1}$	$f(x) = \arctan x + \arctan x^{-1}$
1	0.785392	0.7853982	1.5707963
2	1.1071487	0.4636476	1.5707963
3	1.2490458	0.3217506	1.5707963

49. Since the chain rule gives $h'(x) = n'(m(x))m'(x) = 1$ we must find values a and x such that $a = m(x)$ and $n'(a)m'(x) = 1$.

Calculating slopes from the graph of n gives

$$n'(a) = \begin{cases} 1 & \text{if } 0 < a < 50 \\ 1/2 & \text{if } 50 < a < 100. \end{cases}$$

Calculating slopes from the graph of m gives

$$m'(x) = \begin{cases} -2 & \text{if } 0 < x < 50 \\ 2 & \text{if } 50 < x < 100. \end{cases}$$

The only values of the derivative n' are 1 and 1/2 and the only values of the derivative m' are 2 and -2. In order to have $n'(a)m'(x) = 1$ we must therefore have $n'(a) = 1/2$ and $m'(x) = 2$. Thus $50 < a < 100$ and $50 < x < 100$.

Now $a = m(x)$ and from the graph of m we see that $50 < m(x) < 100$ for $0 < x < 25$ or $75 < x < 100$.

The two conditions on x we have found are both satisfied when $75 < x < 100$. Thus $h'(x) = 1$ for all x in the interval $75 < x < 100$. The question asks for just one of these x values, for example $x = 80$.

53. Since the point $(2, 5)$ is on the curve, we know $f(2) = 5$. The point $(2.1, 5.3)$ is on the tangent line, so

$$\text{Slope tangent} = \frac{5.3 - 5}{2.1 - 2} = \frac{0.3}{0.1} = 3.$$

Thus, $f'(2) = 3$. Since g is the inverse function of f and $f(2) = 5$, we know $f^{-1}(5) = 2$, so $g(5) = 2$.

Differentiating, we have

$$g'(2) = \frac{1}{f'(g(5))} = \frac{1}{f'(2)} = \frac{1}{3}.$$

57. To find $(f^{-1})'(3)$, we first look in the table to find that $3 = f(9)$, so $f^{-1}(3) = 9$. Thus,

$$(f^{-1})'(3) = \frac{1}{f'(f^{-1}(3))} = \frac{1}{f'(9)} = \frac{1}{5}.$$

61. We must have

$$(f^{-1})'(5) = \frac{1}{f'(f^{-1}(5))} = \frac{1}{f'(10)} = \frac{1}{8}.$$

Solutions for Section 3.7

Exercises

1. We differentiate implicitly both sides of the equation with respect to x.

$$2x + 2y\frac{dy}{dx} = 0 \,,$$

$$\frac{dy}{dx} = -\frac{2x}{2y} = -\frac{x}{y}.$$

5. We differentiate implicitly both sides of the equation with respect to x.

$$x^{1/2} = 5y^{1/2}$$
$$\frac{1}{2}x^{-1/2} = \frac{5}{2}y^{-1/2}\frac{dy}{dx}$$
$$\frac{dy}{dx} = \frac{\frac{1}{2}x^{-1/2}}{\frac{5}{2}y^{-1/2}} = \frac{1}{5}\sqrt{\frac{y}{x}} = \frac{1}{25}.$$

We can also obtain this answer by realizing that the original equation represents part of the line $x = 25y$ which has slope $1/25$.

9.

$$2ax - 2by\frac{dy}{dx} = 0$$
$$\frac{dy}{dx} = \frac{-2ax}{-2by} = \frac{ax}{by}$$

13. Using the relation $\cos^2 y + \sin^2 y = 1$, the equation becomes:
$1 = y + 2$ or $y = -1$. Hence, $\frac{dy}{dx} = 0$.

17. We differentiate implicitly both sides of the equation with respect to x.

$$(x - a)^2 + y^2 = a^2$$

$$2(x - a) + 2y\frac{dy}{dx} = 0$$
$$2y\frac{dy}{dx} = 2a - 2x$$
$$\frac{dy}{dx} = \frac{2a - 2x}{2y} = \frac{a - x}{y}.$$

21. Differentiating with respect to x gives
$$3x^2 + 2xy' + 2y + 2yy' = 0$$

so that
$$y' = -\frac{3x^2 + 2y}{2x + 2y}$$

At the point $(1, 1)$ the slope is $-\frac{5}{4}$.

25. First, we must find the slope of the tangent, $\left.\frac{dy}{dx}\right|_{(4,2)}$. Implicit differentiation yields:
$$2y\frac{dy}{dx} = \frac{2x(xy - 4) - x^2\left(x\frac{dy}{dx} + y\right)}{(xy - 4)^2}.$$

Given the complexity of the above equation, we first want to substitute 4 for x and 2 for y (the coordinates of the point where we are constructing our tangent line), then solve for $\dfrac{dy}{dx}$. Substitution yields:

$$2 \cdot 2 \frac{dy}{dx} = \frac{(2 \cdot 4)(4 \cdot 2 - 4) - 4^2 \left(4\frac{dy}{dx} + 2\right)}{(4 \cdot 2 - 4)^2} = \frac{8(4) - 16(4\frac{dy}{dx} + 2)}{16} = -4\frac{dy}{dx}.$$

$$4\frac{dy}{dx} = -4\frac{dy}{dx},$$

Solving for $\dfrac{dy}{dx}$, we have:

$$\frac{dy}{dx} = 0.$$

The tangent is a horizontal line through $(4, 2)$, hence its equation is $y = 2$.

Problems

29. (a) Taking derivatives implicitly, we get

$$\frac{2}{25}x + \frac{2}{9}y\frac{dy}{dx} = 0$$

$$\frac{dy}{dx} = \frac{-9x}{25y}.$$

(b) The slope is not defined anywhere along the line $y = 0$. This ellipse intersects that line in two places, $(-5, 0)$ and $(5, 0)$. (These are the "ends" of the ellipse where the tangent is vertical.)

33. (a) Differentiating both sides of the equation with respect to P gives

$$\frac{d}{dP}\left(\frac{4f^2P}{1 - f^2}\right) = \frac{dK}{dP} = 0.$$

By the product rule

$$\frac{d}{dP}\left(\frac{4f^2P}{1 - f^2}\right) = \frac{d}{dP}\left(\frac{4f^2}{1 - f^2}\right)P + \left(\frac{4f^2}{1 - f^2}\right) \cdot 1$$

$$= \left(\frac{(1 - f^2)(8f) - 4f^2(-2f)}{(1 - f^2)^2}\right)\frac{df}{dP}P + \left(\frac{4f^2}{1 - f^2}\right)$$

$$= \left(\frac{8f}{(1 - f^2)^2}\right)\frac{df}{dP}P + \left(\frac{4f^2}{1 - f^2}\right) = 0.$$

So

$$\frac{df}{dP} = \frac{-4f^2/(1 - f^2)}{8fP/(1 - f^2)^2} = \frac{-1}{2P}f(1 - f^2).$$

(b) Since f is a fraction of a gas, $0 \leq f \leq 1$. Also, in the equation relating f and P we can't have $f = 0$, since that would imply $K = 0$, and we can't have $f = 1$, since the left side is undefined there. So $0 < f < 1$. Thus $1 - f^2 > 0$. Also, pressure can't be negative, and from the equation relating f and P, we see that P can't be zero either, so $P > 0$. Therefore $df/dP = -(1/2P)f(1 - f^2) < 0$ always. This means that at larger pressures less of the gas decomposes.

Solutions for Section 3.8

Exercises

1. Using the chain rule, $\dfrac{d}{dx}(\cosh(2x)) = (\sinh(2x)) \cdot 2 = 2\sinh(2x)$.

5. Using the chain rule,

$$\frac{d}{dt}\left(\cosh^2 t\right) = 2\cosh t \cdot \sinh t.$$

9. Using the chain rule twice,

$$\frac{d}{dy}\left(\sinh\left(\sinh(3y)\right)\right) = \cosh\left(\sinh(3y)\right) \cdot \cosh(3y) \cdot 3$$

$$= 3\cosh(3y) \cdot \cosh\left(\sinh(3y)\right).$$

13. Substituting $-x$ for x in the formula for $\sinh x$ gives

$$\sinh(-x) = \frac{e^{-x} - e^{-(-x)}}{2} = \frac{e^{-x} - e^{x}}{2} = -\frac{e^{x} - e^{-x}}{2} = -\sinh x.$$

Problems

17. The graph of $\sinh x$ in the text suggests that

$$\text{As } x \to \infty, \quad \sinh x \to \frac{1}{2}e^{x}.$$

$$\text{As } x \to -\infty, \quad \sinh x \to -\frac{1}{2}e^{-x}.$$

Using the facts that

$$\text{As } x \to \infty, \quad e^{-x} \to 0,$$

$$\text{As } x \to -\infty, \quad e^{x} \to 0,$$

we can obtain the same results analytically:

$$\text{As } x \to \infty, \quad \sinh x = \frac{e^{x} - e^{-x}}{2} \to \frac{1}{2}e^{x}.$$

$$\text{As } x \to -\infty, \quad \sinh x = \frac{e^{x} - e^{-x}}{2} \to -\frac{1}{2}e^{-x}.$$

21. Recall that

$$\sinh A = \frac{1}{2}(e^{A} - e^{-A}) \quad \text{and} \quad \cosh A = \frac{1}{2}(e^{A} + e^{-A}).$$

Now substitute, expand and collect terms:

$$\cosh A \cosh B + \sinh B \sinh A = \frac{1}{2}(e^{A} + e^{-A}) \cdot \frac{1}{2}(e^{B} + e^{-B}) + \frac{1}{2}(e^{B} - e^{-B}) \cdot \frac{1}{2}(e^{A} - e^{-A})$$

$$= \frac{1}{4}\left(e^{A+B} + e^{A-B} + e^{-A+B} + e^{-(A+B)}\right.$$

$$\left. + e^{B+A} - e^{B-A} - e^{-B+A} + e^{-A-B}\right)$$

$$= \frac{1}{2}\left(e^{A+B} + e^{-(A+B)}\right)$$

$$= \cosh(A + B).$$

25. Note that

$$\frac{\sinh kx}{\cosh 2x} = \frac{e^{kx} - e^{-kx}}{e^{2x} + e^{-2x}}$$

$$= \frac{e^{2x}\left(e^{(k-2)x} - e^{-(k+2)x}\right)}{e^{2x}\left(1 + e^{-4x}\right)}$$

$$= \frac{e^{(k-2)x} - e^{-(k+2)x}}{1 + e^{-4x}}.$$

If $k = 2$, then the limit as $x \to \infty$ is 1.
If $|k| > 2$, then the limit as $x \to \infty$ does not exist.
If $|k| < 2$, then the limit as $x \to \infty$ is 0.

29. (a) Since the cosh function is even, the height, y, is the same at $x = -T/w$ and $x = T/w$. The height at these endpoints is

$$y = \frac{T}{w} \cosh \left(\frac{w}{T} \cdot \frac{T}{w} \right) = \frac{T}{w} \cosh 1 = \frac{T}{w} \left(\frac{e^1 + e^{-1}}{2} \right).$$

At the lowest point, $x = 0$, and the height is

$$y = \frac{T}{w} \cosh 0 = \frac{T}{w}.$$

Thus the "sag" in the cable is given by

$$\text{Sag} = \frac{T}{w} \left(\frac{e + e^{-1}}{2} \right) - \frac{T}{w} = \frac{T}{w} \left(\frac{e + e^{-1}}{2} - 1 \right) \approx 0.54 \frac{T}{w}.$$

(b) To show that the differential equation is satisfied, take derivatives

$$\frac{dy}{dx} = \frac{T}{w} \cdot \frac{w}{T} \sinh \left(\frac{wx}{T} \right) = \sinh \left(\frac{wx}{T} \right)$$

$$\frac{d^2y}{dx^2} = \frac{w}{T} \cosh \left(\frac{wx}{T} \right).$$

Therefore, using the fact that $1 + \sinh^2 a = \cosh^2 a$ and that cosh is always positive, we have:

$$\frac{w}{T} \sqrt{1 + \left(\frac{dy}{dx} \right)^2} = \frac{w}{T} \sqrt{1 + \sinh^2 \left(\frac{wx}{T} \right)} = \frac{w}{T} \sqrt{\cosh^2 \left(\frac{wx}{T} \right)}$$

$$= \frac{w}{T} \cosh \left(\frac{wx}{T} \right).$$

So

$$\frac{w}{T} \sqrt{1 + \left(\frac{dy}{dx} \right)^2} = \frac{d^2y}{dx^2}.$$

Solutions for Section 3.9

Exercises

1. With $f(x) = \sqrt{1+x}$, the chain rule gives $f'(x) = 1/(2\sqrt{1+x})$, so $f(0) = 1$ and $f'(0) = 1/2$. Therefore the tangent line approximation of f near $x = 0$,

$$f(x) \approx f(0) + f'(0)(x - 0),$$

becomes

$$\sqrt{1+x} \approx 1 + \frac{x}{2}.$$

This means that, near $x = 0$, the function $\sqrt{1+x}$ can be approximated by its tangent line $y = 1 + x/2$. (See Figure 3.1.)

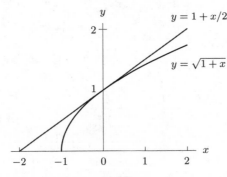

Figure 3.1

5. Let $f(x) = e^{-x}$. Then $f'(x) = -e^{-x}$. So $f(0) = 1$, $f'(0) = -e^0 = -1$. Therefore, $e^{-x} \approx f(0) + f'(0)x = 1 - x$.

Problems

9. (a) Since

$$\frac{d}{dx}(\cos x) = -\sin x,$$

the slope of the tangent line is $-\sin(\pi/4) = -1/\sqrt{2}$. Since the tangent line passes through the point $(\pi/4, \cos(\pi/4)) = (\pi/4, 1/\sqrt{2})$, its equation is

$$y - \frac{1}{\sqrt{2}} = -\frac{1}{\sqrt{2}}\left(x - \frac{\pi}{4}\right)$$

$$y = -\frac{1}{\sqrt{2}}x + \frac{1}{\sqrt{2}}\left(\frac{\pi}{4} + 1\right).$$

Thus, the tangent line approximation to $\cos x$ is

$$\cos x \approx -\frac{1}{\sqrt{2}}x + \frac{1}{\sqrt{2}}\left(\frac{\pi}{4} + 1\right).$$

(b) From Figure 3.2, we see that the tangent line approximation is an overestimate.

(c) From Figure 3.2, we see that the maximum error for $0 \le x \le \pi/2$ is either at $x = 0$ or at $x = \pi/2$. The error can either be estimated from the graph, or as follows. At $x = 0$,

$$|\text{Error}| = \left|\cos 0 - \frac{1}{\sqrt{2}}\left(\frac{\pi}{4} + 1\right)\right| = 0.262 < 0.3.$$

At $x = \pi/2$,

$$|\text{Error}| = \left|\cos\frac{\pi}{2} + \frac{1}{\sqrt{2}}\frac{\pi}{2} - \frac{1}{\sqrt{2}}\left(\frac{\pi}{4} + 1\right)\right| = 0.152 < 0.2.$$

Thus, for $0 \le x \le \pi/2$, we have

$$|\text{Error}| < 0.3.$$

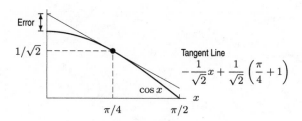

Figure 3.2

13. We have $f(1) = 1$ and $f'(1) = 4$. Thus

$$E(x) = x^4 - (1 + 4(x - 1)).$$

Values of $E(x)/(x - 1)$ near $x = 1$ are in Table 3.3.

Table 3.3

x	1.1	1.01	1.001
$E(x)/(x - 1)$	0.641	0.060401	0.006004

From the table, we can see that

$$\frac{E(x)}{(x - 1)} \approx 6(x - 1),$$

so $k = 6$ and

$$E(x) \approx 6(x - 1)^2.$$

In addition, $f''(1) = 12$, so

$$E(x) \approx 6(x-1)^2 = \frac{f''(1)}{2}(x-1)^2.$$

The same result can be obtained by rewriting the function x^4 using $x = 1 + (x-1)$ and expanding:

$$x^4 = (1 + (x-1))^4 = 1 + 4(x-1) + 6(x-1)^2 + 4(x-1)^3 + (x-1)^4.$$

Thus,

$$E(x) = x^4 - (1 + 4(x-1)) = 6(x-1)^2 + 4(x-1)^3 + (x-1)^4.$$

For x near 1, the value of $x-1$ is small, so we ignore powers of $x-1$ higher than the first, giving

$$E(x) \approx 6(x-1)^2.$$

17. We have $f(1) = 0$ and $f'(1) = 1$. Thus

$$E(x) = \ln x - (x-1).$$

Values of $E(x)/(x-1)$ near $x = 1$ are in Table 3.4.

Table 3.4

x	1.1	1.01	1.001
$E(x)/(x-1)$	-0.047	-0.0050	-0.00050

From the table, we see that

$$\frac{E(x)}{(x-1)} \approx -0.5(x-1),$$

so $k = -1/2$ and

$$E(x) \approx -\frac{1}{2}(x-1)^2.$$

In addition, $f''(1) = -1$, so

$$E(x) \approx -\frac{1}{2}(x-1)^2 = \frac{f''(1)}{2}(x-1)^2.$$

21. (a) Suppose

$$g = f(r) = \frac{GM}{r^2}.$$

Then

$$f'(r) = \frac{-2GM}{r^3}.$$

So

$$f(r + \Delta r) \approx f(r) - \frac{2GM}{r^3}(\Delta r).$$

Since $f(r + \Delta r) - f(r) = \Delta g$, and $g = GM/r^2$, we have

$$\Delta g \approx -2\frac{GM}{r^3}(\Delta r) = -2g\frac{\Delta r}{r}.$$

(b) The negative sign tells us that the acceleration due to gravity decreases as the distance from the center of the earth increases.

(c) The fractional change in g is given by

$$\frac{\Delta g}{g} \approx -2\frac{\Delta r}{r}.$$

So, since $\Delta r = 4.315$ km and $r = 6400$ km, we have

$$\frac{\Delta g}{g} \approx -2\left(\frac{4.315}{6400}\right) = -0.00135 = -0.135\%.$$

25. (a) Since f' is decreasing, $f'(5)$ is larger.
    (b) Since f' is decreasing, its derivative, f'', is negative. Thus, $f''(5)$ is negative, so 0 is larger.
    (c) Since $f''(x)$ is negative for all x, the graph of f is concave down. Thus the graph of $f(x)$ is below its tangent line. From Figure 3.3, we see that $f(5 + \Delta x)$ is below $f(5) + f'(5)\Delta x$. Thus, $f(5) + f'(5)\Delta x$ is larger.

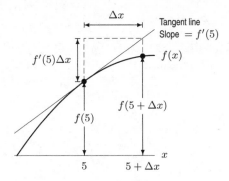

Figure 3.3

Solutions for Section 3.10

Exercises

1. False. The derivative, $f'(x)$, is not equal to zero everywhere, because the function is not continuous at integral values of x, so $f'(x)$ does not exist there. Thus, the Constant Function Theorem does not apply.

5. False. Let $f(x) = x^3$ on $[-1, 1]$. Then $f(x)$ is increasing but $f'(x) = 0$ for $x = 0$.

9. No. This function does not satisfy the hypotheses of the Mean Value Theorem, as it is not continuous.
    However, the function has a point c such that

$$f'(c) = \frac{f(b) - f(a)}{b - a}.$$

Thus, this satisfies the conclusion of the theorem.

Problems

13. The Decreasing Function Theorem is: Suppose that f is continuous on $[a, b]$ and differentiable on (a, b). If $f'(x) < 0$ on (a, b), then f is decreasing on $[a, b]$. If $f'(x) \leq 0$ on (a, b), then f is nonincreasing on $[a, b]$.
    To prove the theorem, we note that if f is decreasing then $-f$ is increasing and vice-versa. Similarly, if f is nonincreasing, then $-f$ is nondecreasing. Thus if $f'(x) < 0$, then $-f'(x) > 0$, so $-f$ is increasing, which means f is decreasing. And if $f'(x) \leq 0$, then $-f'(x) \geq 0$, so $-f$ is nondecreasing, which means f is nonincreasing.

17. By the Mean Value Theorem, Theorem 3.7, there is a number c, with $0 < c < 1$, such that

$$f'(c) = \frac{f(1) - f(0)}{1 - 0}.$$

Since $f(1) - f(0) > 0$, we have $f'(c) > 0$.
    Alternatively if $f'(c) \leq 0$ for all c in $(0, 1)$, then by the Increasing Function Theorem, $f(0) \geq f(1)$.

21. Let $h(x) = f(x) - g(x)$. Then $h'(x) = f'(x) - g'(x) = 0$ for all x in (a, b). Hence, by the Constant Function Theorem, there is a constant C such that $h(x) = C$ on (a, b). Thus $f(x) = g(x) + C$.

Solutions for Chapter 3 Review

Exercises

1. $f'(t) = \dfrac{d}{dt}\left(2te^t - \dfrac{1}{\sqrt{t}}\right) = 2e^t + 2te^t + \dfrac{1}{2t^{3/2}}$.

5. Since $h(\theta) = \theta(\theta^{-1/2} - \theta^{-2}) = \theta\theta^{-1/2} - \theta\theta^{-2} = \theta^{1/2} - \theta^{-1}$, we have $h'(\theta) = \dfrac{1}{2}\theta^{-1/2} + \theta^{-2}$.

9. $y' = 0$

13. $s'(\theta) = \dfrac{d}{d\theta}\sin^2(3\theta - \pi) = 6\cos(3\theta - \pi)\sin(3\theta - \pi)$.

17. $g'(x) = \dfrac{d}{dx}\left(x^{\frac{1}{2}} + x^{-1} + x^{-\frac{3}{2}}\right) = \dfrac{1}{2}x^{-\frac{1}{2}} - x^{-2} - \dfrac{3}{2}x^{-\frac{5}{2}}$.

21. Using the chain rule, we get:

$$m'(n) = \cos(e^n) \cdot (e^n)$$

25. $\dfrac{d}{dx}xe^{\tan x} = e^{\tan x} + xe^{\tan x}\dfrac{1}{\cos^2 x}$.

29. $h(x) = ax \cdot \ln e = ax$, so $h'(x) = a$.

33. Using the product rule gives

$$H'(t) = 2ate^{-ct} - c(at^2 + b)e^{-ct}$$
$$= (-cat^2 + 2at - bc)e^{-ct}.$$

37. Using the quotient rule gives

$$w'(r) = \dfrac{2ar(b + r^3) - 3r^2(ar^2)}{(b + r^3)^2}$$
$$= \dfrac{2abr - ar^4}{(b + r^3)^2}.$$

41. Since $g(w) = 5(a^2 - w^2)^{-2}$, $g'(w) = -10(a^2 - w^2)^{-3}(-2w) = \dfrac{20w}{(a^2 - w^2)^3}$

45. Using the quotient rule gives

$$g'(t) = \dfrac{\left(\frac{k}{kt} + 1\right)(\ln(kt) - t) - (\ln(kt) + t)\left(\frac{k}{kt} - 1\right)}{(\ln(kt) - t)^2}$$

$$g'(t) = \dfrac{\left(\frac{1}{t} + 1\right)(\ln(kt) - t) - (\ln(kt) + t)\left(\frac{1}{t} - 1\right)}{(\ln(kt) - t)^2}$$

$$g'(t) = \dfrac{\ln(kt)/t - 1 + \ln(kt) - t - \ln(kt)/t - 1 + \ln(kt) + t}{(\ln(kt) - t)^2}$$

$$g'(t) = \dfrac{2\ln(kt) - 2}{(\ln(kt) - t)^2}.$$

49. $g'(x) = -\dfrac{1}{2}(5x^4 + 2)$.

53. $g'(x) = \dfrac{d}{dx}(2x - x^{-1/3} + 3^x - e) = 2 + \dfrac{1}{3x^{\frac{4}{3}}} + 3^x \ln 3$.

57. $r'(\theta) = \dfrac{d}{d\theta}\sin[(3\theta - \pi)^2] = \cos[(3\theta - \pi)^2] \cdot 2(3\theta - \pi) \cdot 3 = 6(3\theta - \pi)\cos[(3\theta - \pi)^2]$.

61. $f'(x) = \dfrac{d}{dx}(2 - 4x - 3x^2)(6x^e - 3\pi) = (-4 - 6x)(6x^e - 3\pi) + (2 - 4x - 3x^2)(6ex^{e-1})$.

65.

$$h'(x) = \left(-\frac{1}{x^2} + \frac{2}{x^3}\right)(2x^3 + 4) + \left(\frac{1}{x} - \frac{1}{x^2}\right)(6x^2)$$

$$= -2x + 4 - \frac{4}{x^2} + \frac{8}{x^3} + 6x - 6$$

$$= 4x - 2 - 4x^{-2} + 8x^{-3}$$

69. We wish to find the slope $m = dy/dx$. To do this, we can implicitly differentiate the given formula in terms of x:

$$x^2 + 3y^2 = 7$$

$$2x + 6y\frac{dy}{dx} = \frac{d}{dx}(7) = 0$$

$$\frac{dy}{dx} = \frac{-2x}{6y} = \frac{-x}{3y}.$$

Thus, at $(2, -1)$, $m = -(2)/3(-1) = 2/3$.

Problems

73. Since $r(x) = s(t(x))$, the chain rule gives $r'(x) = s'(t(x)) \cdot t'(x)$. Thus,

$$r'(0) = s'(t(0)) \cdot t'(0) \approx s'(2) \cdot (-2) \approx (-2)(-2) = 4.$$

Note that since $t(x)$ is a linear function whose slope looks like -2 from the graph, $t'(x) \approx -2$ everywhere. To find $s'(2)$, draw a line tangent to the curve at the point $(2, s(2))$, and estimate the slope.

77. We have $r(1) = s(t(1)) \approx s(0) \approx 2$. By the chain rule, $r'(x) = s'(t(x)) \cdot t'(x)$, so

$$r'(1) = s'(t(1)) \cdot t'(1) \approx s'(0) \cdot (-2) \approx 2(-2) = -4.$$

Thus the equation of the tangent line is

$$y - 2 = -4(x - 1)$$

$$y = -4x + 6.$$

Note that since $t(x)$ is a linear function whose slope looks like -2 from the graph, $t'(x) \approx -2$ everywhere. To find $s'(0)$, draw a line tangent to the curve at the point $(0, s(0))$, and estimate the slope.

81. Since W is proportional to r^3, we have $W = kr^3$ for some constant k. Thus, $dW/dr = k(3r^2) = 3kr^2$. Thus, dW/dr is proportional to r^2.

85. (a) $f(x) = x^2 - 4g(x)$
$f(2) = 4 - 4(3) = -8$
$f'(2) = 20$
Thus, we have a point $(2, -8)$ and slope $m = 20$. This gives

$$-8 = 2(20) + b$$

$$b = -48, \quad \text{so}$$

$$y = 20x - 48.$$

(b) $f(x) = \dfrac{x}{g(x)}$
$f(2) = \dfrac{2}{3}$
$f'(2) = \dfrac{11}{9}$
Thus, we have point $(2, \frac{2}{3})$ and slope $m = \frac{11}{9}$. This gives

$$\frac{2}{3} = (\frac{11}{9})(2) + b$$

$$b = \frac{2}{3} - \frac{22}{9} = \frac{-16}{9}, \quad \text{so}$$

$$y = \frac{11}{9}x - \frac{16}{9}.$$

(c) $f(x) = x^2 g(x)$
$f(2) = 4 \cdot g(2) = 4(3) = 12$
$f'(2) = -4$
Thus, we have point $(2, 12)$ and slope $m = -4$. This gives

$$12 = 2(-4) + b$$
$$b = 20, \quad \text{so}$$
$$y = -4x + 20.$$

(d) $f(x) = (g(x))^2$
$f(2) = (g(2))^2 = (3)^2 = 9$
$f'(2) = -24$
Thus, we have point $(2, 9)$ and slope $m = -24$. This gives

$$9 = 2(-24) + b$$
$$b = 57, \quad \text{so}$$
$$y = -24x + 57.$$

(e) $f(x) = x \sin(g(x))$
$f(2) = 2 \sin(g(2)) = 2 \sin 3$
$f'(2) = \sin 3 - 8 \cos 3$
We will use a decimal approximation for $f(2)$ and $f'(2)$, so the point $(2, 2\sin 3) \approx (2, 0.28)$ and $m \approx 8.06$. Thus,

$$0.28 = 2(8.06) + b$$
$$b = -15.84, \quad \text{so}$$
$$y = 8.06x - 15.84.$$

(f) $f(x) = x^2 \ln g(x)$
$f(2) = 4 \ln g(2) = 4 \ln 3 \approx 4.39$
$f'(2) = 4 \ln 3 - \dfrac{16}{3} \approx -0.94.$
Thus, we have point $(2, 4.39)$ and slope $m = -0.94$. This gives

$$4.39 = 2(-0.94) + b$$
$$b = 6.27, \quad \text{so}$$
$$y = -0.94x + 6.27.$$

89. The curves meet when $1 - x^3/3 = x - 1$, that is when $x^3 + 3x - 6 = 0$. So the roots of this equation give us the x-coordinates of the intersection point. By numerical methods, we see there is one solution near $x = 1.3$. See Figure 3.4. Let

$$y_1(x) = 1 - \frac{x^3}{3} \quad \text{and} \quad y_2(x) = x - 1.$$

So we have

$$y_1' = -x^2 \quad \text{and} \quad y_2' = 1.$$

However, $y_2'(x) = +1$, so if the curves are to be perpendicular when they cross, then y_1' must be -1. Since $y_1' = -x^2$, $y_1' = -1$ only at $x = \pm 1$ which is not the point of intersection. The curves are therefore not perpendicular when they cross.

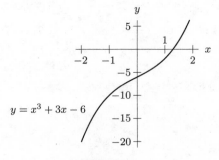

$y = x^3 + 3x - 6$

Figure 3.4

93. Using the definition of $\cosh x$ and $\sinh x$, we have $\cosh x^2 = \dfrac{e^{x^2} + e^{-x^2}}{2}$ and $\sinh x^2 = \dfrac{e^{x^2} - e^{-x^2}}{2}$. Therefore

$$
\begin{aligned}
\lim_{x \to -\infty} \frac{\sinh(x^2)}{\cosh(x^2)} &= \lim_{x \to -\infty} \frac{e^{x^2} - e^{-x^2}}{e^{x^2} + e^{-x^2}} \\
&= \lim_{x \to -\infty} \frac{e^{x^2}(1 - e^{-2x^2})}{e^{x^2}(1 + e^{-2x^2})} \\
&= \lim_{x \to -\infty} \frac{1 - e^{-2x^2}}{1 + e^{-2x^2}} \\
&= 1.
\end{aligned}
$$

97. (a)

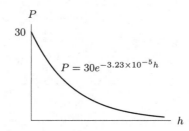

$$P = 30e^{-3.23 \times 10^{-5}h}$$

(b)

$$\frac{dP}{dh} = 30e^{-3.23 \times 10^{-5}h}(-3.23 \times 10^{-5})$$

so

$$\left.\frac{dP}{dh}\right|_{h=0} = -30(3.23 \times 10^{-5}) = -9.69 \times 10^{-4}$$

Hence, at $h = 0$, the slope of the tangent line is -9.69×10^{-4}, so the equation of the tangent line is

$$
\begin{aligned}
y - 30 &= (-9.69 \times 10^{-4})(h - 0) \\
y &= (-9.69 \times 10^{-4})h + 30.
\end{aligned}
$$

(c) The rule of thumb says

$$\text{Drop in pressure from sea level to height } h = \frac{h}{1000}$$

But since the pressure at sea level is 30 inches of mercury, this drop in pressure is also $(30 - P)$, so

$$30 - P = \frac{h}{1000}$$

giving

$$P = 30 - 0.001h.$$

(d) The equations in (b) and (c) are almost the same: both have P intercepts of 30, and the slopes are almost the same $(9.69 \times 10^{-4} \approx 0.001)$. The rule of thumb calculates values of P which are very close to the tangent lines, and therefore yields values very close to the curve.

(e) The tangent line is slightly below the curve, and the rule of thumb line, having a slightly more negative slope, is slightly below the tangent line (for $h > 0$). Thus, the rule of thumb values are slightly smaller.

101. (a) Since $\lim_{t \to \infty} e^{-0.1t} = 0$, we see that $\lim_{t \to \infty} \dfrac{1000000}{1 + 5000e^{-0.1t}} = 1000000$. Thus, in the long run, close to 1,000,000 people will have had the disease. This can be seen in the figure below.

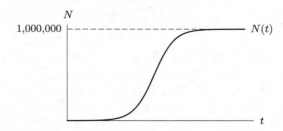

(b) The rate at which people fall sick is given by the first derivative $N'(t)$.
$N'(t) \approx \frac{\Delta N}{\Delta t}$, where $\Delta t = 1$ day.

$$N'(t) = \frac{500,000,000}{e^{0.1t}(1 + 5000e^{-0.1t})^2} = \frac{500,000,000}{e^{0.1t} + 25,000,000e^{-0.1t} + 10^4}$$

In Figure 3.5, we see that the maximum value of $N'(t)$ is approximately 25,000. Therefore the maximum number of people to fall sick on any given day is 25,000. Thus there are no days on which a quarter million or more get sick.

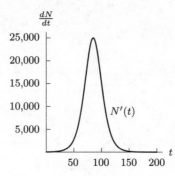

Figure 3.5

105. (a) If $y = \ln x$, then

$$y' = \frac{1}{x}$$
$$y'' = -\frac{1}{x^2}$$
$$y''' = \frac{2}{x^3}$$
$$y'''' = -\frac{3 \cdot 2}{x^4}$$

and so

$$y^{(n)} = (-1)^{n+1}(n-1)!x^{-n}.$$

(b) If $y = xe^x$, then

$$y' = xe^x + e^x$$
$$y'' = xe^x + 2e^x$$
$$y''' = xe^x + 3e^x$$

so that

$$y^{(n)} = xe^x + ne^x.$$

(c) If $y = e^x \cos x$, then

$$y' = e^x(\cos x - \sin x)$$
$$y'' = -2e^x \sin x$$
$$y''' = e^x(-2\cos x - 2\sin x)$$
$$y^{(4)} = -4e^x \cos x$$
$$y^{(5)} = e^x(-4\cos x + 4\sin x)$$
$$y^{(6)} = 8e^x \sin x.$$

Combining these results we get

$$\begin{array}{ll} y^{(n)} = (-4)^{(n-1)/4}e^x(\cos x - \sin x), & n = 4m+1, \quad m = 0, 1, 2, 3, \ldots \\ y^{(n)} = -2(-4)^{(n-2)/4}e^x \sin x, & n = 4m+2, \quad m = 0, 1, 2, 3, \ldots \\ y^{(n)} = -2(-4)^{(n-3)/4}e^x(\cos x + \sin x), & n = 4m+3, \quad m = 0, 1, 2, 3, \ldots \\ y^{(n)} = (-4)^{(n/4)}e^x \cos x, & n = 4m, \quad m = 1, 2, 3, \ldots . \end{array}$$

CAS Challenge Problems

109. (a) A CAS gives $f'(x) = 1$.

(b) By the chain rule,

$$f'(x) = \cos(\arcsin x) \cdot \frac{1}{\sqrt{1 - x^2}}.$$

Now $\cos t = \pm\sqrt{1 - \sin^2 t}$. Furthermore, if $-\pi/2 \leq t \leq \pi/2$ then $\cos t \geq 0$, so we take the positive square root and get $\cos t = \sqrt{1 - \sin^2 t}$. Since $-\pi/2 \leq \arcsin x \leq \pi/2$ for all x in the domain of arcsin, we have

$$\cos(\arcsin x) = \sqrt{1 - (\sin(\arcsin x))^2} = \sqrt{1 - x^2},$$

so

$$\frac{d}{dx}\sin(\arcsin(x)) = \sqrt{1 - x^2} \cdot \frac{1}{\sqrt{1 - x^2}} = 1.$$

(c) Since $\sin(\arcsin(x)) = x$, its derivative is 1.

CHECK YOUR UNDERSTANDING

1. True. Since $d(x^n)/dx = nx^{n-1}$, the derivative of a power function is a power function, so the derivative of a polynomial is a polynomial.

5. True. Since $f'(x)$ is the limit

$$f'(x) = \lim_{h \to 0} \frac{f(x + h) - f(x)}{h},$$

the function f must be defined for all x.

9. True; differentiating the equation with respect to x, we get

$$2y\frac{dy}{dx} + y + x\frac{dy}{dx} = 0.$$

Solving for dy/dx, we get that

$$\frac{dy}{dx} = \frac{-y}{2y + x}.$$

Thus dy/dx exists where $2y + x \neq 0$. Now if $2y + x = 0$, then $x = -2y$. Substituting for x in the original equation, $y^2 + xy - 1 = 0$, we get

$$y^2 - 2y^2 - 1 = 0.$$

This simplifies to $y^2 + 1 = 0$, which has no solutions. Thus dy/dx exists everywhere.

13. False. Since $(\sinh x)' = \cosh x > 0$, the function $\sinh x$ is increasing everywhere so can never repeat any of its values.

17. False; the fourth derivative of $\cos t + C$, where C is any constant, is indeed $\cos t$. But any function of the form $\cos t + p(t)$, where $p(t)$ is a polynomial of degree less than or equal to 3, also has its fourth derivative equal to $\cos t$. So $\cos t + t^2$ will work.

21. False; for example, if both f and g are constant functions, then the derivative of $f(g(x))$ is zero, as is the derivative of $f(x)$. Another example is $f(x) = 5x + 7$ and $g(x) = x + 2$.

25. False. Let $f(x) = e^{-x}$ and $g(x) = x^2$. Let $h(x) = f(g(x)) = e^{-x^2}$. Then $h'(x) = -2xe^{-x^2}$ and $h''(x) = (-2 + 4x^2)e^{-x^2}$. Since $h''(0) < 0$, clearly h is not concave up for all x.

29. False. For example, let $f(x) = x + 5$, and $g(x) = 2x - 3$. Then $f'(x) \leq g'(x)$ for all x, but $f(0) > g(0)$.

33. Let f be defined by

$$f(x) = \begin{cases} x & \text{if } 0 \leq x < 2 \\ 19 & \text{if } x = 2 \end{cases}$$

Then f is differentiable on $(0, 2)$ and $f'(x) = 1$ for all x in $(0, 2)$. Thus there is no c in $(0, 2)$ such that

$$f'(c) = \frac{f(2) - f(0)}{2 - 0} = \frac{19}{2}.$$

The reason that this function does not satisfy the conclusion of the Mean Value Theorem is that it is not continuous at $x = 2$.

CHAPTER FOUR

Solutions for Section 4.1

Exercises

1. There are many possible answers. One possible graph is shown in Figure 4.1.

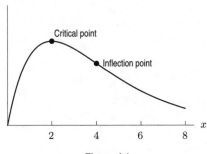

Figure 4.1

5. From the graph of $f(x)$ in the figure below, we see that the function must have two inflection points. We calculate $f'(x) = 4x^3 + 3x^2 - 6x$, and $f''(x) = 12x^2 + 6x - 6$. Solving $f''(x) = 0$ we find that:

$$x_1 = -1 \quad \text{and} \quad x_2 = \frac{1}{2}.$$

Since $f''(x) > 0$ for $x < x_1$, $f''(x) < 0$ for $x_1 < x < x_2$, and $f''(x) > 0$ for $x_2 < x$, it follows that both points are inflection points.

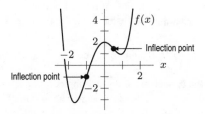

9. The graph of f in Figure 4.2 appears to be increasing for all x, with no critical points. Since $f'(x) = 3x^2 + 6$ and $x^2 \geq 0$ for all x, we have $f'(x) > 0$ for all x. That explains why f is increasing for all x.

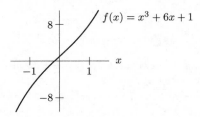

Figure 4.2

13. The graph of f in Figure 4.3 looks like $\sin x$ for $x < 0$ and e^x for $x > 0$. In particular, there are no waves for $x > 0$. We have $f'(x) = \cos x + e^x$, and so the critical points of f occur at those values of x for which $\cos x = -e^x$. Since $e^x > 1$ for all $x > 0$, we know immediately that there are no critical points at positive values of x. The specific locations of the critical points at $x < 0$ must be determined numerically; the first few are $x \approx -1.7, -4.7, -7.9$. For $x < 0$, the quantity e^x is small so that the graph looks like the graph of $\sin x$. For $x > 0$, we have $f'(x) > 0$ since $-1 \le \cos x$ and $e^x > 1$. Thus, the graph is increasing for all $x > 0$ and there are no such waves.

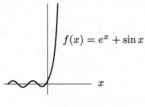

$f(x) = e^x + \sin x$

x

Figure 4.3

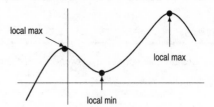

local max

local max

local min

Figure 4.4

17. See Figure 4.4.

Problems

21. Differentiating using the product rule gives

$$f'(x) = 3x^2(1-x)^4 - 4x^3(1-x)^3 = x^2(1-x)^3(3(1-x) - 4x) = x^2(1-x)^3(3 - 7x).$$

The critical points are the solutions to

$$f'(x) = x^2(1-x)^3(3 - 7x) = 0$$
$$x = 0, 1, \frac{3}{7}.$$

For $x < 0$, since $1 - x > 0$ and $3 - 7x > 0$, we have $f'(x) > 0$.
For $0 < x < \frac{3}{7}$, since $1 - x > 0$ and $3 - 7x > 0$, we have $f'(x) > 0$.
For $\frac{3}{7} < x < 1$, since $1 - x > 0$ and $3 - 7x < 0$, we have $f'(x) < 0$.
For $1 < x$, since $1 - x < 0$ and $3 - 7x < 0$, we have $f'(x) > 0$.
Thus, $x = 0$ is neither a local maximum nor a local minimum; $x = 3/7$ is a local maximum; $x = 1$ is a local minimum.

25. (a) It appears that this function has a local maximum at about $x = 1$, a local minimum at about $x = 4$, and a local maximum at about $x = 8$.

(b) The table now gives values of the derivative, so critical points occur where $f'(x) = 0$. Since f' is continuous, this occurs between 2 and 3, so there is a critical point somewhere around 2.5. Since f' is positive for values less than 2.5 and negative for values greater than 2.5, it appears that f has a local maximum at about $x = 2.5$. Similarly, it appears that f has a local minimum at about $x = 6.5$ and another local maximum at about $x = 9.5$.

29. Using the product rule on the function $f(x) = axe^{bx}$, we have $f'(x) = ae^{bx} + abxe^{bx} = ae^{bx}(1 + bx)$. We want $f(\frac{1}{3}) = 1$, and since this is to be a maximum, we require $f'(\frac{1}{3}) = 0$. These conditions give

$$f(1/3) = a(1/3)e^{b/3} = 1,$$
$$f'(1/3) = ae^{b/3}(1 + b/3) = 0.$$

Since $ae^{(1/3)b}$ is non-zero, we can divide both sides of the second equation by $ae^{(1/3)b}$ to obtain $0 = 1 + \frac{b}{3}$. This implies $b = -3$. Plugging $b = -3$ into the first equation gives us $a(\frac{1}{3})e^{-1} = 1$, or $a = 3e$. How do we know we have a maximum at $x = \frac{1}{3}$ and not a minimum? Since $f'(x) = ae^{bx}(1 + bx) = (3e)e^{-3x}(1 - 3x)$, and $(3e)e^{-3x}$ is always positive, it follows that $f'(x) > 0$ when $x < \frac{1}{3}$ and $f'(x) < 0$ when $x > \frac{1}{3}$. Since f' is positive to the left of $x = \frac{1}{3}$ and negative to the right of $x = \frac{1}{3}$, $f(\frac{1}{3})$ is a local maximum.

33. See Figure 4.5.

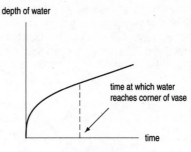

depth of water

time at which water
reaches corner of vase

time

Figure 4.5

37. (a)

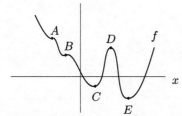

(b)

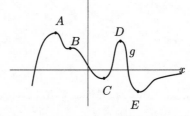

41. Since the derivative of an even function is odd and the derivative of an odd function is even, f and f'' are either both odd or both even, and f' is the opposite. Graphs I and II represent odd functions; III represents an even function, so III is f'. Since the maxima and minima of III occur where I crosses the x-axis, I must be the derivative of f', that is, f''. In addition, the maxima and minima of II occur where III crosses the x-axis, so II is f.

45. See Figure 4.6.

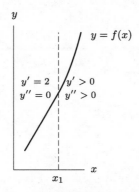

$y = f(x)$

$y' = 2$ $y' > 0$
$y'' = 0$ $y'' > 0$

x_1

Figure 4.6

Solutions for Section 4.2

Exercises

1. The line of slope m through the point (x_0, y_0) has equation

$$y - y_0 = m(x - x_0),$$

so the line we want is

$$y - 0 = 2(x - 5)$$
$$y = 2x - 10.$$

5. The equation of the whole circle is

$$x^2 + y^2 = (\sqrt{2})^2,$$

so the bottom half is

$$y = -\sqrt{2 - x^2}.$$

9. Since the maximum is $y = 2$ and the minimum is $y = 1.5$, the amplitude is $A = (2 - 1.5)/2 = 0.25$. Between the maximum and the minimum, the x-value changes by 10. There is half a period between a maximum and the next minimum, so the period is 20. Thus

$$\frac{2\pi}{B} = 20 \qquad \text{so} \qquad B = \frac{\pi}{10}.$$

The mid-line is $y = C = (2 + 1.5)/2 = 1.75$. Figure 4.7 shows a graph of the function

$$y = 0.25 \sin\left(\frac{\pi x}{10}\right) + 1.75.$$

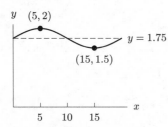

Figure 4.7: Graph of
$y = 0.25 \sin(\pi x/10) + 1.75$

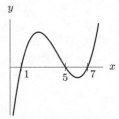

Figure 4.8: Graph of $y = (x - 1)(x - 5)(x - 7)$

13. A cubic polynomial of the form $y = a(x - 1)(x - 5)(x - 7)$ has the correct intercepts for any value of $a \neq 0$. Figure 4.8 shows the graph with $a = 1$, namely $y = (x - 1)(x - 5)(x - 7)$.

Problems

17. (a) We have $p'(x) = 3x^2 - a$, so

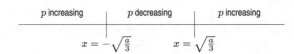

Local maximum: $p(-\sqrt{\frac{a}{3}}) = \frac{-a\sqrt{a}}{\sqrt{27}} + \frac{a\sqrt{a}}{\sqrt{3}} = +\frac{2a\sqrt{a}}{3\sqrt{3}}$

Local minimum: $p(\sqrt{\frac{a}{3}}) = -p(-\sqrt{\frac{a}{3}}) = -\frac{2a\sqrt{a}}{3\sqrt{3}}$

(b) Increasing the value of a moves the critical points of p away from the y-axis, and moves the critical values away from the x-axis. Thus, the "bumps" get further apart and higher. At the same time, increasing the value of a spreads the zeros of p further apart (while leaving the one at the origin fixed).

(c) See Figure 4.9

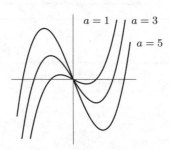

Figure 4.9

21. Since $\lim_{t \to \infty} N = a$, we have $a = 200{,}000$. Note that while $N(t)$ will never actually reach 200,000, it will become arbitrarily close to 200,000. Since N represents the number of people, it makes sense to round up long before $t \to \infty$. When $t = 1$, we have $N = 0.1(200{,}000) = 20{,}000$ people, so plugging into our formula gives

$$N(1) = 20{,}000 = 200{,}000 \left(1 - e^{-k(1)}\right).$$

Solving for k gives

$$0.1 = 1 - e^{-k}$$
$$e^{-k} = 0.9$$
$$k = -\ln 0.9 \approx 0.105.$$

25. Since

$$\frac{dy}{dx} = (1 - bx)e^{-bx},$$

we see

$$\frac{dy}{dx} = 0 \quad \text{at} \quad x = \frac{1}{b}.$$

The critical point has coordinates $(1/b, 1/(be))$. If b is small, the x and y-coordinates of the critical point are both large, indicating a higher maximum further to the right. See figure below.

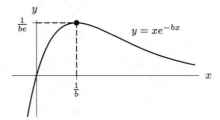

29. (a) The x-intercept occurs where $f(x) = 0$, so

$$ax - x \ln x = 0$$
$$x(a - \ln x) = 0.$$

Since $x > 0$, we must have

$$a - \ln x = 0$$
$$\ln x = a$$
$$x = e^a.$$

(b) See Figures 4.10 and 4.11.

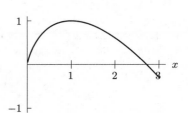

Figure 4.10: Graph of $f(x)$ with $a = -1$

Figure 4.11: Graph of $f(x)$ with $a = 1$

(c) Differentiating gives $f'(x) = a - \ln x - 1$. Critical points are obtained by solving

$$a - \ln x - 1 = 0$$
$$\ln x = a - 1$$
$$x = e^{a-1}.$$

Since $e^{a-1} > 0$ for all a, there is no restriction on a. Now,

$$f(e^{a-1}) = ae^{a-1} - e^{a-1}\ln(e^{a-1}) = ae^{a-1} - (a-1)e^{a-1} = e^{a-1},$$

so the coordinates of the critical point are (e^{a-1}, e^{a-1}). From the graphs, we see that this critical point is a local maximum; this can be confirmed using the second derivative:

$$f''(x) = -\frac{1}{x} < 0 \qquad \text{for } x = e^{a-1}.$$

33. (a) Let $f(x) = axe^{-bx}$. To find the local maxima and local minima of f, we solve

$$f'(x) = ae^{-bx} - abxe^{-bx} = ae^{-bx}(1 - bx) \begin{cases} = 0 & \text{if } x = 1/b \\ < 0 & \text{if } x > 1/b \\ > 0 & \text{if } x < 1/b. \end{cases}$$

Therefore, f is increasing ($f' > 0$) for $x < 1/b$ and decreasing ($f' > 0$) for $x > 1/b$. A local maximum occurs at $x = 1/b$. There are no local minima. To find the points of inflection, we write

$$\begin{aligned} f''(x) &= -abe^{-bx} + ab^2xe^{-bx} - abe^{-bx} \\ &= -2abe^{-bx} + ab^2xe^{-bx} \\ &= ab(bx - 2)e^{-bx}, \end{aligned}$$

so $f'' = 0$ at $x = 2/b$. Therefore, f is concave up for $x < 2/b$ and concave down for $x > 2/b$, and the inflection point is $x = 2/b$.

(b) Varying a stretches or flattens the graph but does not affect the critical point $x = 1/b$ and the inflection point $x = 2/b$. Since the critical and inflection points are depend on b, varying b will change these points, as well as the maximum $f(1/b) = a/be$. For example, an increase in b will shift the critical and inflection points to the left, and also lower the maximum value of f.

(c)

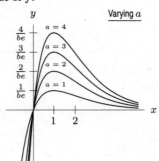

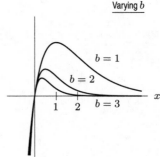

37. (a) Since

$$U = b\left(\frac{a^2 - ax}{x^2}\right) = 0 \quad \text{when} \quad x = a,$$

the x-intercept is $x = a$. There is a vertical asymptote at $x = 0$ and a horizontal asymptote at $U = 0$.

(b) Setting $dU/dx = 0$, we have

$$\frac{dU}{dx} = b\left(-\frac{2a^2}{x^3} + \frac{a}{x^2}\right) = b\left(\frac{-2a^2 + ax}{x^3}\right) = 0.$$

So the critical point is

$$x = 2a.$$

When $x = 2a$,

$$U = b\left(\frac{a^2}{4a^2} - \frac{a}{2a}\right) = -\frac{b}{4}.$$

The second derivative of U is

$$\frac{d^2U}{dx^2} = b\left(\frac{6a^2}{x^4} - \frac{2a}{x^3}\right).$$

When we evaluate this at $x = 2a$, we get

$$\frac{d^2U}{dx^2} = b\left(\frac{6a^2}{(2a)^4} - \frac{2a}{(2a)^3}\right) = \frac{b}{8a^2} > 0.$$

Since $d^2U/dx^2 > 0$ at $x = 2a$, we see that the point $(2a, -b/4)$ is a local minimum.

(c)

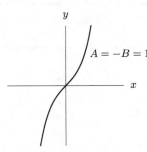

41. (a) The graphs are shown in Figures 4.12–4.17.

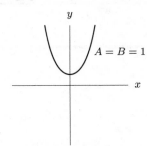

Figure 4.12: $A > 0, B > 0$

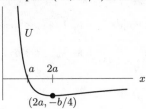

Figure 4.13: $A > 0, B < 0$

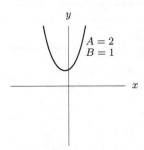

Figure 4.14: $A > 0, B > 0$

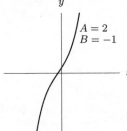

Figure 4.15: $A > 0, B < 0$

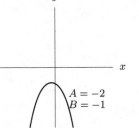

Figure 4.16: $A < 0, B < 0$

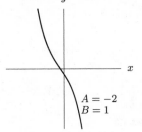

Figure 4.17: $A < 0, B > 0$

(b) If A and B have the same sign, the graph is U-shaped. If A and B are both positive, the graph opens upward. If A and B are both negative, the graph opens downward.

(c) If A and B have different signs, the graph appears to be everywhere increasing (if $A > 0, B < 0$) or decreasing (if $A < 0, B > 0$).

(d) The function appears to have a local maximum if $A < 0$ and $B < 0$, and a local minimum if $A > 0$ and $B > 0$.

To justify this, calculate the derivative

$$\frac{dy}{dx} = Ae^x - Be^{-x}.$$

Setting $dy/dx = 0$ gives

$$Ae^x - Be^{-x} = 0$$
$$Ae^x = Be^{-x}$$
$$e^{2x} = \frac{B}{A}.$$

This equation has a solution only if B/A is positive, that is, if A and B have the same sign. In that case,

$$2x = \ln\left(\frac{B}{A}\right)$$
$$x = \frac{1}{2}\ln\left(\frac{B}{A}\right).$$

This value of x gives the only critical point.

To determine whether the critical point is a local maximum or minimum, we use the first derivative test. Since

$$\frac{dy}{dx} = Ae^x - Be^{-x},$$

we see that:

If $A > 0, B > 0$, we have $dy/dx > 0$ for large positive x and $dy/dx < 0$ for large negative x, so there is a local minimum.

If $A < 0, B < 0$, we have $dy/dx < 0$ for large positive x and $dy/dx > 0$ for large negative x, so there is a local maximum.

Solutions for Section 4.3

Exercises

1. See Figure 4.18.

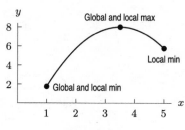

Figure 4.18

5. (a) $f'(x) = 1 - 1/x$. This is zero only when $x = 1$. Now $f'(x)$ is positive when $1 < x \leq 2$, and negative when $0.1 < x < 1$. Thus $f(1) = 1$ is a local minimum. The endpoints $f(0.1) \approx 2.4026$ and $f(2) \approx 1.3069$ are local maxima.

   (b) Comparing values of f shows that $x = 0.1$ gives the global maximum and $x = 1$ gives the global minimum.

9. Differentiating using the product rule gives

$$g'(t) = 1 \cdot e^{-t} - te^{-t} = (1 - t)e^{-t},$$

so the critical point is $t = 1$.

Since $g'(t) > 0$ for $0 < t < 1$ and $g'(t) < 0$ for $t > 1$, the critical point is a local maximum.

As $t \to \infty$, the value of $g(t) \to 0$, and as $t \to 0^+$, the value of $g(t) \to 0$. Thus, the local maximum at $x = 1$ is a global maximum of $g(1) = 1e^{-1} = 1/e$. In addition, the value of $g(t)$ is positive for all $t > 0$; it tends to 0 but never reaches 0. Thus, there is no global minimum. See Figure 4.19.

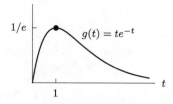

Figure 4.19

13. Let $y = x^3 - 4x^2 + 4x$. To locate the critical points, we solve $y' = 0$. Since $y' = 3x^2 - 8x + 4 = (3x - 2)(x - 2)$, the critical points are $x = 2/3$ and $x = 2$. To find the global minimum and maximum on $0 \leq x \leq 4$, we check the critical points and the endpoints: $y(0) = 0$; $y(2/3) = 32/27$; $y(2) = 0$; $y(4) = 16$. Thus, the global minimum is at $x = 0$ and $x = 2$, the global maximum is at $x = 4$, and $0 \leq y \leq 16$.

17. Let $y = \ln(1 + x)$. Since $y' = 1/(1 + x)$, y is increasing for all $x \geq 0$. The lower bound is at $x = 0$, so, $\ln(1) = 0 \leq y$. There is no upper bound.

Problems

21. (a) We have

$$T(D) = \left(\frac{C}{2} - \frac{D}{3}\right)D^2 = \frac{CD^2}{2} - \frac{D^3}{3},$$

and

$$\frac{dT}{dD} = CD - D^2 = D(C - D).$$

Since, by this formula, dT/dD is zero when $D = 0$ or $D = C$, negative when $D > C$, and positive when $D < C$, we have (by the first derivative test) that the temperature change is maximized when $D = C$.

(b) The sensitivity is $dT/dD = CD - D^2$; its derivative is $d^2T/dD^2 = C - 2D$, which is zero if $D = C/2$, negative if $D > C/2$, and positive if $D < C/2$. Thus by the first derivative test the sensitivity is maximized at $D = C/2$.

25. (a) To show that R is an increasing function of r_1, we show that $dR/dr_1 > 0$ for all values of r_1. We first solve for R:

$$\frac{1}{R} = \frac{1}{r_1} + \frac{1}{r_2}$$

$$\frac{1}{R} = \frac{r_2 + r_1}{r_1 r_2}$$

$$R = \frac{r_1 r_2}{r_2 + r_1}.$$

We use the quotient rule (and remember that r_2 is a constant) to find dR/dr_1:

$$\frac{dR}{dr_1} = \frac{(r_2 + r_1)(r_2) - (r_1 r_2)(1)}{(r_2 + r_1)^2} = \frac{(r_2)^2}{(r_2 + r_1)^2}.$$

Since dR/dr_1 is the square of a number, we have $dR/dr_1 > 0$ for all values of r_1, and thus R is increasing for all r_1.

(b) Since R is increasing on any interval $a \leq r_1 \leq b$, the maximum value of R occurs at the right endpoint $r_1 = b$.

29. Suppose the points are given by x and $-x$, where $x \geq 0$. The function is odd, since

$$y = \frac{(-x)^3}{1 + (-x)^4} = -\frac{x^3}{1 + x^4},$$

so the corresponding y-coordinates are also opposite. See Figure 4.20. For $x > 0$, we have

$$m = \frac{\frac{x^3}{1+x^4} - \left(-\frac{x^3}{1+x^4}\right)}{x - (-x)} = \frac{1}{2x} \cdot \frac{2x^3}{1 + x^4} = \frac{x^2}{1 + x^4}.$$

For the maximum slope,

$$\frac{dm}{dx} = \frac{2x}{1 + x^4} - \frac{x^2(4x^3)}{(1 + x^4)^2} = 0$$

$$\frac{2x(1 + x^4) - 4x^5}{(1 + x^4)^2} = 0$$

$$\frac{2x(1 - x^4)}{(1 + x^4)^2} = 0$$

$$x\left(1 - x^4\right) = 0$$

$$x = 0, \pm 1.$$

For $x > 0$, there is one critical point, $x = 1$. Since m tends to 0 when $x \to 0$ and when $x \to \infty$, the critical point $x = 1$ gives the maximum slope. Thus, the maximum slope occurs when the line has endpoints

$$\left(-1, -\frac{1}{2}\right) \quad \text{and} \quad \left(1, \frac{1}{2}\right).$$

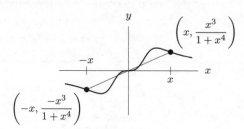

Figure 4.20

33. (a) To obtain $g(v)$, which is in gallons per mile, we need to divide $f(v)$ (in gallons per hour) by v (in miles per hour). Thus, $g(v) = f(v)/v$.
    (b) By inspecting the graph, we see that $f(v)$ is minimized at approximately 220 mph.
    (c) Note that a point on the graph of $f(v)$ has the coordinates $(v, f(v))$. The line passing through this point and the origin $(0, 0)$ has

$$\text{Slope} = \frac{f(v) - 0}{v - 0} = \frac{f(v)}{v} = g(v).$$

So minimizing $g(v)$ corresponds to finding the line of minimum slope from the family of lines which pass through the origin $(0, 0)$ and the point $(v, f(v))$ on the graph of $f(v)$. This line is the unique member of the family which is tangent to the graph of $f(v)$. The value of v corresponding to the point of tangency will minimize $g(v)$. This value of v will satisfy $f(v)/v = f'(v)$. From the graph in Figure 4.21, we see that $v \approx 300$ mph.

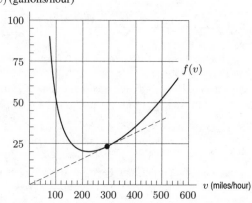

Figure 4.21

    (d) The pilot's goal with regard to $f(v)$ and $g(v)$ would depend on the purpose of the flight, and might even vary within a given flight. For example, if the mission involved aerial surveillance or banner-towing over some limited area, or if the plane was flying a holding pattern, then the pilot would want to minimize $f(v)$ so as to remain aloft as long as possible. In a more normal situation where the purpose was economical travel between two fixed points, then the minimum net fuel expenditure for the trip would result from minimizing $g(v)$.

37. False. For example, if $f(x) = x^3$, then $f'(0) = 0$, so $x = 0$ is a critical point, but $x = 0$ is neither a local maximum nor a local minimum.

41. (a) If both the global minimum and the global maximum are at the endpoints, then $f(x) = 0$ everywhere in $[a, b]$, since $f(a) = f(b) = 0$. In that case $f'(x) = 0$ everywhere as well, so any point in (a, b) will do for c.
    (b) Suppose that either the global maximum or the global minimum occurs at an interior point of the interval. Let c be that point. Then c must be a local extremum of f, so, by the theorem concerning local extrema on page 168, we have $f'(c) = 0$, as required.

Solutions for Section 4.4

Exercises

1. The fixed costs are \$5000, the marginal cost per item is \$2.40, and the price per item is \$4.

5. The profit function is positive when $R(q) > C(q)$, and negative when $C(q) > R(q)$. It's positive for $5.5 < q < 12.5$, and negative for $0 < q < 5.5$ and $12.5 < q$. Profit is maximized when $R(q) > C(q)$ and $R'(q) = C'(q)$ which occurs at about $q = 9.5$. See Figure 4.22.

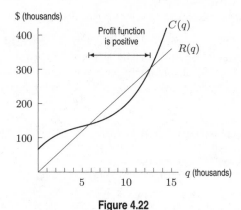

Figure 4.22

9. (a) Profit is maximized when $R(q) - C(q)$ is as large as possible. This occurs at $q = 2500$, where profit $= 7500 - 5500 = \$2000$.

(b) We see that $R(q) = 3q$ and so the price is $p = 3$, or \$3 per unit.

(c) Since $C(0) = 3000$, the fixed costs are \$3000.

Problems

13. (a) We know that Profit = Revenue − Cost, so differentiating with respect to q gives:

$$\text{Marginal Profit} = \text{Marginal Revenue} - \text{Marginal Cost}.$$

We see from the figure in the problem that just to the left of $q = a$, marginal revenue is less than marginal cost, so marginal profit is negative there. To the right of $q = a$ marginal revenue is greater than marginal cost, so marginal profit is positive there. At $q = a$ marginal profit changes from negative to positive. This means that profit is decreasing to the left of a and increasing to the right. The point $q = a$ corresponds to a local minimum of profit, and does not maximize profit. It would be a terrible idea for the company to set its production level at $q = a$.

(b) We see from the figure in the problem that just to the left of $q = b$ marginal revenue is greater than marginal cost, so marginal profit is positive there. Just to the right of $q = b$ marginal revenue is less than marginal cost, so marginal profit is negative there. At $q = b$ marginal profit changes from positive to negative. This means that profit is increasing to the left of b and decreasing to the right. The point $q = b$ corresponds to a local maximum of profit. In fact, since the area between the MC and MR curves in the figure in the text between $q = a$ and $q = b$ is bigger than the area between $q = 0$ and $q = a$, $q = b$ is in fact a global maximum.

17. For each month,

$$\text{Profit} = \text{Revenue} - \text{Cost}$$
$$\pi = pq - wL = pcK^\alpha L^\beta - wL$$

The variable on the right is L, so at the maximum

$$\frac{d\pi}{dL} = \beta pcK^\alpha L^{\beta-1} - w = 0$$

Now $\beta - 1$ is negative, since $0 < \beta < 1$, so $1 - \beta$ is positive and we can write

$$\frac{\beta pc K^\alpha}{L^{1-\beta}} = w$$

giving

$$L = \left(\frac{\beta pc K^\alpha}{w} \right)^{\frac{1}{1-\beta}}$$

Since $\beta - 1$ is negative, when L is just above 0, the quantity $L^{\beta-1}$ is huge and positive, so $d\pi/dL > 0$. When L is large, $L^{\beta-1}$ is small, so $d\pi/dL < 0$. Thus the value of L we have found gives a global maximum, since it is the only critical point.

21. (a) $a(q) = C(q)/q$, so $C(q) = 0.01q^3 - 0.6q^2 + 13q$.

(b) Taking the derivative of $C(q)$ gives an expression for the marginal cost:

$$C'(q) = MC(q) = 0.03q^2 - 1.2q + 13.$$

To find the smallest MC we take its derivative and find the value of q that makes it zero. So: $MC'(q) = 0.06q - 1.2 = 0$ when $q = 1.2/0.06 = 20$. This value of q must give a minimum because the graph of $MC(q)$ is a parabola opening upward. Therefore the minimum marginal cost is $MC(20) = 1$. So the marginal cost is at a minimum when the additional cost per item is \$1.

(c) $a'(q) = 0.02q - 0.6$

Setting $a'(q) = 0$ and solving for q gives $q = 30$ as the quantity at which the average is minimized, since the graph of a is a parabola which opens upward. The minimum average cost is $a(30) = 4$ dollars per item.

(d) The marginal cost at $q = 30$ is $MC(30) = 0.03(30)^2 - 1.2(30) + 13 = 4$. This is the same as the average cost at this quantity. Note that since $a(q) = C(q)/q$, we have $a'(q) = (qC'(q) - C(q))/q^2$. At a critical point, q_0, of $a(q)$, we have

$$0 = a'(q_0) = \frac{q_0 C'(q_0) - C(q_0)}{q_0^2},$$

so $C'(q_0) = C(q_0)/q_0 = a(q_0)$. Therefore $C'(30) = a(30) = 4$ dollars per item.

Another way to see why the marginal cost at $q = 30$ must equal the minimum average cost $a(30) = 4$ is to view $C'(30)$ as the approximate cost of producing the 30th or 31st good. If $C'(30) < a(30)$, then producing the 31st good would lower the average cost, i.e. $a(31) < a(30)$. If $C'(30) > a(30)$, then producing the 30th good would raise the average cost, i.e. $a(30) > a(29)$. Since $a(30)$ is the global minimum, we must have $C'(30) = a(30)$.

Solutions for Section 4.5

Exercises

1. We look for critical points of M:

$$\frac{dM}{dx} = \frac{1}{2}wL - wx.$$

Now $dM/dx = 0$ when $x = L/2$. At this point $d^2M/dx^2 = -w$ so this point is a local maximum. The graph of $M(x)$ is a parabola opening downward, so the local maximum is also the global maximum.

5. (a) If we expect the rate to be nonnegative, we must have $0 \le y \le a$ and $0 \le y \le b$. Since we assume $a < b$, we restrict y to $0 \le y \le a$.

In fact, the expression for the rate is nonnegative for y greater than b, but these values of y are not meaningful for the reaction. See Figure 4.23.

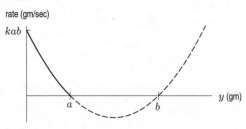

Figure 4.23

(b) From the graph, we see that the maximum rate occurs when $y = 0$; that is, at the start of the reaction.

9. The domain for E is all real x. Note $E \to 0$ as $x \to \pm\infty$. The critical points occur where $dE/dx = 0$. The derivative is

$$\frac{dE}{dx} = \frac{k}{(x^2 + r_0^2)^{3/2}} - \frac{3}{2} \cdot \frac{kx(2x)}{(x^2 + r_0^2)^{5/2}}$$

$$= \frac{k\left(x^2 + r_0^2 - 3x^2\right)}{(x^2 + r_0^2)^{5/2}}$$

$$= \frac{k\left(r_0^2 - 2x^2\right)}{(x^2 + r_0^2)^{5/2}}.$$

So $dE/dx = 0$ where

$$r_0^2 - 2x^2 = 0$$

$$x = \pm\frac{r_0}{\sqrt{2}}.$$

Looking at the formula for dE/dx shows

$$\frac{dE}{dx} > 0 \text{ for } -\frac{r_0}{\sqrt{2}} < x < \frac{r_0}{\sqrt{2}}$$

$$\frac{dE}{dx} < 0 \text{ for } x < -\frac{r_0}{\sqrt{2}}$$

$$\frac{dE}{dx} < 0 \text{ for } x > \frac{r_0}{\sqrt{2}}.$$

Therefore, $x = -r_0/\sqrt{2}$ gives the minimum value of E and $x = r_0/\sqrt{2}$ gives the maximum value of E.

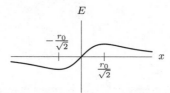

Problems

13. The rectangle in Figure 4.24 has area, A, given by

$$A = 2xy = \frac{2x}{1 + x^2} \qquad \text{for } x \geq 0.$$

At a critical point,

$$\frac{dA}{dx} = \frac{2}{1 + x^2} + 2x\left(\frac{-2x}{(1 + x^2)^2}\right) = 0$$

$$\frac{2(1 + x^2 - 2x^2)}{(1 + x^2)^2} = 0$$

$$1 - x^2 = 0$$

$$x = \pm 1.$$

Since $A = 0$ for $x = 0$ and $A \to 0$ as $x \to \infty$, the critical point $x = 1$ is a local and global maximum for the area. Then $y = 1/2$, so the vertices are

$$(-1, 0),\ (1, 0),\ \left(1, \frac{1}{2}\right),\ \left(-1, \frac{1}{2}\right).$$

$$y = \frac{1}{1 + x^2}$$

Figure 4.24

17. Let w and l be the width and length, respectively, of the rectangular area you wish to enclose. Then

$$w + w + l = 100 \text{ feet}$$
$$l = 100 - 2w$$
$$\text{Area} = w \cdot l = w(100 - 2w) = 100w - 2w^2$$

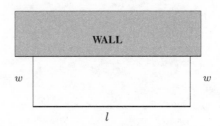

To maximize area, we solve $A' = 0$ to find critical points. This gives $A' = 100 - 4w = 0$, so $w = 25$, $l = 50$. So the area is $25 \cdot 50 = 1250$ square feet. This is a local maximum by the second derivative test because $A'' = -4 < 0$. Since the graph of A is a parabola, the local maximum is in fact a global maximum.

21. Volume: $V = x^2 y$,
Surface: $S = x^2 + 4xy = x^2 + 4xV/x^2 = x^2 + 4V/x$.
To find the dimensions which minimize the area, find x such that $dS/dx = 0$.

$$\frac{dS}{dx} = 2x - \frac{4V}{x^2} = 0,$$

so

$$x^3 = 2V,$$

and solving for x gives $x = \sqrt[3]{2V}$. To see that this gives a minimum, note that for small x, $S \approx 4V/x$ is decreasing. For large x, $S \approx x^2$ is increasing. Since there is only one critical point, this must give a global minimum. Using x to find y gives $y = V/x^2 = V/(2V)^{2/3} = \sqrt[3]{V/4}$.

25. We see that the width of the tunnel is $2r$. The area of the rectangle is then $(2r)h$. The area of the semicircle is $(\pi r^2)/2$. The cross-sectional area, A, is then

$$A = 2rh + \frac{1}{2}\pi r^2$$

and the perimeter, P, is

$$P = 2h + 2r + \pi r.$$

From $A = 2rh + (\pi r^2)/2$ we get

$$h = \frac{A}{2r} - \frac{\pi r}{4}.$$

Thus,

$$P = 2\left(\frac{A}{2r} - \frac{\pi r}{4}\right) + 2r + \pi r = \frac{A}{r} + 2r + \frac{\pi r}{2}.$$

We now have the perimeter in terms of r and the constant A. Differentiating, we obtain

$$\frac{dP}{dr} = -\frac{A}{r^2} + 2 + \frac{\pi}{2}.$$

To find the critical points we set $P' = 0$:

$$-\frac{A}{r^2} + \frac{\pi}{2} + 2 = 0$$

$$\frac{r^2}{A} = \frac{2}{4+\pi}$$

$$r = \sqrt{\frac{2A}{4+\pi}}.$$

Substituting this back into our expression for h, we have

$$h = \frac{A}{2} \cdot \frac{\sqrt{4+\pi}}{\sqrt{2A}} - \frac{\pi}{4} \cdot \frac{\sqrt{2A}}{\sqrt{4+\pi}}.$$

Since $P \to \infty$ as $r \to 0^+$ and as $r \to \infty$, this critical point must be a global minimum. Notice that the h-value simplifies to

$$h = \sqrt{\frac{2A}{4+\pi}} = r.$$

29. (a) We have

$$x^{1/x} = e^{\ln\left(x^{1/x}\right)} = e^{(1/x)\ln x}.$$

Thus

$$\frac{d(x^{1/x})}{dx} = \frac{d(e^{(1/x)\ln x})}{dx} = \frac{d(\frac{1}{x}\ln x)}{dx}e^{(1/x)\ln x}$$

$$= \left(-\frac{\ln x}{x^2} + \frac{1}{x^2}\right)x^{1/x}$$

$$= \frac{x^{1/x}}{x^2}(1 - \ln x) \begin{cases} = 0 & \text{when } x = e \\ < 0 & \text{when } x > e \\ > 0 & \text{when } x < e. \end{cases}$$

Hence $e^{1/e}$ is the global maximum for $x^{1/x}$, by the first derivative test.

(b) Since $x^{1/x}$ is increasing for $0 < x < e$ and decreasing for $x > e$, and 2 and 3 are the closest integers to e, either $2^{1/2}$ or $3^{1/3}$ is the maximum for $n^{1/n}$. We have $2^{1/2} \approx 1.414$ and $3^{1/3} \approx 1.442$, so $3^{1/3}$ is the maximum.

(c) Since $e < 3 < \pi$, and $x^{1/x}$ is decreasing for $x > e$, $3^{1/3} > \pi^{1/\pi}$.

33. (a) The distance the pigeon flies over water is

$$\overline{BP} = \frac{\overline{AB}}{\sin\theta} = \frac{500}{\sin\theta},$$

and over land is

$$\overline{PL} = \overline{AL} - \overline{AP} = 2000 - \frac{500}{\tan\theta} = 2000 - \frac{500\cos\theta}{\sin\theta}.$$

Therefore the energy required is

$$E = 2e\left(\frac{500}{\sin\theta}\right) + e\left(2000 - \frac{500\cos\theta}{\sin\theta}\right)$$

$$= 500e\left(\frac{2 - \cos\theta}{\sin\theta}\right) + 2000e, \quad \text{for} \quad \arctan\left(\frac{500}{2000}\right) \le \theta \le \frac{\pi}{2}.$$

(b) Notice that E and the function $f(\theta) = \dfrac{2 - \cos\theta}{\sin\theta}$ must have the same critical points since the graph of E is just a stretch and a vertical shift of the graph of f. The graph of $\dfrac{2 - \cos\theta}{\sin\theta}$ for $\arctan(\frac{500}{2000}) \le \theta \le \frac{\pi}{2}$ in Figure 4.25 shows that E has precisely one critical point, and that a minimum for E occurs at this point.

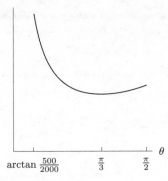

Figure 4.25: Graph of $f(\theta) = \frac{2 - \cos\theta}{\sin\theta}$ for $\arctan\left(\frac{500}{2000}\right) \leq \theta \leq \frac{\pi}{2}$

To find the critical point θ, we solve $f'(\theta) = 0$ or

$$E' = 0 = 500e\left(\frac{\sin\theta \cdot \sin\theta - (2 - \cos\theta) \cdot \cos\theta}{\sin^2\theta}\right)$$

$$= 500e\left(\frac{1 - 2\cos\theta}{\sin^2\theta}\right).$$

Therefore $1 - 2\cos\theta = 0$ and so $\theta = \pi/3$.

(c) Letting $a = \overline{AB}$ and $b = \overline{AL}$, our formula for E becomes

$$E = 2e\left(\frac{a}{\sin\theta}\right) + e\left(b - \frac{a\cos\theta}{\sin\theta}\right)$$

$$= ea\left(\frac{2 - \cos\theta}{\sin\theta}\right) + eb, \quad \text{for} \quad \arctan\left(\frac{a}{b}\right) \leq \theta \leq \frac{\pi}{2}.$$

Again, the graph of E is just a stretch and a vertical shift of the graph of $\dfrac{2 - \cos\theta}{\sin\theta}$. Thus, the critical point $\theta = \pi/3$ is independent of e, a, and b. But the maximum of E *on the domain* $\arctan(a/b) \leq \theta \leq \frac{\pi}{2}$ is dependent on the ratio $a/b = \dfrac{\overline{AB}}{\overline{AL}}$. In other words, the optimal angle is $\theta = \pi/3$ provided $\arctan(a/b) \leq \frac{\pi}{3}$; otherwise, the optimal angle is $\arctan(a/b)$, which means the pigeon should fly over the lake for the entire trip—this occurs when $a/b > 1.733$.

37. We know that the time taken is given by

$$T = \frac{\sqrt{a^2 + (c - x)^2}}{v_1} + \frac{\sqrt{b^2 + x^2}}{v_2}$$

$$\frac{dT}{dx} = \frac{-(c - x)}{v_1\sqrt{a^2 + (c - x)^2}} + \frac{x}{v_2\sqrt{b^2 + x^2}}.$$

Differentiating again gives

$$\frac{d^2T}{dx^2} = \frac{1}{v_1\sqrt{a^2 + (c - x)^2}} + \frac{(c - x)(-2(c - x))}{2v_1(a^2 + (c - x)^2)^{3/2}} + \frac{1}{v_2\sqrt{b^2 + x^2}} - \frac{x(2x)}{2v_2(b^2 + x^2)^{3/2}}$$

$$= \frac{a^2 + (c - x)^2 - (c - x)^2}{v_1(a^2 + (c - x)^2)^{3/2}} + \frac{b^2 + x^2 - x^2}{v_2(b^2 + x^2)^{3/2}}$$

$$= \frac{a^2}{v_1(a^2 + (c - x)^2)^{3/2}} + \frac{b^2}{v_2(b^2 + x^2)^{3/2}}.$$

This expression for d^2T/dx^2 shows that for any value of x, a, c, v_1, and v_2 with v_1, $v_2 > 0$, we have $d^2T/dx^2 > 0$. Thus, any critical point must be a local minimum. Since there is only one critical point, it must be a global minimum.

Solutions for Section 4.6

Exercises

1. The rate of growth, in billions of people per year, is

$$\frac{dP}{dt} = 6.342(0.011)e^{0.011t}.$$

On January 1, 2004, we have $t = 0$, so

$$\frac{dP}{dt} = 6.342(0.011)e^0 = 0.0698 \text{ billion/year} = 69.8 \text{ million/year}.$$

5. (a) The rate of change of thickness of ice is

$$\frac{dy}{dt} = 0.2(1.5)t^{0.5} = 0.3t^{0.5} \text{ cm/hr}.$$

Thus, at $t = 1$

$$\left.\frac{dy}{dt}\right|_{t=1} = 0.3(1)^{0.5} = 0.3 \text{ cm/hr}.$$

At $t = 2$,

$$\left.\frac{dy}{dt}\right|_{t=2} = 0.3(2)^{0.5} = 0.424 \text{ cm/hr}.$$

(b) Since both $y = 0.2t^{1.5}$ and $dy/dt = 0.3t^{0.5}$ increase as t increase on the interval $0 \le t \le 3$, the thickness, y, and the rate, dy/dt, are both greatest when $t = 3$.

9. (a) The rate of change of force with respect to distance is

$$\frac{dF}{dr} = \frac{2A}{r^3} - \frac{3B}{r^4}.$$

The units are units of force per units of distance.

(b) We are told that $dr/dt = k$ and we want dF/dt. By the chain rule

$$\frac{dF}{dt} = \frac{dF}{dr} \cdot \frac{dr}{dt} = \left(\frac{2A}{r^3} - \frac{3B}{r^4}\right)k.$$

The units are units of force per unit time.

Problems

13. (a) From the second figure in the problem, we see that $\theta \approx 3.3$ when $t = 2$. The coordinates of P are given by $x = \cos\theta$, $y = \sin\theta$. When $t = 2$, the coordinates of P are

$$(x, y) \approx (\cos 3.3, \sin 3.3) = (-0.99, -0.16).$$

(b) Using the chain rule, the velocity in the x-direction is given by

$$v_x = \frac{dx}{dt} = \frac{dx}{d\theta} \cdot \frac{d\theta}{dt} = -\sin\theta \cdot \frac{d\theta}{dt}.$$

From Figure 4.26, we estimate that when $t = 2$,

$$\left.\frac{d\theta}{dt}\right|_{t=2} \approx 2.$$

So

$$v_x = \frac{dx}{dt} \approx -(-0.16) \cdot (2) = 0.32.$$

Similarly, the velocity in the y-direction is given by

$$v_y = \frac{dy}{dt} = \frac{dy}{d\theta} \cdot \frac{d\theta}{dt} = \cos\theta \cdot \frac{d\theta}{dt}.$$

When $t = 2$

$$v_y = \frac{dy}{dt} \approx (-0.99) \cdot (2) = -1.98.$$

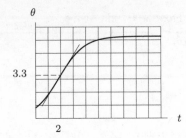

Figure 4.26

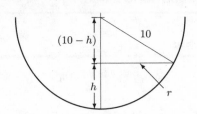

Figure 4.27

17. (a) The surface of the water is circular with radius r cm. Applying Pythagoras' Theorem to the triangle in Figure 4.27 shows that

$$(10 - h)^2 + r^2 = 10^2$$

so

$$r = \sqrt{10^2 - (10 - h)^2} = \sqrt{20h - h^2} \text{ cm.}$$

(b) We know $dh/dt = -0.1$ cm/hr and we want to know dr/dt when $h = 5$ cm. Differentiating

$$r = \sqrt{20h - h^2}$$

gives

$$\frac{dr}{dt} = \frac{1}{2}(20h - h^2)^{-1/2}\left(20\frac{dh}{dt} - 2h\frac{dh}{dt}\right) = \frac{10 - h}{\sqrt{20h - h^2}} \cdot \frac{dh}{dt}.$$

Substituting $dh/dt = -0.1$ and $h = 5$ gives

$$\left.\frac{dr}{dt}\right|_{h=5} = \frac{5}{\sqrt{20 \cdot 5 - 5^2}} \cdot (-0.1) = -\frac{1}{2\sqrt{75}} = -0.0577 \text{ cm/hr.}$$

Thus, the radius is decreasing at 0.0577 cm per hour.

21. Let the volume of clay be V. The clay is in the shape of a cylinder, so $V = \pi r^2 L$. We know $dL/dt = 0.1$ cm/sec and we want to know dr/dt when $r = 1$ cm and $L = 5$ cm. Differentiating with respect to time t gives

$$\frac{dV}{dt} = \pi 2rL\frac{dr}{dt} + \pi r^2\frac{dL}{dt}.$$

However, the amount of clay is unchanged, so $dV/dt = 0$ and

$$2rL\frac{dr}{dt} = -r^2\frac{dL}{dt},$$

therefore

$$\frac{dr}{dt} = -\frac{r}{2L}\frac{dL}{dt}.$$

When the radius is 1 cm and the length is 5 cm, and the length is increasing at 0.1 cm per second, the rate at which the radius is changing is

$$\frac{dr}{dt} = -\frac{1}{2 \cdot 5} \cdot 0.1 = -0.01 \text{ cm/sec.}$$

Thus, the radius is decreasing at 0.01 cm/sec.

25. Using the triangle OSL in Figure 4.28, we label the distance x.

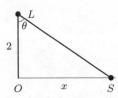

Figure 4.28

We want to calculate $dx/d\theta$. First we must find x as a function of θ. From the triangle, we see

$$\frac{x}{2} = \tan\theta \quad \text{so} \quad x = 2\tan\theta.$$

Thus,

$$\frac{dx}{d\theta} = \frac{2}{\cos^2\theta}.$$

29. Let r be the radius of the raindrop. Then its volume $V = \frac{4}{3}\pi r^3$ cm^3 and its surface area is $S = 4\pi r^2$ cm^2. It is given that

$$\frac{dV}{dt} = 2S = 8\pi r^2.$$

Furthermore,

$$\frac{dV}{dr} = 4\pi r^2,$$

so from the chain rule,

$$\frac{dV}{dt} = \frac{dV}{dr} \cdot \frac{dr}{dt} \quad \text{and thus} \quad \frac{dr}{dt} = \frac{dV/dt}{dV/dr} = 2.$$

Since dr/dt is a constant, $dr/dt = 2$, the radius is increasing at a constant rate of 2 cm/sec.

33. (a) Since the elevator is descending at 30 ft/sec, its height from the ground is given by $h(t) = 300 - 30t$, for $0 \le t \le 10$.

(b) From the triangle in the figure,

$$\tan\theta = \frac{h(t) - 100}{150} = \frac{300 - 30t - 100}{150} = \frac{200 - 30t}{150}.$$

Therefore

$$\theta = \arctan\left(\frac{200 - 30t}{150}\right)$$

and

$$\frac{d\theta}{dt} = \frac{1}{1 + \left(\frac{200-30t}{150}\right)^2} \cdot \left(\frac{-30}{150}\right) = -\frac{1}{5}\left(\frac{150^2}{150^2 + (200 - 30t)^2}\right).$$

Notice that $\frac{d\theta}{dt}$ is always negative, which is reasonable since θ decreases as the elevator descends.

(c) If we want to know when θ changes (decreases) the fastest, we want to find out when $d\theta/dt$ has the largest magnitude. This will occur when the denominator, $150^2 + (200 - 30t)^2$, in the expression for $d\theta/dt$ is the smallest, or when $200 - 30t = 0$. This occurs when $t = \frac{200}{30}$ seconds, and so $h(\frac{200}{30}) = 100$ feet, i.e., when the elevator is at the level of the observer.

Solutions for Section 4.7

Exercises

1. Since $f'(a) > 0$ and $g'(a) < 0$, l'Hopital's rule tells us that

$$\lim_{x \to a} \frac{f(x)}{g(x)} = \frac{f'(a)}{g'(a)} < 0.$$

5. The denominator approaches zero as x goes to zero and the numerator goes to zero even faster, so you should expect that the limit to be 0. You can check this by substituting several values of x close to zero. Alternatively, using l'Hopital's rule, we have

$$\lim_{x \to 0} \frac{x^2}{\sin x} = \lim_{x \to 0} \frac{2x}{\cos x} = 0.$$

9. The larger power dominates. Using l'Hopital's rule

$$\lim_{x \to \infty} \frac{x^5}{0.1x^7} = \lim_{x \to \infty} \frac{5x^4}{0.7x^6} = \lim_{x \to \infty} \frac{20x^3}{4.2x^5}$$

$$= \lim_{x \to \infty} \frac{60x^2}{21x^4} = \lim_{x \to \infty} \frac{120x}{84x^3} = \lim_{x \to \infty} \frac{120}{252x^2} = 0$$

so $0.1x^7$ dominates.

13. Let $f(x) = \ln x$ and $g(x) = 1/x$ so $f'(x) = 1/x$ and $g'(x) = -1/x^2$ and

$$\lim_{x \to 0^+} \frac{\ln x}{1/x} = \lim_{x \to 0^+} \frac{1/x}{-1/x^2} = \lim_{x \to 0^+} \frac{x}{-1} = 0.$$

Problems

17. Let $f(x) = \ln x$ and $g(x) = x^2 - 1$, so $f(1) = 0$ and $g(1) = 0$ and l'Hopital's rule can be used. To apply l'Hopital's rule, we first find $f'(x) = 1/x$ and $g'(x) = 2x$, then

$$\lim_{x \to 1} \frac{\ln x}{x^2 - 1} = \lim_{x \to 1} \frac{1/x}{2x} = \lim_{x \to 1} \frac{1}{2x^2} = \frac{1}{2}.$$

21. To get this expression in a form in which l'Hopital's rule applies, we rewrite it as a fraction:

$$x^a \ln x = \frac{\ln x}{x^{-a}}.$$

Letting $f(x) = \ln x$ and $g(x) = x^{-a}$, we have

$$\lim_{x \to 0^+} f(x) = \lim_{x \to 0^+} \ln x = -\infty, \quad \text{and} \quad \lim_{x \to 0^+} g(x) = \lim_{x \to 0^+} \frac{1}{x^a} = \infty.$$

So l'Hopital's rule can be used. To apply l'Hopital's rule we differentiate to get $f'(x) = 1/x$ and $g'(x) = -ax^{-a-1}$. Then

$$\lim_{x \to 0^+} x^a \ln x = \lim_{x \to 0^+} \frac{\ln x}{x^{-a}}$$
$$= \lim_{x \to 0^+} \frac{1/x}{-ax^{-a-1}}$$
$$= -\frac{1}{a} \lim_{x \to 0^+} x^a$$
$$= 0.$$

25. Let $y = (1 + \sin 3/x)^x$. Taking logs gives

$$\ln y = x \ln \left(1 + \sin \frac{3}{x}\right).$$

To use l'Hopital's rule, we rewrite $\ln y$ as a fraction:

$$\lim_{x \to \infty} \ln y = \lim_{x \to \infty} x \ln \left(1 + \sin \left(\frac{3}{x}\right)\right)$$
$$= \lim_{x \to \infty} \frac{\ln(1 + \sin(3/x))}{1/x}.$$

Let $f(x) = \ln(1 + \sin(3/x))$ and $g(x) = 1/x$ then

$$f'(x) = \frac{\cos(3/x)(-3/x^2)}{(1 + \sin(3/x))} \quad \text{and} \quad g'(x) = -\frac{1}{x^2}.$$

Now apply l'Hopital's rule to get

$$\lim_{x \to \infty} \ln y = \lim_{x \to \infty} \frac{f(x)}{g(x)}$$
$$= \lim_{x \to \infty} \frac{f'(x)}{g'(x)}$$

$$= \lim_{x \to \infty} \frac{\cos(3/x)(-3/x^2)/(1 + \sin(3/x))}{-1/x^2}$$

$$= \lim_{x \to \infty} \frac{3 \cos(3/x)}{1 + \sin(3/x)} = \frac{3 \cos 0}{1 + \sin 0}$$

$$= 3.$$

Since $\lim_{x \to \infty} \ln y = 3$, we have

$$\lim_{x \to \infty} y = e^3.$$

Thus,

$$\lim_{x \to \infty} \left(1 + \sin \frac{3}{x}\right)^x = e^3.$$

29. (a) Since $a + b = 1$, as $p \to 0$, we have

$$\ln(ax^p + by^p) \to \ln(ax^0 + by^0) = \ln(a + b) = \ln 1 = 0.$$

Thus, the limit is of the form $0/0$, so l'Hopital's rule applies. Since $d(x^p)/dp = (\ln x)x^p$,

$$\frac{d \ln(ax^p + by^p)}{dp} = \frac{a(\ln x)x^p + b(\ln y)y^p}{ax^p + by^p}$$

and $dp/dp = 1$ we have

$$\lim_{p \to 0} \frac{\ln(ax^p + by^p)}{p} = \lim_{p \to 0} \frac{a(\ln x)x^p + b(\ln y)y^p}{(ax^p + by^p) \cdot 1} = \frac{a(\ln x)x^0 + b(\ln y)y^0}{ax^0 + by^0} = \frac{a(\ln x) + b(\ln y)}{a + b} = a \ln x + b \ln y.$$

(b) Because

$$(ax^p + by^p)^{1/p} = e^{(\ln(ax^p + by^p))/p}$$

and the exponential function is continuous, using part (a), we have

$$\lim_{p \to 0} (ax^p + by^p)^{1/p} = \lim_{p \to 0} e^{(\ln(ax^p + by^p))/p} = e^{\lim_{p \to 0}(\ln(ax^p + by^p))/p} = e^{a \ln x + b \ln y} = x^a y^b.$$

33. Let $n = 1/(kx)$, so $n \to \infty$ as $x \to 0^+$. Thus

$$\lim_{x \to 0^+} (1 + kx)^{t/x} = \lim_{n \to \infty} \left(1 + \frac{1}{n}\right)^{nkt} = \lim_{n \to \infty} \left(\left(1 + \frac{1}{n}\right)^n\right)^{kt} = e^{kt}.$$

37. This limit is of the form ∞^0 so we apply l'Hopital's rule to

$$\ln f(t) = \frac{\ln\left((3^t + 5^t)/2\right)}{t}.$$

We have

$$\lim_{t \to +\infty} \ln f(t) = \lim_{t \to +\infty} \frac{\left((\ln 3)3^t + (\ln 5)5^t\right) / \left(3^t + 5^t\right)}{1}$$

$$= \lim_{t \to +\infty} \frac{(\ln 3)3^t + (\ln 5)5^t}{3^t + 5^t}$$

$$= \lim_{t \to +\infty} \frac{(\ln 3)(3/5)^t + \ln 5}{(3/5)^t + 1}$$

$$= \lim_{t \to +\infty} \frac{0 + \ln 5}{0 + 1} = \ln 5.$$

Thus

$$\lim_{t \to -\infty} f(t) = \lim_{t \to -\infty} e^{\ln f(t)} = e^{\lim_{t \to -\infty} \ln f(t)} = e^{\ln 5} = 5.$$

41. (a) Let $f(r) = x^r - 1$ and $g(r) = r$, where we are to compute $\lim_{r \to 0} f(r)/g(r)$. Since $f(0) = g(0) = 0$, the limit is of the form $0/0$ and l'Hopital's rule applies. Then $f'(r) = (\ln x)x^r$ and $g'(r) = 1$. Thus

$$h_0(x) = \lim_{r \to 0} \frac{x^r - 1}{r} = \frac{(\ln x)x^0}{1} = \ln x.$$

(b) See Figure 4.29.

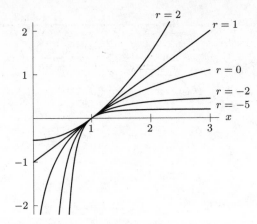

Figure 4.29

Solutions for Section 4.8

Exercises

1. Between times $t = 0$ and $t = 1$, x goes at a constant rate from 0 to 1 and y goes at a constant rate from 1 to 0. So the particle moves in a straight line from $(0, 1)$ to $(1, 0)$. Similarly, between times $t = 1$ and $t = 2$, it goes in a straight line to $(0, -1)$, then to $(-1, 0)$, then back to $(0, 1)$. So it traces out the diamond shown in Figure 4.30.

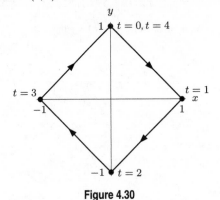

Figure 4.30

5. For $0 \le t \le \frac{\pi}{2}$, we have $x = \sin t$ increasing and $y = \cos t$ decreasing, so the motion is clockwise for $0 \le t \le \frac{\pi}{2}$. Similarly, we see that the motion is clockwise for the time intervals $\frac{\pi}{2} \le t \le \pi, \pi \le t \le \frac{3\pi}{2}$, and $\frac{3\pi}{2} \le t \le 2\pi$.

9. Let $f(t) = \ln t$. Then $f'(t) = \frac{1}{t}$. The particle is moving counterclockwise when $f'(t) > 0$, that is, when $t > 0$. Any other time, when $t \le 0$, the position is not defined.

13. We have
$$\frac{dx}{dt} = -2\sin 2t, \frac{dy}{dt} = \cos t.$$

The speed is
$$v = \sqrt{4\sin^2(2t) + \cos^2 t}.$$

Thus, $v = 0$ when $\sin(2t) = \cos t = 0$, and so the particle stops when $t = \pm\pi/2, \pm 3\pi/2, \ldots$ or $t = (2n+1)\frac{\pi}{2}$, for any integer n.

17. One possible answer is $x = -2, y = t$.

21. The ellipse $x^2/25 + y^2/49 = 1$ can be parameterized by $x = 5\cos t, y = 7\sin t, 0 \le t \le 2\pi$.

25. We have
$$\frac{dy}{dx} = \frac{dy/dt}{dx/dt} = \frac{4\cos(4t)}{3\cos(3t)}.$$

Thus when $t = \pi$, the slope of the tangent line is $-4/3$. Since $x = 0$ and $y = 0$ when $t = \pi$, the equation of the tangent line is $y = -(4/3)x$.

Problems

29. In all three cases, $y = x^2$, so that the motion takes place on the parabola $y = x^2$.

In case (a), the x-coordinate always increases at a constant rate of one unit distance per unit time, so the equations describe a particle moving to the right on the parabola at constant horizontal speed.

In case (b), the x-coordinate is never negative, so the particle is confined to the right half of the parabola. As t moves from $-\infty$ to $+\infty$, $x = t^2$ goes from ∞ to 0 to ∞. Thus the particle first comes down the right half of the parabola, reaching the origin $(0, 0)$ at time $t = 0$, where it reverses direction and goes back up the right half of the parabola.

In case (c), as in case (a), the particle traces out the entire parabola $y = x^2$ from left to right. The difference is that the horizontal speed is not constant. This is because a unit change in t causes larger and larger changes in $x = t^3$ as t approaches $-\infty$ or ∞. The horizontal motion of the particle is faster when it is farther from the origin.

33. (a) The curve is a spiral as shown in Figure 4.31.

(b) At $t = 2$, the position is $(2\cos 2, 2\sin 2) = (-0.8323, 1.8186)$, and at $t = 2.01$ the position is $(2.01\cos 2.01, 2.01\sin 2.01) = (-0.8546, 1.8192)$. The distance between these points is

$$\sqrt{(-0.8546 - (-0.8323))^2 + (1.8192 - 1.8186)^2} \approx 0.022.$$

Thus the speed is approximately $0.022/0.01 \approx 2.2$. See Figure 4.32.

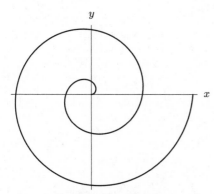

Figure 4.31: The spiral
$x = t\cos t, y = t\sin t$ for $0 \le t \le 4\pi$

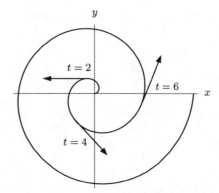

Figure 4.32: The spiral $x = t\cos t, y = t\sin t$ and three velocity vectors

(c) Evaluating the exact formula

$$v = \sqrt{(\cos t - t\sin t)^2 + (\sin t + t\cos t)^2}$$

gives :

$$v(2) = \sqrt{(-2.235)^2 + (0.077)^2} = 2.2363.$$

37. (a) Figure 4.33 shows the path and the clockwise direction of motion. (The curve is an ellipse.)

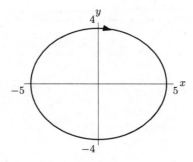

Figure 4.33

(b) At $t = \pi/4$, the position is given by

$$x(\pi/4) = 5\sin\frac{2\pi}{4} = 5\sin\frac{\pi}{2} = 5 \quad \text{and} \quad y(\pi/4) = 4\cos\frac{2\pi}{4} = 4\cos\frac{\pi}{2} = 0.$$

Differentiating, we get $x'(t) = 10\cos(2t)$ and $y'(t) = -8\sin(2t)$. At $t = \pi/4$, the velocity is given by

$$x'(\pi/4) = 10\cos\frac{2\pi}{4} = 10\cos\frac{\pi}{2} = 0 \quad \text{and} \quad y'(\pi/4) = -8\sin\frac{2\pi}{4} = -8\sin\frac{\pi}{2} = -8.$$

(c) As t increases from 0 to 2π, the ellipse is traced out twice. Thus, the particle passes through the point $(5, 0)$ twice.

(d) Since $x'(\pi/4) = 0$ and $y'(\pi/4) = -8$, when $t = \pi/4$, the particle is moving in the negative y-direction, parallel to the y-axis.

(e) At time t,

$$\text{Speed} = \sqrt{(x'(t))^2 + (y'(t))^2} = \sqrt{(10\cos(2t))^2 + (-8\sin(2t))^2}.$$

When $t = \pi$,

$$\text{Speed} = \sqrt{(10\cos(2\pi))^2 + (-8\sin(2\pi))^2} = \sqrt{(10 \cdot 1)^2 + (-8 \cdot 0)^2} = 10.$$

41. (a) The x and y-coordinates of the point on the graph when $t = \pi/3$ are given by

$$x = 3 \cdot \frac{\pi}{3} = \pi \quad \text{and} \quad y = \cos\left(\frac{2\pi}{3}\right) = -\frac{1}{2}.$$

Thus when $t = \pi/3$, the particle is at the point $(\pi, -1/2)$.

To find the slope, we find dy/dx

$$\frac{dy}{dx} = \frac{dy/dt}{dx/dt} = \frac{-2\sin(2t)}{3}.$$

When $t = \pi/3$,

$$\frac{dy}{dx} = \frac{-2\sin(2\pi/3)}{3} = -\frac{\sqrt{3}}{3}.$$

The equation of the tangent line when $t = \pi/3$ is:

$$y + \frac{1}{2} = -\frac{\sqrt{3}}{3}(x - \pi).$$

(b) To find the smallest positive value of t for which the y-coordinate is a local maximum, we set $dy/dt = 0$. We have

$$\frac{dy}{dt} = -2\sin(2t) = 0$$

$$2t = \pi \quad \text{or} \quad 2t = 2\pi$$

$$t = \frac{\pi}{2} \quad \text{or} \quad t = \pi.$$

There is a minimum of $y = \cos(2t)$ at $t = \pi/2$, and a maximum at $t = \pi$.

(c) To find d^2y/dx^2 when $t = 2$, we use the formula:

$$\frac{d^2y}{dx^2} = \frac{dw/dt}{dx/dt} \quad \text{where} \quad w = \frac{dy}{dx}.$$

Since $w = -2\sin(2t)/3$ from part (a), we have

$$\frac{d^2y}{dx^2} = \frac{-4\cos(2t)/3}{3}.$$

When $t = 2$, we have

$$\frac{d^2y}{dx^2} = \frac{-4\cos(4)/3}{3} = 0.291.$$

Since the second derivative is positive, the graph is concave up when $t = 2$.

45. For $0 \leq t \leq 2\pi$, we get Figure 4.34.

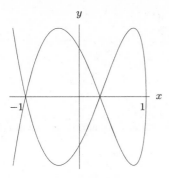

Figure 4.34

Figure 4.35

49. (a) To find the equations of the moon's motion relative to the star, you must first calculate the equation of the planet's motion relative to the star, and then the moon's motion relative to the planet, and then add the two together.

   The distance from the planet to the star is R, and the time to make one revolution is one unit, so the parametric equations for the planet relative to the star are $x = R \cos t$, $y = R \sin t$.

   The distance from the moon to the planet is 1, and the time to make one revolution is twelve units, therefore, the parametric equations for the moon relative to the planet are $x = \cos 12t$, $y = \sin 12t$.

   Adding these together, we get:

$$x = R \cos t + \cos 12t,$$
$$y = R \sin t + \sin 12t.$$

(b) For the moon to stop completely at time t, the velocity of the moon must be equal to zero. Therefore,

$$\frac{dx}{dt} = -R \sin t - 12 \sin 12t = 0,$$
$$\frac{dy}{dt} = R \cos t + 12 \cos 12t = 0.$$

There are many possible values to choose for R and t that make both of these equations equal to zero. We choose $t = \pi$, and $R = 12$.

(c) The graph with $R = 12$ is shown in Figure 4.35.

Solutions for Chapter 4 Review

Exercises

1. See Figure 4.36.

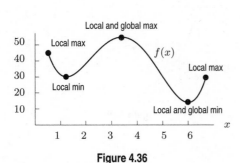

Figure 4.36

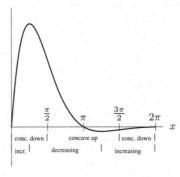

Figure 4.37

5. (a) First we find f' and f'':

$$f'(x) = -e^{-x}\sin x + e^{-x}\cos x$$
$$f''(x) = e^{-x}\sin x - e^{-x}\cos x$$
$$-e^{-x}\cos x - e^{-x}\sin x$$
$$= -2e^{-x}\cos x$$

(b) The critical points are $x = \pi/4, 5\pi/4$, since $f'(x) = 0$ here.

(c) The inflection points are $x = \pi/2, 3\pi/2$, since f'' changes sign at these points.

(d) At the endpoints, $f(0) = 0$, $f(2\pi) = 0$. So we have $f(\pi/4) = (e^{-\pi/4})(\sqrt{2}/2)$ as the global maximum; $f(5\pi/4) = -e^{-5\pi/4}(\sqrt{2}/2)$ as the global minimum.

(e) See Figure 4.37.

9. As $x \to -\infty$, $e^{-x} \to \infty$, so $xe^{-x} \to -\infty$. Thus $\lim_{x\to-\infty} xe^{-x} = -\infty$.

As $x \to \infty$, $\frac{x}{e^x} \to 0$, since e^x grows much more quickly than x. Thus $\lim_{x\to\infty} xe^{-x} = 0$.

Using the product rule,

$$f'(x) = e^{-x} - xe^{-x} = (1-x)e^{-x},$$

which is zero when $x = 1$, negative when $x > 1$, and positive when $x < 1$. Thus $f(1) = 1/e^1 = 1/e$ is a local maximum.

Again, using the product rule,

$$f''(x) = -e^{-x} - e^{-x} + xe^{-x}$$
$$= xe^{-x} - 2e^{-x}$$
$$= (x-2)e^{-x},$$

which is zero when $x = 2$, positive when $x > 2$, and negative when $x < 2$, giving an inflection point at $(2, \frac{2}{e^2})$. With the above, we have the following diagram:

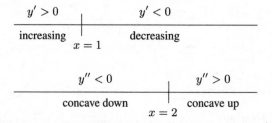

The graph of f is shown in Figure 4.38.   and $f(x)$ has one global maximum at $1/e$ and no local or global minima.

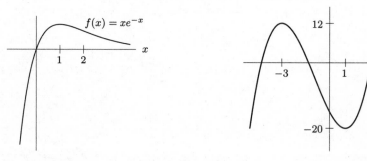

Figure 4.38

Figure 4.39

13. $\lim\limits_{x\to\infty} f(x) = +\infty$, and $\lim\limits_{x\to-\infty} f(x) = -\infty$.

There are no asymptotes.

$f'(x) = 3x^2 + 6x - 9 = 3(x+3)(x-1)$. Critical points are $x = -3$, $x = 1$.

$f''(x) = 6(x+1)$.

x		-3		-1		1	
f'	$+$	0	$-$	$-$	$-$	0	$+$
f''	$-$	$-$	$-$	0	$+$	$+$	$+$
f	⤴		⤵		⤵		⤴

Thus, $x = -1$ is an inflection point. $f(-3) = 12$ is a local maximum; $f(1) = -20$ is a local minimum. There are no global maxima or minima. See Figure 4.39.

17. Since $\lim\limits_{x\to-\infty} f(x) = \lim\limits_{x\to+\infty} f(x) = 0$, $y = 0$ is a horizontal asymptote.

$f'(x) = -2xe^{-x^2}$. So, $x = 0$ is the only critical point.

$f''(x) = -2(e^{-x^2} + x(-2x)e^{-x^2}) = 2e^{-x^2}(2x^2 - 1) = 2e^{-x^2}(\sqrt{2}x - 1)(\sqrt{2}x + 1)$.

Thus, $x = \pm 1/\sqrt{2}$ are inflection points.

Table 4.1

x		$-1/\sqrt{2}$		0		$1/\sqrt{2}$	
f'	$+$	$+$	$+$	0	$-$	$-$	$-$
f''	$+$	0	$-$	$-$	$-$	0	$+$
f	⤴		⤴		⤵		⤵

Thus, $f(0) = 1$ is a local and global maximum.

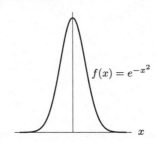

$f(x) = e^{-x^2}$

21. To find the best possible bounds for $f(x) = x^3 - 6x^2 + 9x + 5$ on $0 \le x \le 5$, we find the global maximum and minimum for the function on the interval. First, we find the critical points. Differentiating yields

$$f'(x) = 3x^2 - 12x + 9$$

Letting $f'(x) = 0$ and factoring yields

$$3x^2 - 12x + 9 = 0$$
$$3(x^2 - 4x + 3) = 0$$
$$3(x-3)(x-1) = 0$$

So $x = 1$ and $x = 3$ are critical points for the function on $0 \le x \le 5$. Evaluating the function at the critical points and endpoints gives us

$$f(0) = (0)^3 - 6(0)^2 + 9(0) + 5 = 5$$
$$f(1) = (1)^3 - 6(1)^2 + 9(1) + 5 = 9$$
$$f(3) = (3)^3 - 6(3)^2 + 9(3) + 5 = 5$$
$$f(5) = (5)^3 - 6(5)^2 + 9(5) + 5 = 25$$

So the global minimum on this interval is $f(0) = f(3) = 5$ and the global maximum is $f(5) = 25$. From this we conclude

$$5 \le x^3 - 6x^2 + 9x + 5 \le 25$$

are the best possible bounds for the function on the interval $0 \le x \le 5$.

Problems

25. The critical points of f occur where f' is zero. These two points are indicated in the figure below.

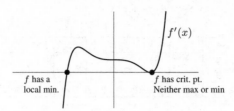

Note that the point labeled as a local minimum of f is not a critical point of f'.

29. (a) Decreasing for $x < 0$, increasing for $0 < x < 4$, and decreasing for $x > 4$.
   (b) $f(0)$ is a local minimum, and $f(4)$ is a local maximum.

33.

$$r(\lambda) = a(\lambda)^{-5}(e^{b/\lambda} - 1)^{-1}$$
$$r'(\lambda) = a(-5\lambda^{-6})(e^{b/\lambda} - 1)^{-1} + a(\lambda^{-5})\left(\frac{b}{\lambda^2}e^{b/\lambda}\right)(e^{b/\lambda} - 1)^{-2}$$

$(0.96, 3.13)$ is a maximum, so $r'(0.96) = 0$ implies that the following holds, with $\lambda = 0.96$:

$$5\lambda^{-6}(e^{b/\lambda} - 1)^{-1} = \lambda^{-5}\left(\frac{b}{\lambda^2}e^{b/\lambda}\right)(e^{b/\lambda} - 1)^{-2}$$
$$5\lambda(e^{b/\lambda} - 1) = be^{b/\lambda}$$
$$5\lambda e^{b/\lambda} - 5\lambda = be^{b/\lambda}$$
$$5\lambda e^{b/\lambda} - be^{b/\lambda} = 5\lambda$$
$$\left(\frac{5\lambda - b}{5\lambda}\right)e^{b/\lambda} = 1$$
$$\frac{4.8 - b}{4.8}e^{b/0.96} - 1 = 0.$$

Using Newton's method, or some other approximation method, we search for a root. The root should be near 4.8. Using our initial guess, we get $b \approx 4.7665$. At $\lambda = 0.96$, $r = 3.13$, so

$$3.13 = \frac{a}{0.96^5(e^{b/0.96} - 1)} \quad \text{or}$$
$$a = 3.13(0.96)^5(e^{b/0.96} - 1)$$
$$\approx 363.23.$$

As a check, we try $r(4) \approx 0.155$, which looks about right on the given graph.

37. The volume is given by $V = x^2y$. The surface area is given by

$$S = 2x^2 + 4xy$$
$$= 2x^2 + 4xV/x^2 = 2x^2 + 4V/x.$$

To find the dimensions which minimize the area, find x such that $dS/dx = 0$:

$$\frac{dS}{dx} = 4x - \frac{4V}{x^2} = 0$$
$$x^3 = V.$$

Solving for x gives $x = \sqrt[3]{V} = y$. To see that this gives a minimum, note that for small x, $S \approx 4V/x$ is decreasing. For large x, $S \approx 2x^2$ is increasing. Since there is only one critical point, it must give a global minimum. Therefore, when the width equals the height, the surface area is minimized.

41. Since the volume is fixed at 200 ml (i.e. 200 cm^3), we can solve the volume expression for h in terms of r to get (with h and r in centimeters)

$$h = \frac{200 \cdot 3}{7\pi r^2}.$$

Using this expression in the surface area formula we arrive at

$$S = 3\pi r \sqrt{r^2 + \left(\frac{600}{7\pi r^2}\right)^2}$$

By plotting $S(r)$ we see that there is a minimum value near $r = 2.7$ cm.

45. (a) The length of the piece of wire made into a circle is x cm, so the length of the piece made into a square is $(L - x)$ cm. See Figure 4.40.

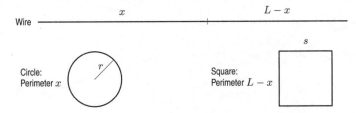

Figure 4.40

The circumference of the circle is x, so its radius, r cm, is given by

$$r = \frac{x}{2\pi} \text{ cm.}$$

The perimeter of the square is $(L - x)$, so the side length, s cm, is given by

$$s = \frac{L - x}{4} \text{ cm.}$$

Thus, the sum of areas is given by

$$A = \pi r^2 + s^2 = \pi \left(\frac{x}{2\pi}\right)^2 + \left(\frac{L - x}{4}\right)^2 = \frac{x^2}{4\pi} + \frac{(L - x)^2}{16}, \quad \text{for } 0 \le x \le L.$$

Setting $dA/dx = 0$ to find the critical points gives

$$\frac{dA}{dx} = \frac{x}{2\pi} - \frac{(L - x)}{8} = 0$$
$$8x = 2\pi L - 2\pi x$$
$$(8 + 2\pi)x = 2\pi L$$
$$x = \frac{2\pi L}{8 + 2\pi} = \frac{\pi L}{4 + \pi} \approx 0.44L.$$

To find the maxima and minima, we substitute the critical point and the endpoints, $x = 0$ and $x = L$, into the area function.

For $x = 0$, we have $A = \dfrac{L^2}{16}$.

For $x = \dfrac{\pi L}{4 + \pi}$, we have $L - x = L - \dfrac{\pi L}{4 + \pi} = \dfrac{4L}{4 + \pi}$. Then

$$A = \frac{\pi^2 L^2}{4\pi(4 + \pi)^2} + \frac{1}{16}\left(\frac{4L}{4 + \pi}\right)^2 = \frac{\pi L^2}{4(4 + \pi)^2} + \frac{L^2}{(4 + \pi)^2}$$
$$= \frac{\pi L^2 + 4L^2}{4(4 + \pi)^2} = \frac{L^2}{4(4 + \pi)} = \frac{L^2}{16 + 4\pi}.$$

For $x = L$, we have $A = \dfrac{L^2}{4\pi}$.

Thus, $x = \dfrac{\pi L}{4 + \pi}$ gives the minimum value of $A = \dfrac{L^2}{16 + 4\pi}$.

Since $4\pi < 16$, we see that $x = L$ gives the maximum value of $A = \dfrac{L^2}{4\pi}$.

This corresponds to the situation in which we do not cut the wire at all and use the single piece to make a circle.

(b) At the maximum, $x = L$, so

$$\frac{\text{Length of wire in square}}{\text{Length of wire in circle}} = \frac{0}{L} = 0.$$

$$\frac{\text{Area of square}}{\text{Area of circle}} = \frac{0}{L^2/4\pi} = 0.$$

At the minimum, $x = \dfrac{\pi L}{4 + \pi}$, so $L - x = L - \dfrac{\pi L}{4 + \pi} = \dfrac{4L}{4 + \pi}$.

$$\frac{\text{Length of wire in square}}{\text{Length of wire in circle}} = \frac{4L/(4 + \pi)}{\pi L/(4 + \pi)} = \frac{4}{\pi}.$$

$$\frac{\text{Area of square}}{\text{Area of circle}} = \frac{L^2/(4 + \pi)^2}{\pi L^2/(4(4 + \pi)^2)} = \frac{4}{\pi}.$$

(c) For a general value of x,

$$\frac{\text{Length of wire in square}}{\text{Length of wire in circle}} = \frac{L - x}{x}.$$

$$\frac{\text{Area of square}}{\text{Area of circle}} = \frac{(L - x)^2/16}{x^2/(4\pi)} = \frac{\pi}{4} \cdot \frac{(L - x)^2}{x^2}.$$

If the ratios are equal, we have

$$\frac{L - x}{x} = \frac{\pi}{4} \cdot \frac{(L - x)^2}{x^2}.$$

So either $L - x = 0$, giving $x = L$, or we can cancel $(L - x)$ and multiply through by $4x^2$, giving

$$4x = \pi(L - x)$$

$$x = \frac{\pi L}{4 + \pi}.$$

Thus, the two values of x found in part (a) are the only values of x in $0 \leq x \leq L$ making the ratios in part (b) equal. (The ratios are not defined if $x = 0$.)

49. If $f(x) = x - \sinh x$ and $g(x) = x^3$, then $f(0) = g(0) = 0$. However, $f'(0) = g'(0) = f''(0) = g''(0) = 0$ also, so we use l'Hopital's Rule three times. Since $f'''(x) = -\cosh x$ and $g'''(x) = 6$:

$$\lim_{x \to 0} \frac{x - \sinh x}{x^3} = \lim_{x \to 0} \frac{1 - \cosh x}{3x^2} = \lim_{x \to 0} \frac{-\sinh x}{6x} = \lim_{x \to 0} \frac{-\cosh x}{6} = -\frac{1}{6}.$$

53. (a) Since $d\theta/dt$ represents the rate of change of θ with time, $d\theta/dt$ represents the angular velocity of the disk.

(b) Suppose P is the point on the rim shown in Figure 4.41.

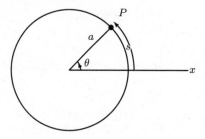

Figure 4.41

Any other point on the rim is moving at the same speed, though in a different direction. We know that since θ is in radians,

$$s = a\theta.$$

Since a is a constant, we know

$$\frac{ds}{dt} = a\frac{d\theta}{dt}.$$

But $ds/dt = v$, the speed of the point on the rim, so

$$v = a\frac{d\theta}{dt}.$$

57. We want to find dP/dV. Solving $PV = k$ for P gives

$$P = k/V$$

so,

$$\frac{dP}{dV} = -\frac{k}{V^2}.$$

CAS Challenge Problems

61. (a) A CAS gives

$$\frac{d}{dx}\operatorname{arcsinh} x = \frac{1}{\sqrt{1+x^2}}$$

(b) Differentiating both sides of $\sinh(\operatorname{arcsinh} x) = x$, we get

$$\cosh(\operatorname{arcsinh} x)\frac{d}{dx}(\operatorname{arcsinh} x) = 1$$

$$\frac{d}{dx}(\operatorname{arcsinh} x) = \frac{1}{\cosh(\operatorname{arcsinh} x)}.$$

Since $\cosh^2 x - \sinh^2 x = 1$, $\cosh x = \pm\sqrt{1+\sinh^2 x}$. Furthermore, since $\cosh x > 0$ for all x, we take the positive square root, so $\cosh x = \sqrt{1+\sinh^2 x}$. Therefore, $\cosh(\operatorname{arcsinh} x) = \sqrt{1+(\sinh(\operatorname{arcsinh} x))^2} = \sqrt{1+x^2}$. Thus

$$\frac{d}{dx}\operatorname{arcsinh} x = \frac{1}{\sqrt{1+x^2}}.$$

65. (a)

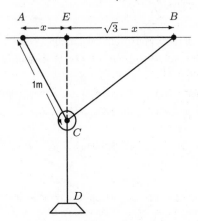

Figure 4.42

We want to maximize the sum of the lengths EC and CD in Figure 4.42. Let x be the distance AE. Then x can be between 0 and 1, the length of the left rope. By the Pythagorean theorem,

$$EC = \sqrt{1-x^2}.$$

The length of the rope from B to C can also be found by the Pythagorean theorem:

$$BC = \sqrt{EC^2 + EB^2} = \sqrt{1-x^2 + (\sqrt{3}-x)^2} = \sqrt{4-2\sqrt{3}x}.$$

Since the entire rope from B to D has length 3 m, the length from C to D is

$$CD = 3 - \sqrt{4-2\sqrt{3}x}.$$

The distance we want to maximize is

$$f(x) = EC + CD = \sqrt{1-x^2} + 3 - \sqrt{4-2\sqrt{3}x}, \quad \text{for} \quad 0 \le x \le 1.$$

Differentiating gives

$$f'(x) = \frac{-2x}{2\sqrt{1-x^2}} - \frac{-2\sqrt{3}}{2\sqrt{4-2\sqrt{3}x}}.$$

Setting $f'(x) = 0$ gives the cubic equation

$$2\sqrt{3}x^3 - 7x^2 + 3 = 0.$$

Using a computer algebra system to solve the equation gives three roots: $x = -1/\sqrt{3}, x = \sqrt{3}/2, x = \sqrt{3}$. We discard the negative root. Since x cannot be larger than 1 meter (the length of the left rope), the only critical point of interest is $x = \sqrt{3}/2$, that is, halfway between A and B.

To find the global maximum, we calculate the distance of the weight from the ceiling at the critical point and at the endpoints:

$$f(0) = \sqrt{1} + 3 - \sqrt{4} = 2$$

$$f\left(\frac{\sqrt{3}}{2}\right) = \sqrt{1 - \frac{3}{4}} + 3 - \sqrt{4 - 2\sqrt{3} \cdot \frac{\sqrt{3}}{2}} = 2.5$$

$$f(1) = \sqrt{0} + 3 - \sqrt{4 - 2\sqrt{3}} = 4 - \sqrt{3} = 2.27.$$

Thus, the weight is at the maximum distance from the ceiling when $x = \sqrt{3}/2$; that is, the weight comes to rest at a point halfway between points A and B.

(b) No, the equilibrium position depends on the length of the rope. For example, suppose that the left-hand rope was 1 cm long. Then there is no way for the pulley at its end to move to a point halfway between the anchor points.

CHECK YOUR UNDERSTANDING

1. True. Since the domain of f is all real numbers, all local minima occur at critical points.

5. False. For example, if $f(x) = x^3$, then $f'(0) = 0$, but $f(x)$ does not have either a local maximum or a local minimum at $x = 0$.

9. Let $f(x) = ax^2$, with $a \neq 0$. Then $f'(x) = 2ax$, so f has a critical point only at $x = 0$.

13. True. If the maximum is not at an endpoint, then it must be at critical point of f. But $x = 0$ is the only critical point of $f(x) = x^2$ and it gives a minimum, not a maximum.

17. False. The circumference A and radius r are related by $A = \pi r^2$, so $dA/dt = 2\pi r \, dr/dt$. Thus dA/dt depends on r and since r is not constant, neither is dA/dt.

21. $f(x) = x^2 + 1$ is positive for all x and concave up.

25. This is impossible. Since f'' exists, so must f', which means that f is differentiable and hence continuous. If $f(x)$ were positive for some values of x and negative for other values, then by the Intermediate Value Theorem, $f(x)$ would have to be zero somewhere, but this is impossible since $f(x)f''(x) < 0$ for all x. Thus either $f(x) > 0$ for all values of x, in which case $f''(x) < 0$ for all values of x, that is f is concave down. But this is impossible by Problem 22. Or else $f(x) < 0$ for all x, in which case $f''(x) > 0$ for all x, that is f is concave up. But this is impossible by Problem 24.

CHAPTER FIVE

Solutions for Section 5.1

Exercises

1. (a) Lower estimate $= (45)(2) + (16)(2) + (0)(2) = 122$ feet.
   Upper estimate $= (88)(2) + (45)(2) + (16)(2) = 298$ feet.

   (b)

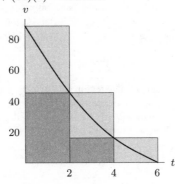

5. (a) With $n = 4$, we have $\Delta t = 4$. Then

$$t_0 = 0, t_1 = 4, t_2 = 8, t_3 = 12, t_4 = 16 \quad \text{and} \quad f(t_0) = 25, f(t_1) = 23, f(t_2) = 22, f(t_3) = 20, f(t_4) = 17$$

   (b)

$$\text{Left sum } = (25)(4) + (23)(4) + (22)(4) + (20)(4) = 360$$
$$\text{Right sum } = (23)(4) + (22)(4) + (20)(4) + (17)(4) = 328.$$

   (c) With $n = 2$, we have $\Delta t = 8$. Then

$$t_0 = 0, t_1 = 8, t_2 = 16 \quad \text{and} \quad f(t_0) = 25, f(t_1) = 22, f(t_2) = 17$$

   (d)

$$\text{Left sum } = (25)(8) + (22)(8) = 376$$
$$\text{Right sum } = (22)(8) + (17)(8) = 312.$$

9. Figure 5.1 shows the graph of $f(t)$. The region under the graph of $f(t)$ from $t = 0$ to $t = 10$ is a triangle of base 10 seconds and height 50 meter/sec. Then

$$\text{Distance traveled } = \text{Area of triangle} = \frac{1}{2} \cdot 10 \cdot 50 = 250 \text{ meters.}$$

Thus the distance traveled is 250 meters.

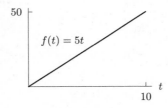

Figure 5.1

13. Since f is decreasing, the right-hand sum is the lower estimate and the left-hand sum is the upper estimate. We have $f(0) = 1$, $f(2) = e^{-2}$ and $\Delta t = (b - a)/n = 2/20 = 1/10$. Thus,

$$|\text{Difference in estimates}| = |f(b) - f(a)|\Delta t$$

$$= |e^{-2} - 1|\frac{1}{10} = 0.086.$$

Problems

17. The velocity is constant and negative, so the change in position is $-3 \cdot 5$ cm, that is 15 cm to the left.

21. The change in position is calculated from the area between the velocity graph and the t-axis, with the region below the axis corresponding to negatives velocities and counting negatively.

Figure 5.2 shows the graph of $f(t)$. From $t = 0$ to $t = 3$ the velocity is positive. The region under the graph of $f(t)$ is a triangle with height 6 cm/sec and base 3 seconds. Thus, from $t = 0$ to $t = 3$, the particle moves

$$\text{Distance moved to right } = \frac{1}{2} \cdot 3 \cdot 6 = 9 \text{ centimeters.}$$

From $t = 3$ to $t = 4$, the velocity is negative. The region between the graph of $f(t)$ and the t-axis is a triangle with height 2 cm/sec and base 1 second, so in this interval the particle moves

$$\text{Distance moved to left } = \frac{1}{2} \cdot 1 \cdot 2 = 1 \text{ centimeter.}$$

Thus, the total change in position is $9 - 1 = 8$ centimeters to the right.

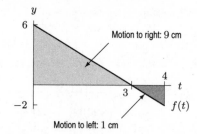

Figure 5.2

25. (a) Car A has the largest maximum velocity because the peak of car A's velocity curve is higher than the peak of B's.
   (b) Car A stops first because the curve representing its velocity hits zero (on the t-axis) first.
   (c) Car B travels farther because the area under car B's velocity curve is the larger.

Solutions for Section 5.2

Exercises

1.

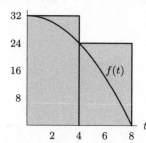

Figure 5.3: Left Sum, $\Delta t = 4$

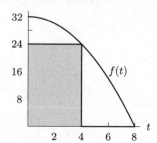

Figure 5.4: Right Sum, $\Delta t = 4$

(a) Left-hand sum $= 32 \cdot 4 + 24 \cdot 4 = 224$.

(b) Right-hand sum = $24 \cdot 4 + 0 \cdot 4 = 96$.

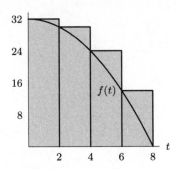

Figure 5.5: Left Sum, $\Delta t = 2$

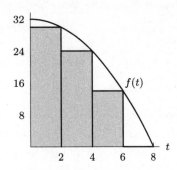

Figure 5.6: Right Sum, $\Delta t = 2$

(c) Left-hand sum = $32 \cdot 2 + 30 \cdot 2 + 24 \cdot 2 + 14 \cdot 2 = 200$.
(d) Right-hand sum = $30 \cdot 2 + 24 \cdot 2 + 14 \cdot 2 + 0 \cdot 2 = 136$.

5. The graph given shows that f is positive for $0 \le t \le 1$. Since the graph is contained within a rectangle of height 100 and length 1, the answers -98.35 and 100.12 are both either too small or too large to represent $\int_0^1 f(t)dt$. Since the graph of f is above the horizontal line $y = 80$ for $0 \le t \le 0.95$, the best estimate is 93.47 and not 71.84.

9. We use a calculator or computer to see that $\displaystyle\int_0^3 2^x\,dx = 10.0989$.

13. A graph of $y = 6x^3 - 2$ shows that this function is nonnegative on the interval $x = 5$ to $x = 10$. Thus,

$$\text{Area} = \int_5^{10} (6x^3 - 2)\,dx = 14{,}052.5.$$

The integral was evaluated on a calculator.

17. The graph of $y = 7 - x^2$ has intercepts $x = \pm\sqrt{7}$. See Figure 5.7. Therefore we have

$$\text{Area} = \int_{-\sqrt{7}}^{\sqrt{7}} (7 - x^2)\,dx = 24.694.$$

The integral was evaluated on a calculator.

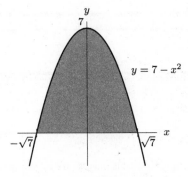

Figure 5.7

Problems

21. Left-hand sum gives: $1^2(1/4) + (1.25)^2(1/4) + (1.5)^2(1/4) + (1.75)^2(1/4) = 1.96875$.
    Right-hand sum gives: $(1.25)^2(1/4) + (1.5)^2(1/4) + (1.75)^2(1/4) + (2)^2(1/4) = 2.71875$.

We estimate the value of the integral by taking the average of these two sums, which is 2.34375. Since x^2 is monotonic on $1 \le x \le 2$, the true value of the integral lies between 1.96875 and 2.71875. Thus the most our estimate could be off is 0.375. We expect it to be much closer. (And it is—the true value of the integral is $7/3 \approx 2.333$.)

25. Looking at the graph of $e^{-x} \sin x$ for $0 \leq x \leq 2\pi$ in Figure 5.8, we see that the area, A_1, below the curve for $0 \leq x \leq \pi$ is much greater than the area, A_2, above the curve for $\pi \leq x \leq 2\pi$. Thus, the integral is

$$\int_0^{2\pi} e^{-x} \sin x \, dx = A_1 - A_2 > 0.$$

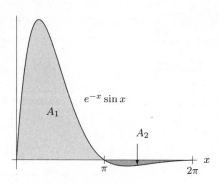

Figure 5.8

29. (a) $\displaystyle\int_{-3}^0 f(x) \, dx = -2.$

(b) $\displaystyle\int_{-3}^4 f(x) \, dx = \int_{-3}^0 f(x) \, dx + \int_0^3 f(x) \, dx + \int_3^4 f(x) \, dx = -2 + 2 - \frac{A}{2} = -\frac{A}{2}.$

33. We have

$$\Delta x = \frac{4}{3} = \frac{b-a}{n} \quad \text{and} \quad n = 3, \quad \text{so} \quad b - a = 4 \quad \text{or} \quad b = a + 4.$$

The function, $f(x)$, is squaring something. Since it is a left-hand sum, $f(x)$ could equal x^2 with $a = 2$ and $b = 6$ (note that $2 + 3(\frac{4}{3})$ gives the right-hand endpoint of the last interval). Or, $f(x)$ could possibly equal $(x+2)^2$ with $a = 0$ and $b = 4$. Other answers are possible.

Solutions for Section 5.3

Exercises

1. The units of measurement are dollars.

5. The integral $\int_0^6 a(t) \, dt$ represents the change in velocity between times $t = 0$ and $t = 6$ seconds; it is measured in km/hr.

9. Average value $= \dfrac{1}{2-0} \displaystyle\int_0^2 (1+t) \, dt = \dfrac{1}{2}(4) = 2.$

13. For any t, consider the interval $[t, t + \Delta t]$. During this interval, oil is leaking out at an approximately constant rate of $f(t)$ gallons/minute. Thus, the amount of oil which has leaked out during this interval can be expressed as

$$\text{Amount of oil leaked} = \text{Rate} \times \text{Time} = f(t) \, \Delta t$$

and the units of $f(t) \, \Delta t$ are gallons/minute $\times$ minutes $=$ gallons. The total amount of oil leaked is obtained by adding all these amounts between $t = 0$ and $t = 60$. (An hour is 60 minutes.) The sum of all these infinitesimal amounts is the integral

$$\begin{array}{c} \text{Total amount of} \\ \text{oil leaked, in gallons} \end{array} = \int_0^{60} f(t) \, dt.$$

Problems

17. (a) The integral $\int_0^{50} f(t)\,dt$ represents the total emissions of nitrogen oxides, in millions of metric tons, during the period 1940 to 1990.

(b) We estimate the integral using left- and right-hand sums:

$$\text{Left sum} = (6.9)(10) + (9.4)(10) + (13.0)(10) + (18.5)(10) + (20.9)(10) = 687.$$

$$\text{Right sum} = (9.4)(10) + (13.0)(10) + (18.5)(10) + (20.9)(10) + (19.6)(10) = 814.$$

We average the left- and right-hand sums to find the best estimate of the integral:

$$\int_0^{50} f(t)\,dt \approx \frac{687 + 814}{2} = 750.5 \text{ million metric tons.}$$

Between 1940 and 1990, about 750.5 million metric tons of nitrogen oxides were emitted.

21. Since W is in tons per week and t is in weeks since January 1, 2000, the integral $\int_0^{52} W\,dt$ gives the amount of waste, in tons, produced during the year 2000.

$$\text{Total waste during the year} = \int_0^{52} 3.75 e^{-0.008t}\,dt = 159.5249 \text{ tons.}$$

Since waste removal costs \$15/ton, the cost of waste removal for the company is $159.5249 \cdot 15 = \$2392.87$.

25. (a) The integral is the area above the x-axis minus the area below the x-axis. Thus, we can see that $\int_{-3}^{3} f(x)\,dx$ is about $-6 + 2 = -4$ (the negative of the area from $t = -3$ to $t = 1$ plus the area from $t = 1$ to $t = 3$.)

(b) Since the integral in part (a) is negative, the average value of $f(x)$ between $x = -3$ and $x = 3$ is negative. From the graph, however, it appears that the average value of $f(x)$ from $x = 0$ to $x = 3$ is positive. Hence (ii) is the larger quantity.

29. (a) Average value $= \int_0^{1} \sqrt{1 - x^2}\,dx = 0.79$.

(b) The area between the graph of $y = 1 - x$ and the x-axis is 0.5. Because the graph of $y = \sqrt{1 - x^2}$ is concave down, it lies above the line $y = 1 - x$, so its average value is above 0.5. See figure below.

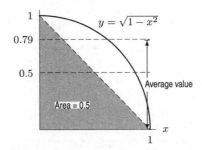

33. (a) Over the interval $[-1, 3]$, we estimate that the total change of the population is about 1.5, by counting boxes between the curve and the x-axis; we count about 1.5 boxes below the x-axis from $x = -1$ to $x = 1$ and about 3 above from $x = 1$ to $x = 3$. So the average rate of change is just the total change divided by the length of the interval, that is $1.5/4 = 0.375$ thousand/hour.

(b) We can estimate the total change of the algae population by counting boxes between the curve and the x-axis. Here, there is about 1 box above the x-axis from $x = -3$ to $x = -2$, about 0.75 of a box below the x-axis from $x = -2$ to $x = -1$, and a total change of about 1.5 boxes thereafter (as discussed in part (a)). So the total change is about $1 - 0.75 + 1.5 = 1.75$ thousands of algae.

37. On the interval $a \le t \le b$, we have

$$\begin{array}{cc} \text{Average value} \\ \text{of } v(t) \end{array} = \frac{1}{b - a} \int_a^b v(t)\,dt.$$

Since $v(t) = s'(t)$, by the Fundamental Theorem of Calculus, we get:

$$\frac{1}{b - a} \int_a^b v(t)\,dt = \frac{1}{b - a}\left(s(b) - s(a)\right) = \text{Average velocity.}$$

Solutions for Section 5.4

Exercises

1. (a) A graph of $f'(x) = \sin(x^2)$ is shown in Figure 5.9. Since the derivative $f'(x)$ is positive between $x = 0$ and $x = 1$, the change in $f(x)$ is positive, so $f(1)$ is larger than $f(0)$. Between $x = 2$ and $x = 2.5$, we see that $f'(x)$ is negative, so the change in $f(x)$ is negative; thus, $f(2)$ is greater than $f(2.5)$.

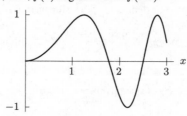

Figure 5.9: Graph of $f'(x) = \sin(x^2)$

(b) The change in $f(x)$ between $x = 0$ and $x = 1$ is given by the Fundamental Theorem of Calculus:

$$f(1) - f(0) = \int_0^1 \sin(x^2)dx = 0.310.$$

Since $f(0) = 2$, we have

$$f(1) = 2 + 0.310 = 2.310.$$

Similarly, since

$$f(2) - f(0) = \int_0^2 \sin(x^2)dx = 0.805,$$

we have

$$f(2) = 2 + 0.805 = 2.805.$$

Since

$$f(3) - f(0) = \int_0^3 \sin(x^2)dx = 0.774,$$

we have

$$f(3) = 2 + 0.774 = 2.774.$$

The results are shown in the table.

x	0	1	2	3
$f(x)$	2	2.310	2.805	2.774

5. The graph of $y = 5\ln(2x)$ is above the line $y = 3$ for $3 \leq x \leq 5$. See Figure 5.10. Therefore

$$\text{Area} = \int_3^5 (5\ln(2x) - 3)\, dx = 14.688.$$

The integral was evaluated on a calculator.

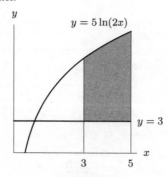

Figure 5.10

9. The graph of $y = \cos t$ is above the graph of $y = \sin t$ for $0 \leq t \leq \pi/4$ and $y = \cos t$ is below $y = \sin t$ for $\pi/4 < t < \pi$. See Figure 5.11. Therefore, we find the area in two pieces:

$$\text{Area} = \int_0^{\pi/4} (\cos t - \sin t)\, dt + \int_{\pi/4}^{\pi} (\sin t - \cos t)\, dt = 2.828.$$

The integral was evaluated on a calculator.

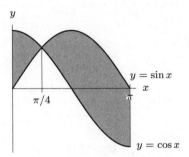

Figure 5.11

13. We have $f(t) = F'(t) = 4t^3$, so by the Fundamental Theorem of Calculus,

$$\int_{-1}^{1} 4t^3\, dt = F(1) - F(-1) = 1 - 1 = 0.$$

Notice in this case the integral is 0 because the function being integrated, $f(t) = 4t^3$, is odd: the negative contribution to the integral from $a = -1$ to $b = 1$ exactly cancels the positive.

Problems

17. We have

$$\int_a^b (f(x))^2\, dx - \left(\int_a^b f(x)\, dx \right)^2 = 12 - 8^2 = -52.$$

21. We have

$$8 = \int_{-2}^{5} f(x)\, dx = \int_{-2}^{2} f(x)\, dx + \int_{2}^{5} f(x)\, dx.$$

Since f is odd, $\int_{-2}^{2} f(x)\, dx = 0$, so $\int_{-2}^{5} f(x)\, dx = 8$.

25. (a) See Figure 5.12. Since the shaded region lies within a rectangle of area 1, the area is less than 1.

(b) Since the area is given by the integral

$$\text{Area} = \int_0^1 e^{-x^2/2}\, dx = 0.856.$$

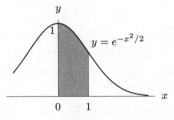

Figure 5.12

29. (a) The integrand is positive, so the integral cannot be negative.

(b) The integrand ≥ 0. If the integral $= 0$, then the integrand must be identically 0, which is not true.

33. (a) Yes.

(b) No, because the sum of the left sums has 20 subdivisions. The result is the left sum approximation with 20 subdivisions to $\int_1^3 f(x)\,dx$.

37. See Figure 5.13.

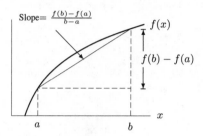

Figure 5.13

41. By the given property, $\displaystyle\int_a^a f(x)\,dx = -\int_a^a f(x)\,dx$, so $2\displaystyle\int_a^a f(x)\,dx = 0$. Thus $\displaystyle\int_a^a f(x)\,dx = 0$.

Solutions for Chapter 5 Review

Exercises

1. (a) Suppose $f(t)$ is the flowrate in m^3/hr at time t. We are only given two values of the flowrate, so in making our estimates of the flow, we use one subinterval, with $\Delta t = 3/1 = 3$:

$$\text{Left estimate} = 3[f(6\text{ am})] = 3\cdot 100 = 300\text{ m}^3 \quad \text{(an underestimate)}$$
$$\text{Right estimate} = 3[f(9\text{ am})] = 3\cdot 280 = 840\text{ m}^3 \quad \text{(an overestimate)}.$$

The best estimate is the average of these two estimates,

$$\text{Best estimate} = \frac{\text{Left} + \text{Right}}{2} = \frac{300 + 840}{2} = 570\text{ m}^3.$$

(b) Since the flowrate is increasing throughout, the error, i.e., the difference between over- and under-estimates, is given by

$$\text{Error} \leq \Delta t\,[f(9\text{ am}) - f(6\text{ am})] = \Delta t[280 - 100] = 180\Delta t.$$

We wish to choose Δt so that the the error $180\Delta t \leq 6$, or $\Delta t \leq 6/180 = 1/30$. So the flowrate gauge should be read every $1/30$ of an hour, or every 2 minutes.

5. We take $\Delta t = 20$. Then:

$$\text{Left-hand sum} = 1.2(20) + 2.8(20) + 4.0(20) + 4.7(20) + 5.1(20)$$
$$= 356.$$
$$\text{Right-hand sum} = 2.8(20) + 4.0(20) + 4.7(20) + 5.1(20) + 5.2(20)$$
$$= 436.$$
$$\int_0^{100} f(t)\,dt \approx \text{Average} = \frac{356 + 436}{2} = 396.$$

9. Since x intercepts are $x = 0, \pi, 2\pi, \ldots,$

$$\text{Area} = \int_0^{\pi} \sin x \, dx = 2.$$

The integral was evaluated on a calculator.

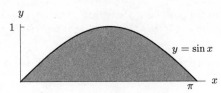

13. The graph of $y = -e^x + e^{2(x-1)}$ has intercepts where $e^x = e^{2(x-1)}$, or where $x = 2(x-1)$, so $x = 2$. See Figure 5.14. Since the region is below the x-axis, the integral is negative, so

$$\text{Area} = -\int_0^2 -e^x + e^{2(x-1)} \, dx = 2.762.$$

The integral was evaluated on a calculator.

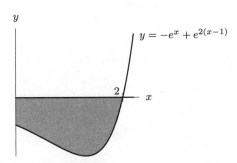

Figure 5.14

17. Distance traveled $= \displaystyle\int_0^{1.1} \sin(t^2) \, dt \approx 0.40$ miles.

Problems

21. On the interval $2 \le x \le 5$,

$$\begin{array}{c} \text{Average value} \\ \text{of } f \end{array} = \frac{1}{5-2} \int_2^5 f(x) \, dx = 4,$$

so

$$\int_2^5 f(x) \, dx = 12.$$

Thus

$$\int_2^5 (3f(x) + 2) \, dx = 3 \int_2^5 f(x) \, dx + 2 \int_2^5 1 \, dx = 3(12) + 2(5-2) = 42.$$

25. (a) Train A starts earlier than Train B, and stops later. At every moment Train A is going faster than Train B. Both trains increase their speed at a constant rate through the first half of their trip and slow down during the second half. Both trains reach their maximum speed at the same time. The area under the velocity graph for Train A is larger than the area under the velocity graph for Train B, meaning that Train A travels farther—as would be expected, given that its speed is always higher than B's.

   (b) (i) The maximum velocity is read off the vertical axis. The graph for Train A appears to go about twice as high as the graph for Train B; see Figure 5.15. So

$$\frac{\text{Maximum velocity of Train } A}{\text{Maximum velocity of Train } B} = \frac{v_A}{v_B} \approx 2.$$

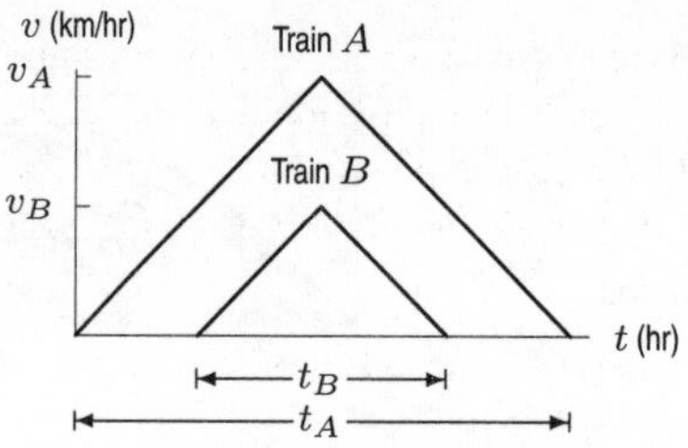

Figure 5.15

(ii) The time of travel is the horizontal distance between the start and stop times (the two t-intercepts). The horizontal distance for Train A appears to be about twice the corresponding distance for Train B; see Figure 5.15. So

$$\frac{\text{Time traveled by Train } A}{\text{Time traveled by Train } B} = \frac{t_A}{t_B} \approx 2.$$

(iii) The distance traveled by each train is given by the area under its graph. Since the area of triangle is $\frac{1}{2} \cdot$ Base $\cdot$ Height, and since the base and height for Train A is approximately twice that for Train B, we have

$$\frac{\text{Distance traveled by Train } A}{\text{Distance traveled by Train } B} = \frac{\frac{1}{2} \cdot v_A \cdot t_A}{\frac{1}{2} \cdot v_B \cdot t_B} \approx 2 \cdot 2 = 4.$$

29. (a) At $t = 20$ minutes, she stops moving toward the lake (with $v > 0$) and starts to move away from the lake (with $v < 0$). So at $t = 20$ minutes the cyclist turns around.

(b) The cyclist is going the fastest when v has the greatest magnitude, either positive or negative. Looking at the graph, we can see that this occurs at $t = 40$ minutes, when $v = -25$ and the cyclist is pedaling at 25 km/hr away from the lake.

(c) From $t = 0$ to $t = 20$ minutes, the cyclist comes closer to the lake, since $v > 0$; thereafter, $v < 0$ so the cyclist moves away from the lake. So at $t = 20$ minutes, the cyclist comes the closest to the lake. To find out how close she is, note that between $t = 0$ and $t = 20$ minutes the distance she has come closer is equal to the area under the graph of v. Each box represents 5/6 of a kilometer, and there are about 2.5 boxes under the graph, giving a distance of about 2 km. Since she was originally 5 km away, she then is about $5 - 2 = 3$ km from the lake.

(d) At $t = 20$ minutes she turns around , since v changes sign then. Since the area below the t-axis is greater than the area above, the farthest she is from the lake is at $t = 60$ minutes. Between $t = 20$ and $t = 60$ minutes, the area under the graph is about 10.8 km. (Since 13 boxes $\cdot 5/6 = 10.8$.) So at $t = 60$ she will be about $3 + 10.8 = 13.8$ km from the lake.

33. The change in the amount of water is the integral of rate of change, so we have

$$\text{Number of liters pumped out} = \int_0^{60} (5 - 5e^{-0.12t}) dt = 258.4 \text{ liters.}$$

Since the tank contained 1000 liters of water initially, we see that

$$\text{Amount in tank after one hour} = 1000 - 258.4 = 741.6 \text{ liters.}$$

37. (a) We know that $\int_2^5 f(x)\, dx = \int_0^5 f(x)\, dx - \int_0^2 f(x)\, dx$. By symmetry, $\int_0^2 f(x)\, dx = \frac{1}{2} \int_{-2}^2 f(x)\, dx$, so $\int_2^5 f(x)\, dx = \int_0^5 f(x)\, dx - \frac{1}{2} \int_{-2}^2 f(x)\, dx$.

(b) $\int_2^5 f(x)\, dx = \int_{-2}^5 f(x)\, dx - \int_{-2}^2 f(x)\, dx = \int_{-2}^5 f(x)\, dx - 2 \int_{-2}^0 f(x)\, dx$.

(c) Using symmetry again, $\int_0^2 f(x)\, dx = \frac{1}{2} \left(\int_{-2}^5 f(x)\, dx - \int_2^5 f(x)\, dx \right)$.

41. (a) Looking at the graph, it appears that the graph of B is above $F = 10$ between $t = 2.3$ and $t = 4.2$, or for about 1.9 seconds.

(b) The total impulse of each rocket is represented by the area under its thrust curve. From $t = 0$ to $t = 2$, the graph of A looks like a triangle with base 2 and height 12, for an area of 12. From $t = 2$ to $t = 4$, the graph of A looks a trapezoid with base 2 and heights 13 and 6, for an area of 19. From $t = 4$ to $t = 16$, A is approximately a rectangle with height 5.8 and width 12, for an area of 69.6. Finally, from $t = 16$ to $t = 17$, A looks like a triangle with base 1 and height 5.8, for an area of 2.9. So

$$A's \text{ total impulse} = \text{Area under } A's \text{ thrust curve} = 12 + 19 + 69.6 + 2.9 = 103.5 \text{ newton-seconds.}$$

(c) Note that when we calculated the impulse in part (b), we multiplied height, measured in newtons, by width, measured in seconds. So the units of impulse are newton-seconds.

(d) The graph of B's thrust looks like a triangle with base 6 and height 22, for a total impulse of about 66 newton-seconds. So rocket A, with total impulse 103.5 newton-seconds, has a larger total impulse than rocket B.

(e) As we can see from the graph, rocket B reaches a maximum thrust of 22, whereas A only reaches a maximum thrust of 13. So rocket B has the largest maximum thrust.

45. (a) When the aircraft is climbing at v ft/min, it takes $1/v$ minutes to climb 1 foot. Therefore

$$\text{Lower estimate} = \left(\frac{1 \text{ min}}{925 \text{ ft}}\right)(1000 \text{ ft}) + \left(\frac{1 \text{ min}}{875 \text{ ft}}\right)(1000 \text{ ft}) + \cdots + \left(\frac{1 \text{ min}}{490 \text{ ft}}\right)(1000 \text{ ft})$$
$$\approx 14.73 \text{ minutes.}$$
$$\text{Upper estimate} = \left(\frac{1 \text{ min}}{875 \text{ ft}}\right)(1000 \text{ ft}) + \left(\frac{1 \text{ min}}{830 \text{ ft}}\right)(1000 \text{ ft}) + \cdots + \left(\frac{1 \text{ min}}{440 \text{ ft}}\right)(1000 \text{ ft})$$
$$\approx 15.93 \text{ minutes.}$$

Note: The Pilot Operating Manual for this aircraft gives 16 minutes as the estimated time required to climb to 10,000 ft.

(b) The difference between upper and lower sums with $\Delta x = 500$ ft would be

$$\text{Difference} = \left(\frac{1 \text{ min}}{440 \text{ ft}} - \frac{1 \text{ min}}{925 \text{ ft}}\right)(500 \text{ ft}) = 0.60 \text{ minutes.}$$

CAS Challenge Problems

49. (a) We have $\Delta x = (1 - 0)/n = 1/n$ and $x_i = 0 + i \cdot \Delta x = i/n$. So we get

$$\text{Right-hand sum} = \sum_{i=1}^{n} (x_i)^4 \Delta x = \sum_{i=1}^{n} \left(\frac{i}{n}\right)^4 \left(\frac{1}{n}\right) = \sum_{i=1}^{n} \frac{i^4}{n^5}.$$

(b) The CAS gives

$$\text{Right-hand sum} = \sum_{i=1}^{n} \frac{i^4}{n^5} = \frac{6n^4 + 15n^3 + 10n^2 - 1}{30n^4}.$$

(The results may look slightly different depending on the CAS you use.)

(c) Using a CAS or by hand, we get

$$\lim_{n \to \infty} \frac{6n^4 + 15n^3 + 10n^2 - 1}{30n^4} = \lim_{n \to \infty} \frac{6n^4}{30n^4} = \frac{1}{5}.$$

The numerator is dominated by the highest power term, which is $6n^4$, so when n is large, the ratio behaves like $6n^4/30n^4 = 1/5$ as $n \to \infty$. Thus we see that

$$\int_0^1 x^4 \, dx = \frac{1}{5}.$$

53. (a) Since the length of the interval of integration is π, the width of each subdivision is $\Delta x = \pi/n$. Thus the endpoints of the subdivision are

$$x_0 = 0, \quad x_1 = 0 + \Delta x = \frac{\pi}{n}, \quad x_2 = 0 + 2\Delta x = \frac{2\pi}{n}, \ldots,$$
$$x_i = 0 + i\Delta x = \frac{i\pi}{n}, \quad \ldots, \quad x_n = 0 + n\Delta x = \frac{n\pi}{n} = \pi.$$

Thus, since the integrand is $f(x) = \sin x$,

$$\text{Right-hand sum} = \sum_{i=1}^{n} f(x_i)\Delta x = \sum_{i=1}^{n} \sin(x_i)\Delta x = \sum_{i=1}^{n} \sin\left(\frac{i\pi}{n}\right)\frac{\pi}{n}.$$

(b) If the CAS can evaluate this sum, we get

$$\sum_{i=1}^{n} \sin\left(\frac{i\pi}{n}\right)\frac{\pi}{n} = \frac{\pi \cot(\pi/2n)}{n} = \frac{\pi \cos(\pi/2n)}{n \sin(\pi/2n)}.$$

(c) Using the computer algebra system, we find that

$$\lim_{n\to\infty} \frac{\pi \cos(\pi/2n)}{n \sin(\pi/2n)} = 2.$$

(d) The computer algebra system gives

$$\int_0^{\pi} \sin x \, dx = 2.$$

CHECK YOUR UNDERSTANDING

1. False. The units of the integral are the product of the units for $f(x)$ times the units for x.

5. False. The integral is the change in position from $t = q$ to $t = b$. If the velocity changes sign in the interval, the total distance traveled and the change in position will not be the same.

9. True, since $\int_0^2 2f(x)dx = 2\int_0^2 f(x)dx$.

13. False. Let $f(x) = 7$ and $g(x) = 9$ for all x.
Then $\int_1^2 f(x)\,dx + \int_2^3 g(x)\,dx = 7 + 9 = 16$, but $\int_1^3 (f(x) + g(x))\,dx = \int_1^3 16\,dx = 32$.

17. False. Let $f(x) = x$ and $g(x) = 5$. Then $\int_2^6 f(x)\,dx = 16$ and $\int_2^6 g(x)\,dx = 20$, so $\int_2^6 f(x)\,dx \le \int_2^6 g(x)\,dx$, but $f(x) > g(x)$ for $5 < x < 6$.

21. True. We have by the properties of integrals in Theorem 5.3,

$$\int_1^9 f(x)dx = \int_1^4 f(x)dx + \int_4^9 f(x)dx.$$

Since $(1/(4-1))\int_1^4 f(x)dx = A$ and $(1/(9-4))\int_4^9 f(x)dx = B$, we have

$$\int_1^9 f(x)dx = 3A + 5B.$$

Dividing this equation through by 8, we get that the average value of f on the interval $[1, 9]$ is $(3/8)A + (5/8)B$.

25. False. A counterexample is given by the functions f and g in Figure 5.16. The function f is decreasing, g is increasing, and we have

$$\int_1^2 f(x)\,dx = \int_1^2 g(x)\,dx,$$

because both integrals equal $1/2$, the area of of the same sized triangle.

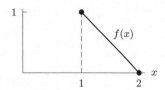
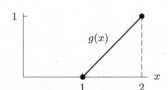

Figure 5.16

CHAPTER SIX

Solutions for Section 6.1

Exercises

1. See Figure 6.1.

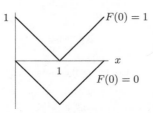

Figure 6.1

5. By the Fundamental Theorem of Calculus, we know that

$$f(2) - f(0) = \int_0^2 f'(x)dx.$$

Using a left-hand sum, we estimate $\int_0^2 f'(x)dx \approx (10)(2) = 20$. Using a right-hand sum, we estimate $\int_0^2 f'(x)dx \approx (18)(2) = 36$. Averaging, we have

$$\int_0^2 f'(x)dx \approx \frac{20 + 36}{2} = 28.$$

We know $f(0) = 100$, so

$$f(2) = f(0) + \int_0^2 f'(x)dx \approx 100 + 28 = 128.$$

Similarly, we estimate

$$\int_2^4 f'(x)dx \approx \frac{(18)(2) + (23)(2)}{2} = 41,$$

so

$$f(4) = f(2) + \int_2^4 f'(x)dx \approx 128 + 41 = 169.$$

Similarly,

$$\int_4^6 f'(x)dx \approx \frac{(23)(2) + (25)(2)}{2} = 48,$$

so

$$f(6) = f(4) + \int_4^6 f'(x)dx \approx 169 + 48 = 217.$$

The values are shown in the table.

x	0	2	4	6
$f(x)$	100	128	169	217

Problems

9. (a) Critical points of $F(x)$ are the zeros of f: $x = 1$ and $x = 3$.
   (b) $F(x)$ has a local minimum at $x = 1$ and a local maximum at $x = 3$.
   (c) See Figure 6.2.

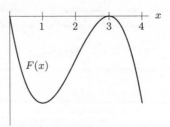

Figure 6.2

Notice that the graph could also be above or below the x-axis at $x = 3$.

13. See Figure 6.3. Note that since $f(x_1) = 0$, $F(x_1)$ is either a local minimum or a point of inflection; it is impossible to tell which from the graph. Since $f'(x_3) = 0$, and f' changes sign around $x = x_3$, $F(x_3)$ is an inflection point. Also, since $f'(x_2) = 0$ and f changes from increasing to decreasing about $x = x_2$, F has another inflection point at $x = x_2$.

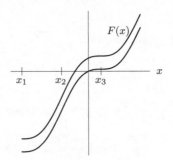

Figure 6.3

17. The critical points are at $(0, 5)$, $(2, 21)$, $(4, 13)$, and $(5, 15)$. A graph is given in Figure 6.4.

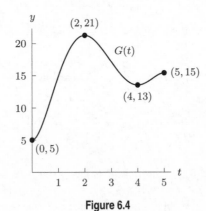

Figure 6.4

21. (a) The total volume emptied must increase with time and cannot decrease. The smooth graph (I) that is always increasing is therefore the volume emptied from the bladder. The jagged graph (II) that increases then decreases to zero is the flow rate.
   (b) The total change in volume is the integral of the flow rate. Thus, the graph giving total change (I) shows an antiderivative of the rate of change in graph (II).

25. (a) Suppose $Q(t)$ is the amount of water in the reservoir at time t. Then

$$Q'(t) = \frac{\text{Rate at which water}}{\text{in reservoir is changing}} = \frac{\text{Inflow}}{\text{rate}} - \frac{\text{Outflow}}{\text{rate}}$$

Thus the amount of water in the reservoir is increasing when the inflow curve is above the outflow, and decreasing when it is below. This means that $Q(t)$ is a maximum where the curves cross in July 1993 (as shown in Figure 6.5), and $Q(t)$ is decreasing fastest when the outflow is farthest above the inflow curve, which occurs about October 1993 (see Figure 6.5).

To estimate values of $Q(t)$, we use the Fundamental Theorem which says that the change in the total quantity of water in the reservoir is given by

$$Q(t) - Q(\text{Jan'93}) = \int_{\text{Jan93}}^{t} (\text{inflow rate} - \text{outflow rate})\, dt$$

or $$Q(t) = Q(\text{Jan'93}) + \int_{\text{Jan93}}^{t} (\text{inflow rate} - \text{outflow rate})\, dt.$$

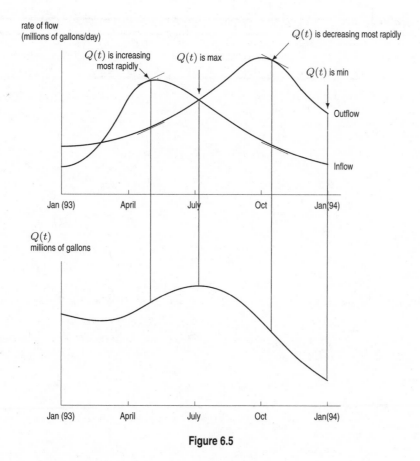

Figure 6.5

(b) See Figure 6.5. Maximum in July 1993. Minimum in Jan 1994.
(c) See Figure 6.5. Increasing fastest in May 1993. Decreasing fastest in Oct 1993.
(d) In order for the water to be the same as Jan '93 the total amount of water which has flowed into the reservoir must be 0. Referring to Figure 6.6, we have

$$\int_{\text{Jan93}}^{\text{July94}} (\text{inflow} - \text{outflow})\, dt = -A_1 + A_2 - A_3 + A_4 = 0$$

giving $A_1 + A_3 = A_2 + A_4$

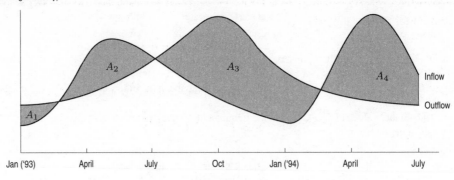

rate of flow
(millions of gallons/day)

A_1 A_2 A_3 A_4 Inflow   Outflow

Jan ('93)   April   July   Oct   Jan ('94)   April   July

Figure 6.6

Solutions for Section 6.2

Exercises

1. $5x$

5. $\sin t$

9. $-\dfrac{1}{2z^2}$

13. $\dfrac{t^4}{4} - \dfrac{t^3}{6} - \dfrac{t^2}{2}$

17. $-\cos 2\theta$

21. $\dfrac{5}{2}x^2 - \dfrac{2}{3}x^{\frac{3}{2}}$

25. $R(t) = \displaystyle\int (t^3 + 5t - 1)\, dt = \dfrac{t^4}{4} + \dfrac{5}{2}t^2 - t + C$

29. $H(x) = \displaystyle\int (4x^3 - 7)\, dx = x^4 - 7x + C$

33. $f(x) = 3$, so $F(x) = 3x + C$. $F(0) = 0$ implies that $3 \cdot 0 + C = 0$, so $C = 0$. Thus $F(x) = 3x$ is the only possibility.

37. $f(x) = x^2$, so $F(x) = \dfrac{x^3}{3} + C$. $F(0) = 0$ implies that $\dfrac{0^3}{3} + C = 0$, so $C = 0$. Thus $F(x) = \dfrac{x^3}{3}$ is the only possibility.

41. $\displaystyle\int 5x\, dx = \dfrac{5}{2}x^2 + C$.

45. $\displaystyle\int \left(t^2 + \dfrac{1}{t^2}\right) dt = \dfrac{t^3}{3} - \dfrac{1}{t} + C$

49. $2t^2 + 7t + C$

53. $-\cos t + C$

57. $\frac{1}{2}e^{2r} + C$

61. $\displaystyle\int_1^3 \dfrac{1}{t}\, dt = \ln|t| \Big|_1^3 = \ln|3| - \ln|1| = \ln 3 \approx 1.0986$.

65. $\displaystyle\int_0^1 \sin\theta\, d\theta = -\cos\theta \Big|_0^1 = 1 - \cos 1 \approx 0.460$.

69. $\displaystyle\int_0^1 2e^x\, dx = 2e^x \Big|_0^1 = 2e - 2 \approx 3.437$.

Problems

73. The graph crosses the x-axis where

$$7 - 8x + x^2 = 0$$
$$(x - 7)(x - 1) = 0;$$

so $x = 1$ and $x = 7$. See Figure 6.7. The parabola opens upward and the region is below the x-axis, so

$$\text{Area} = -\int_1^7 (7 - 8x + x^2)\, dx$$

$$= -\left(7x - 4x^2 + \frac{x^3}{3} \right)\Big|_1^7 = 36.$$

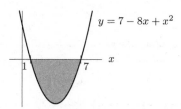

$$y = 7 - 8x + x^2$$

Figure 6.7

77. Since the graph of $y = e^x$ is above the graph of $y = \cos x$ (see the figure below), we have

$$\text{Area} = \int_0^1 (e^x - \cos x)\, dx$$

$$= \int_0^1 e^x\, dx - \int_0^1 \cos x\, dx$$

$$= e^x\Big|_0^1 - \sin x\Big|_0^1$$

$$= e^1 - e^0 - \sin 1 + \sin 0$$

$$= e - 1 - \sin 1.$$

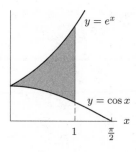

$$y = e^x$$

$$y = \cos x$$

81. We have

$$\text{Average value} = \frac{1}{10 - 0}\int_0^{10} (x^2 + 1)dx = \frac{1}{10}\left(\frac{x^3}{3} + x \right)\Big|_0^{10} = \frac{1}{10}\left(\frac{10^3}{3} + 10 - 0 \right) = \frac{103}{3}.$$

We see in Figure 6.8 that the average value of $103/3 \approx 34.33$ for $f(x)$ looks right.

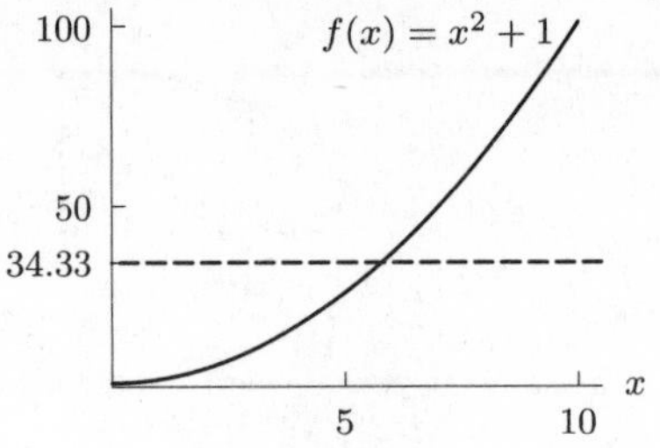

Figure 6.8

85. The curves $y = x$ and $y = x^n$ cross at $x = 0$ and $x = 1$. For $0 < x < 1$, the curve $y = x$ is above $y = x^n$. Thus the area is given by

$$A_n = \int_0^1 (x - x^n)\,dx = \left[\frac{x^2}{2} - \frac{x^{n+1}}{n+1}\right]_0^1 = \frac{1}{2} - \frac{1}{n+1} \to \frac{1}{2}.$$

Since $x^n \to 0$ for $0 \le x < 1$, as $n \to \infty$, the area between the curves approaches the area under the line $y = x$ between $x = 0$ and $x = 1$.

Solutions for Section 6.3

Exercises

1. $y = \displaystyle\int (x^3 + 5)\,dx = \dfrac{x^4}{4} + 5x + C$

5. Since $y = x + \sin x - \pi$, we differentiate to see that $dy/dx = 1 + \cos x$, so y satisfies the differential equation. To show that it also satisfies the initial condition, we check that $y(\pi) = 0$:

$$y = x + \sin x - \pi$$
$$y(\pi) = \pi + \sin \pi - \pi = 0.$$

9. Integrating gives

$$\int \frac{dq}{dz}\,dz = \int (2 + \sin z)\,dz = 2z - \cos z + C.$$

If $q = 5$ when $z = 0$, then $2(0) - \cos(0) + C = 5$ so $C = 6$. Thus $q = 2z - \cos z + 6$.

Problems

13.

$$\frac{dy}{dt} = k\sqrt{t} = kt^{1/2}$$
$$y = \frac{2}{3}kt^{3/2} + C.$$

Since $y = 0$ when $t = 0$, we have $C = 0$, so

$$y = \frac{2}{3}kt^{3/2}.$$

17. (a)

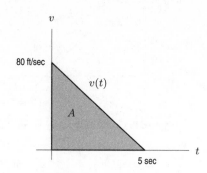

(b) The total distance is represented by the shaded region A, the area under the graph of $v(t)$.

(c) The area A, a triangle, is given by

$$A = \frac{1}{2}(\text{base})(\text{height}) = \frac{1}{2}(5\,\text{sec})(80\,\text{ft/sec}) = 200\,\text{ft}.$$

(d) Using integration and the Fundamental Theorem of Calculus, we have $A = \int_0^5 v(t)\,dt$ or $A = s(5) - s(0)$, where $s(t)$ is an antiderivative of $v(t)$.

We have that $a(t)$, the acceleration, is constant: $a(t) = k$ for some constant k. Therefore $v(t) = kt + C$ for some constant C. We have $80 = v(0) = k(0) + C = C$, so that $v(t) = kt + 80$. Putting in $t = 5, 0 = v(5) = (k)(5) + 80$, or $k = -80/5 = -16$.

Thus $v(t) = -16t + 80$, and an antiderivative for $v(t)$ is $s(t) = -8t^2 + 80t + C$. Since the total distance traveled at $t = 0$ is 0, we have $s(0) = 0$ which means $C = 0$. Finally, $A = \int_0^5 v(t)\,dt = s(5) - s(0) = (-8(5)^2 + (80)(5)) - (-8(0)^2 + (80)(0)) = 200\,\text{ft}$, which agrees with the previous part.

21. The equation of motion is $y = -\frac{gt^2}{2} + v_0 t + y_0 = -16t^2 + 128t + 320$. Taking the first derivative, we get $v = -32t + 128$. The second derivative gives us $a = -32$.

(a) At its highest point, the stone's velocity is zero:
$v = 0 = -32t + 128$, so $t = 4$.

(b) At $t = 4$, the height is $y = -16(4)^2 + 128(4) + 320 = 576$ ft

(c) When the stone hits the beach,

$$y = 0 = -16t^2 + 128t + 320$$
$$0 = -t^2 + 8t + 20 = (10 - t)(2 + t).$$

So $t = 10$ seconds.

(d) Impact is at $t = 10$. The velocity, v, at this time is $v(10) = -32(10) + 128 = -192$ ft/sec. Upon impact, the stone's velocity is 192 ft/sec downward.

25. The first thing we should do is convert our units. We'll bring everything into feet and seconds. Thus, the initial speed of the car is

$$\frac{70\,\text{miles}}{\text{hour}}\left(\frac{1\,\text{hour}}{3600\,\text{sec}}\right)\left(\frac{5280\,\text{feet}}{1\,\text{mile}}\right) \approx 102.7\,\text{ft/sec}.$$

We assume that the acceleration is constant as the car comes to a stop. A graph of its velocity versus time is given in Figure 6.9. We know that the area under the curve represents the distance that the car travels before it comes to a stop, 157 feet. But this area is a triangle, so it is easy to find t_0, the time the car comes to rest. We solve

$$\frac{1}{2}(102.7)t_0 = 157,$$

which gives

$$t_0 \approx 3.06\,\text{sec}.$$

Since acceleration is the rate of change of velocity, the car's acceleration is given by the slope of the line in Figure 6.9. Thus, the acceleration, k, is given by

$$k = \frac{102.7 - 0}{0 - 3.06} \approx -33.56\ \text{ft/sec}^2.$$

Notice that k is negative because the car is slowing down.

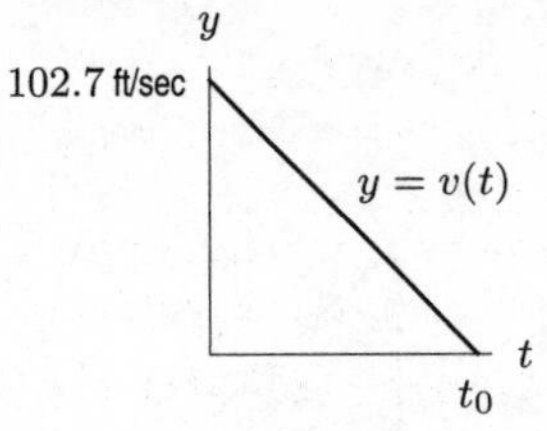

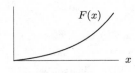

Figure 6.9: Graph of velocity versus time

Solutions for Section 6.4

Exercises

1. By the Fundamental Theorem, $f(x) = F'(x)$. Since f is positive and increasing, F is increasing and concave up. Since $F(0) = \int_0^0 f(t)dt = 0$, the graph of F must start from the origin. See Figure 6.10.

$F(x)$

x

Figure 6.10

5.

Table 6.1

x	0	0.5	1	1.5	2
$I(x)$	0	0.50	1.09	2.03	3.65

9. If $f'(x) = \dfrac{\sin x}{x}$, then $f(x)$ is of the form

$$f(x) = C + \int_a^x \frac{\sin t}{t}\, dt.$$

Since $f(1) = 5$, we take $a = 1$ and $C = 5$, giving

$$f(x) = 5 + \int_1^x \frac{\sin t}{t}\, dt.$$

Problems

13. Since $F'(x) = e^{-x^2}$ and $F(0) = 2$, we have

$$F(x) = F(0) + \int_0^x e^{-t^2}\, dt = 2 + \int_0^x e^{-t^2}\, dt.$$

Substituting $x = 1$ and evaluating the integral numerically gives

$$F(1) = 2 + \int_0^1 e^{-t^2}\, dt = 2.747.$$

17. $\arctan(x^2)$.

21. (a) The definition of g gives $g(0) = \int_0^0 f(t)\,dt = 0$.

(b) The Fundamental Theorem gives $g'(1) = f(1) = -2$.

(c) The function g is concave upward where g'' is positive. Since $g'' = f'$, we see that g is concave up where f is increasing. This occurs on the interval $1 \le x \le 6$.

(d) The function g decreases from $x = 0$ to $x = 3$ and increases for $3 < x \le 8$, and the magnitude of the increase is more than the magnitude of the decrease. Thus g takes its maximum value at $x = 8$.

25. If we let $f(t) = \int_1^t \cos(x^2)\,dx$ and $g(t) = \sin t$, using the chain rule gives

$$\frac{d}{dt}\int_1^{\sin t}\cos(x^2)\,dx = f'(g(t))\cdot g'(t) = \cos((\sin t)^2)\cdot\cos t = \cos(\sin^2 t)(\cos t).$$

29. By the Fundamental Construction Theorem, $s'(t) = v(t)$. Hence, by the Fundamental Theorem of Calculus, $\int_a^b v(t)\,dt = s(b) - s(a)$.

33.

$$\frac{d}{dx}[x\,\mathrm{erf}(x)] = \mathrm{erf}(x)\frac{d}{dx}(x) + x\frac{d}{dx}[\mathrm{erf}(x)]$$

$$= \mathrm{erf}(x) + x\frac{d}{dx}\left(\frac{2}{\sqrt{\pi}}\int_0^x e^{-t^2}\,dt\right)$$

$$= \mathrm{erf}(x) + \frac{2}{\sqrt{\pi}}xe^{-x^2}.$$

Solutions for Section 6.5

Exercises

1. (a) The object is thrown from an initial height of $y = 1.5$ meters.

(b) The velocity is obtained by differentiating, which gives $v = -9.8t + 7$ m/sec. The initial velocity is $v = 7$ m/sec upward.

(c) The acceleration due to gravity is obtained by differentiating again, giving $g = -9.8$ m/sec^2, or 9.8 m/sec^2 downward.

Problems

5. $a(t) = -32$. Since $v(t)$ is the antiderivative of $a(t)$, $v(t) = -32t + v_0$. But $v_0 = 0$, so $v(t) = -32t$. Since $s(t)$ is the antiderivative of $v(t)$, $s(t) = -16t^2 + s_0$, where s_0 is the height of the building. Since the ball hits the ground in 5 seconds, $s(5) = 0 = -400 + s_0$. Hence $s_0 = 400$ feet, so the window is 400 feet high.

9. (a) Since $s(t) = -\frac{1}{2}gt^2$, the distance a body falls in the first second is

$$s(1) = -\frac{1}{2}\cdot g\cdot 1^2 = -\frac{g}{2}.$$

In the second second, the body travels

$$s(2) - s(1) = -\frac{1}{2}\left(g\cdot 2^2 - g\cdot 1^2\right) = -\frac{1}{2}(4g - g) = -\frac{3g}{2}.$$

In the third second, the body travels

$$s(3) - s(2) = -\frac{1}{2}\left(g\cdot 3^2 - g\cdot 2^2\right) = -\frac{1}{2}(9g - 4g) = -\frac{5g}{2},$$

and in the fourth second, the body travels

$$s(4) - s(3) = -\frac{1}{2}\left(g\cdot 4^2 - g\cdot 3^2\right) = -\frac{1}{2}(16g - 9g) = -\frac{7g}{2}.$$

(b) Galileo seems to have been correct. His observation follows from the fact that the differences between consecutive squares are consecutive odd numbers. For, if n is any number, then $n^2 - (n-1)^2 = 2n - 1$, which is the nth odd number (where 1 is the first).

Solutions for Chapter 6 Review

Exercises

1. See Figure 6.11

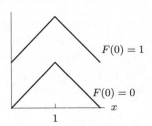

$F(0) = 1$

$F(0) = 0$

x

1

Figure 6.11

5. $\int (2 + \cos t)\, dt = 2t + \sin t + C$

9. $\int \dfrac{8}{\sqrt{x}}\, dx = 16x^{1/2} + C$

13. $\tan x + C$

17. $\frac{1}{10}(x+1)^{10} + C$

21. $3\sin x + 7\cos x + C$

25. $F(x) = \int \dfrac{1}{x^2}\, dx = -\dfrac{1}{x} + C$

29. $F(x) = \int 5e^x\, dx = 5e^x + C$

33. We have $F(x) = \dfrac{x^4}{4} + 2x^3 - 4x + C$. Since $F(0) = 4$, we have $4 = 0 + C$, so $C = 4$. So $F(x) = \dfrac{x^4}{4} + 2x^3 - 4x + 4$.

37. $F(x) = \int \cos x\, dx = \sin x + C$. If $F(0) = 4$, then $F(0) = 0 + C = 4$ and thus $C = 4$. So $F(x) = \sin x + 4$.

Problems

41. The area we want (the shaded area in Figure 6.12) is symmetric about the y-axis and so is given by

$$
\begin{aligned}
\text{Area} &= 2\int_0^{\pi/3} \left(\cos x - \frac{1}{2}\left(\frac{3}{\pi}x\right)^2 \right) dx \\
&= 2\int_0^{\pi/3} \cos x\, dx - \int_0^{\pi/3} \frac{9}{\pi^2} x^2\, dx \\
&= 2\sin x \Big|_0^{\pi/3} - \frac{9}{\pi^2} \cdot \frac{x^3}{3} \Big|_0^{\pi/3} \\
&= 2 \cdot \frac{\sqrt{3}}{2} - \frac{3}{\pi^2} \cdot \frac{\pi^3}{3^3} = \sqrt{3} - \frac{\pi}{9}.
\end{aligned}
$$

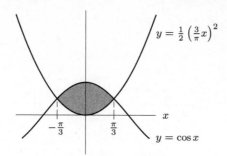

Figure 6.12

45. The graph of $y = c(1 - x^2)$ has x-intercepts of $x = \pm 1$. See Figure 6.13. Since it is symmetric about the y-axis, we have

$$\text{Area} = \int_{-1}^{1} c(1 - x^2)\, dx = 2c \int_{0}^{1} (1 - x^2)\, dx$$

$$= 2c \left(x - \frac{x^3}{3} \right) \Bigg|_{0}^{1} = \frac{4c}{3}.$$

We want the area to be 1, so

$$\frac{4c}{3} = 1, \quad \text{giving} \quad c = \frac{3}{4}.$$

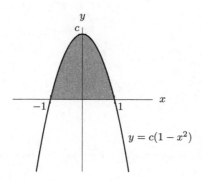

Figure 6.13

49. (a) Since $f'(t)$ is positive on the interval $0 < t < 2$ and negative on the interval $2 < t < 5$, the function $f(t)$ is increasing on $0 < t < 2$ and decreasing on $2 < t < 5$. Thus $f(t)$ attains its maximum at $t = 2$. Since the area under the t-axis is greater than the area above the t-axis, the function $f(t)$ decreases more than it increases. Thus, the minimum is at $t = 5$.

(b) To estimate the value of f at $t = 2$, we see that the area under $f'(t)$ between $t = 0$ and $t = 2$ is about 1 box, which has area 5. Thus,

$$f(2) = f(0) + \int_{0}^{2} f'(t)dt \approx 50 + 5 = 55.$$

The maximum value attained by the function is $f(2) \approx 55$.

The area between $f'(t)$ and the t-axis between $t = 2$ and $t = 5$ is about 3 boxes, each of which has an area of 5. Thus

$$f(5) = f(2) + \int_{2}^{5} f'(t)dt \approx 55 + (-15) = 40.$$

The minimum value attained by the function is $f(5) = 40$.

(c) Using part (b), we have $f(5) - f(0) = 40 - 50 = -10$. Alternately, we can use the Fundamental Theorem:

$$f(5) - f(0) = \int_{0}^{5} f'(t)dt \approx 5 - 15 = -10.$$

53. We have
$$\frac{d}{dx}\int_2^x \arccos(t^7)\,dt = \arccos x^7.$$

57. We split the integral at $x = 1$ (or any other point we choose):
$$\int_{t^5}^{\cos t} 4^{7x}\,dx = \int_1^{\cos t} 4^{7x}\,dx + \int_{t^5}^1 4^{7x}\,dx = \int_1^{\cos t} 4^{7x}\,dx - \int_1^{t^5} 4^{7x}\,dx.$$

Differentiating each part separately and using the chain rule gives
$$\frac{d}{dt}\int_{t^5}^{\cos t} 4^{7x}\,dx = \frac{d}{dt}\int_1^{\cos t} 4^{7x}\,dx - \frac{d}{dt}\int_1^{t^5} 4^{7x}\,dx$$
$$= 4^{7\cos t}(-\sin t) - 4^{7t^5}(5t^4)$$
$$= -(\sin t)4^{7\cos t} - 5t^4 4^{7t^5}.$$

61. A function whose derivative is e^{x^2} is of the form
$$f(x) = C + \int_a^x e^{t^2}\,dt \qquad \text{for some value of } C.$$

(a) To ensure that the function goes through the point $(0, 3)$, we take $a = 0$ and $C = 3$:
$$f(x) = 3 + \int_0^x e^{t^2}\,dt.$$

(b) To ensure that the function goes through $(-1, 5)$, we take $a = -1$ and $C = 5$:
$$f(x) = 5 + \int_{-1}^x e^{t^2}\,dt.$$

65. (a) Since 6 sec $= 1/10$ min,
$$\text{Angular acceleration} = \frac{2500 - 1100}{1/10} = 14{,}000 \text{ revs/min}^2.$$

(b) We know angular acceleration is the derivative of angular velocity. Since
$$\text{Angular acceleration} = 14{,}000,$$
we have
$$\text{Angular velocity} = 14{,}000t + C.$$
Measuring time from the moment at which the angular velocity is 1100 revs/min, we have $C = 1100$. Thus,
$$\text{Angular velocity} = 14{,}000t + 1100.$$
Thus the total number of revolutions performed during the period from $t = 0$ to $t = 1/10$ min is given by
$$\begin{array}{l}\text{Number of} \\ \text{revolutions}\end{array} = \int_0^{1/10} (14000t + 1100)dt = 7000t^2 + 1100t\Big|_0^{1/10} = 180 \text{ revolutions}.$$

69. (a) In the beginning, both birth and death rates are small; this is consistent with a very small population. Both rates begin climbing, the birth rate faster than the death rate, which is consistent with a growing population. The birth rate is then high, but it begins to decrease as the population increases.

(b)

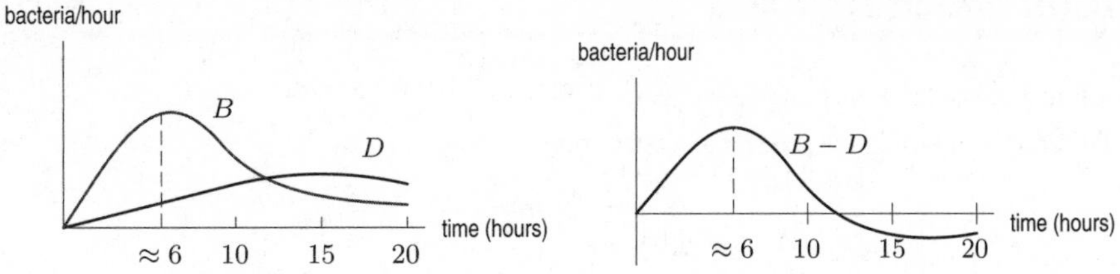

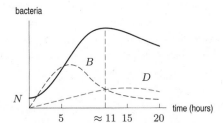

Figure 6.14: Difference between B and D is greatest at $t \approx 6$

The bacteria population is growing most quickly when $B - D$, the rate of change of population, is maximal; that happens when B is farthest above D, which is at a point where the slopes of both graphs are equal. That point is $t \approx 6$ hours.

(c) Total number born by time t is the area under the B graph from $t = 0$ up to time t. See Figure 6.15.

Total number alive at time t is the number born minus the number that have died, which is the area under the B graph minus the area under the D graph, up to time t. See Figure 6.16.

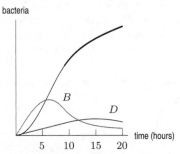

Figure 6.15: Number born by time t is $\int_0^t B(x)\, dx$

Figure 6.16: Number alive at time t is $\int_0^t (B(x) - D(x))\, dx$

From Figure 6.16, we see that the population is at a maximum when $B = D$, that is, after about 11 hours. This stands to reason, because $B - D$ is the rate of change of population, so population is maximized when $B - D = 0$, that is, when $B = D$.

CAS Challenge Problems

73. (a) A CAS gives

$$\int e^{2x}\, dx = \frac{1}{2} e^{2x} \qquad \int e^{3x}\, dx = \frac{1}{3} e^{3x} \qquad \int e^{3x+5}\, dx = \frac{1}{3} e^{3x+5}.$$

(b) The three integrals in part (a) obey the rule

$$\int e^{ax+b}\, dx = \frac{1}{a} e^{ax+b}.$$

(c) Checking the formula by calculating the derivative

$$\frac{d}{dx} \left(\frac{1}{a} e^{ax+b} \right) = \frac{1}{a} \frac{d}{dx} e^{ax+b} \quad \text{by the constant multiple rule}$$

$$= \frac{1}{a} e^{ax+b} \frac{d}{dx} (ax + b) \quad \text{by the chain rule}$$

$$= \frac{1}{a} e^{ax+b} \cdot a = e^{ax+b}.$$

CHECK YOUR UNDERSTANDING

1. True. A function can have only one derivative.

5. False. Differentiating using the product and chain rules gives

$$\frac{d}{dx}\left(\frac{-1}{2x}e^{-x^2}\right) = \frac{1}{2x^2}e^{-x^2} + e^{-x^2}.$$

9. True. If $y = F(x)$ is a solution to the differential equation $dy/dx = f(x)$, then $F'(x) = f(x)$, so $F(x)$ is an antiderivative of $f(x)$.

13. False. The solution of the initial value problem $dy/dx = 1$ with $y(0) = -5$ is a solution of the differential equation that is not positive at $x = 0$.

17. True. All solutions of the differential equation $dy/dt = 3t^2$ are in the family $y(t) = t^3 + C$ of antiderivatives of $3t^2$. The initial condition $y(1) = \pi$ tells us that $y(1) = \pi = 1^3 + C$, so $C = \pi - 1$. Thus $y(t) = t^3 + \pi - 1$ is the only solution of the initial value problem.

21. True. Suppose t is measured in seconds from when the ball was thrown. The acceleration $a = dv/dt$ is -32 ft/sec^2, so the velocity of the ball is $v = -32t + C$ feet/second at time t. At $t = 0$ the velocity is -10, so $v = -32t - 10$. Since $v = ds/dt$, an antiderivative gives the height $s = -16t^2 - 10t + K$ feet of the ball at time t. Since the ball starts at the top of the building, $s = 100$ when $t = 0$. Substituting gives $s = -16t^2 - 10t + 100$. The ball hits the ground when $s = 0$, so we solve $0 = -16t^2 - 10t + 100$. The positive solution $t = 2.2$ tells us that the ball hits the ground after 2.2 seconds.

25. True. Since F and G are both antiderivatives of f, they must differ by a constant. In fact, we can see that the constant C is equal to $\int_0^2 f(t)dt$ since

$$F(x) = \int_0^x f(t)dt = \int_2^x f(t)dt + \int_0^2 f(t)dt = G(x) + C.$$

CHAPTER SEVEN

Solutions for Section 7.1

Exercises

1. (a) We substitute $w = 1 + x^2$, $dw = 2x\,dx$.

$$\int_{x=0}^{x=1} \frac{x}{1+x^2}\,dx = \frac{1}{2}\int_{w=1}^{w=2}\frac{1}{w}\,dw = \frac{1}{2}\ln|w|\Big|_1^2 = \frac{1}{2}\ln 2.$$

(b) We substitute $w = \cos x$, $dw = -\sin x\,dx$.

$$\int_{x=0}^{x=\frac{\pi}{4}} \frac{\sin x}{\cos x}\,dx = -\int_{w=1}^{w=\sqrt{2}/2}\frac{1}{w}\,dw$$

$$= -\ln|w|\Big|_1^{\sqrt{2}/2} = -\ln\frac{\sqrt{2}}{2} = \frac{1}{2}\ln 2.$$

5. We use the substitution $w = -0.2t$, $dw = -0.2\,dt$.

$$\int 25e^{-0.2t}\,dt = \frac{25}{-0.2}\int e^w\,dw = -125e^w + C = -125e^{-0.2t} + C.$$

Check: $\frac{d}{dt}(-125e^{-0.2t} + C) = -125e^{-0.2t}(-0.2) = 25e^{-0.2t}$.

9. We use the substitution $w = -x^2$, $dw = -2x\,dx$.

$$\int xe^{-x^2}\,dx = -\frac{1}{2}\int e^{-x^2}(-2x\,dx) = -\frac{1}{2}\int e^w\,dw$$

$$= -\frac{1}{2}e^w + C = -\frac{1}{2}e^{-x^2} + C.$$

Check: $\frac{d}{dx}(-\frac{1}{2}e^{-x^2} + C) = (-2x)(-\frac{1}{2}e^{-x^2}) = xe^{-x^2}$.

13. We use the substitution $w = x^2 + 3$, $dw = 2x\,dx$.

$$\int x(x^2+3)^2\,dx = \int w^2(\frac{1}{2}\,dw) = \frac{1}{2}\frac{w^3}{3} + C = \frac{1}{6}(x^2+3)^3 + C.$$

Check: $\frac{d}{dx}\left[\frac{1}{6}(x^2+3)^3 + C\right] = \frac{1}{6}\left[3(x^2+3)^2(2x)\right] = x(x^2+3)^2$.

17. We use the substitution $w = y + 5$, $dw = dy$, to get

$$\int \frac{dy}{y+5} = \int \frac{dw}{w} = \ln|w| + C = \ln|y+5| + C.$$

Check: $\frac{d}{dy}(\ln|y+5| + C) = \frac{1}{y+5}$.

21. We use the substitution $w = \cos\theta + 5$, $dw = -\sin\theta\,d\theta$.

$$\int \sin\theta(\cos\theta + 5)^7\,d\theta = -\int w^7\,dw = -\frac{1}{8}w^8 + C$$

$$= -\frac{1}{8}(\cos\theta + 5)^8 + C.$$

Check:

$$\frac{d}{d\theta}\left[-\frac{1}{8}(\cos\theta + 5)^8 + C\right] = -\frac{1}{8}\cdot 8(\cos\theta + 5)^7\cdot(-\sin\theta)$$

$$= \sin\theta(\cos\theta + 5)^7$$

25. We use the substitution $w = \sin 5\theta$, $dw = 5\cos 5\theta\, d\theta$.

$$\int \sin^6 5\theta \cos 5\theta\, d\theta = \frac{1}{5}\int w^6\, dw = \frac{1}{5}\left(\frac{w^7}{7}\right) + C = \frac{1}{35}\sin^7 5\theta + C.$$

Check: $\dfrac{d}{d\theta}\left(\dfrac{1}{35}\sin^7 5\theta + C\right) = \dfrac{1}{35}[7\sin^6 5\theta](5\cos 5\theta) = \sin^6 5\theta \cos 5\theta.$

Note that we could also use Problem 23 to solve this problem, substituting $w = 5\theta$ and $dw = 5\, d\theta$ to get:

$$\int \sin^6 5\theta \cos 5\theta\, d\theta = \frac{1}{5}\int \sin^6 w \cos w\, dw$$

$$= \frac{1}{5}\left(\frac{\sin^7 w}{7}\right) + C = \frac{1}{35}\sin^7 5\theta + C.$$

29. We use the substitution $w = y^2 + 4$, $dw = 2y\, dy$.

$$\int \frac{y}{y^2+4}\, dy = \frac{1}{2}\int \frac{dw}{w} = \frac{1}{2}\ln|w| + C = \frac{1}{2}\ln(y^2+4) + C.$$

(We can drop the absolute value signs since $y^2 + 4 \geq 0$ for all y.)

Check: $\dfrac{d}{dy}\left[\dfrac{1}{2}\ln(y^2+4) + C\right] = \dfrac{1}{2}\cdot\dfrac{1}{y^2+4}\cdot 2y = \dfrac{y}{y^2+4}.$

33. We use the substitution $w = 2 + e^x$, $dw = e^x\, dx$.

$$\int \frac{e^x}{2+e^x}\, dx = \int \frac{dw}{w} = \ln|w| + C = \ln(2+e^x) + C.$$

(We can drop the absolute value signs since $2 + e^x \geq 0$ for all x.)

Check: $\dfrac{d}{dx}[\ln(2+e^x) + C] = \dfrac{1}{2+e^x}\cdot e^x = \dfrac{e^x}{2+e^x}.$

37. It seems easier not to substitute.

$$\int \frac{(t+1)^2}{t^2}\, dt = \int \frac{(t^2+2t+1)}{t^2}\, dt$$

$$= \int\left(1 + \frac{2}{t} + \frac{1}{t^2}\right) dt = t + 2\ln|t| - \frac{1}{t} + C.$$

Check: $\dfrac{d}{dt}\left(t + 2\ln|t| - \dfrac{1}{t} + C\right) = 1 + \dfrac{2}{t} + \dfrac{1}{t^2} = \dfrac{(t+1)^2}{t^2}.$

41. Since $d(\cosh z)/dz = \sinh z$, the chain rule shows that

$$\frac{d}{dz}(e^{\cosh z}) = (\sinh z)e^{\cosh z}.$$

Thus,

$$\int (\sinh z)e^{\cosh z}\, dz = e^{\cosh z} + C.$$

45. The general antiderivative is $\int(\pi t^3 + 4t)\, dt = (\pi/4)t^4 + 2t^2 + C.$

49. Make the substitution $w = 2 - 5x$, then $dw = -5dx$. We have

$$\int \sin(2-5x)dx = \int \sin w\left(-\frac{1}{5}\right)dw = -\frac{1}{5}(-\cos w) + C = \frac{1}{5}\cos(2-5x) + C.$$

53. $\displaystyle\int_0^\pi \cos(x+\pi)\, dx = \sin(x+\pi)\Big|_0^\pi = \sin(2\pi) - \sin(\pi) = 0 - 0 = 0$

57. We substitute $w = \sqrt[3]{x} = x^{\frac{1}{3}}$. Then $dw = \dfrac{1}{3}x^{-\frac{2}{3}}\, dx = \dfrac{1}{3\sqrt[3]{x^2}}\, dx.$

$$\int_1^8 \frac{e^{\sqrt[3]{x}}}{\sqrt[3]{x^2}}\, dx = \int_{x=1}^{x=8} e^w(3\, dw) = 3e^w\Big|_{x=1}^{x=8} = 3e^{\sqrt[3]{x}}\Big|_1^8 = 3(e^2 - e).$$

61.
$$\int_{-1}^{3} (x^3 + 5x)\, dx = \frac{x^4}{4}\bigg|_{-1}^{3} + \frac{5x^2}{2}\bigg|_{-1}^{3} = 40.$$

65. $\int_{-1}^{2} \sqrt{x+2}\, dx = \frac{2}{3}(x+2)^{3/2}\bigg|_{-1}^{2} = \frac{2}{3}\left[(4)^{3/2} - (1)^{3/2}\right] = \frac{2}{3}(7) = \frac{14}{3}$

69. Let $w = \sqrt{t+1}$, so $t = w^2 - 1$ and $dt = 2w\, dw$. Thus

$$\int \frac{t^2 + t}{\sqrt{t+1}}\, dt = \int \frac{(w^2 - 1)^2 + (w^2 - 1)}{w} 2w\, dw = 2\int w^4 - w^2\, dw$$

$$= \frac{2}{5}w^5 - \frac{2}{3}w^3 + C = \frac{2}{5}(t+1)^{5/2} - \frac{2}{3}(t+1)^{3/2} + C.$$

73. Let $w = \sqrt{t+1}$, so $t = w^2 - 1$ and $dt = 2w\, dw$. Thus

$$\int \frac{t}{\sqrt{t+1}}\, dt = \int \frac{w^2 - 1}{w} 2w\, dw = 2\int w^2 - 1\, dw$$

$$= \frac{2}{3}w^3 - 2w + C = \frac{2}{3}(t+1)^{3/2} - 2(t+1)^{1/2} + C.$$

Problems

77. (a) If $w = 2t$, then $dw = 2dt$. When $t = 0$, $w = 0$; when $t = 0.5$, $w = 1$. Thus,

$$\int_{0}^{0.5} f(2t)\, dt = \int_{0}^{1} f(w)\frac{1}{2}dw = \frac{1}{2}\int_{0}^{1} f(w)\, dw = \frac{3}{2}.$$

(b) If $w = 1 - t$, then $dw = -dt$. When $t = 0$, $w = 1$; when $t = 1$, $w = 0$. Thus,

$$\int_{0}^{1} f(1 - t)\, dt = \int_{1}^{0} f(w)(-dw) = +\int_{0}^{1} f(w)\, dw = 3.$$

(c) If $w = 3 - 2t$, then $dw = -2dt$. When $t = 1$, $w = 1$; when $t = 1.5$, $w = 0$. Thus,

$$\int_{1}^{1.5} f(3 - 2t)\, dt = \int_{1}^{0} f(w)\left(-\frac{1}{2}dw\right) = +\frac{1}{2}\int_{0}^{1} f(w)\, dw = \frac{3}{2}.$$

81. See Figure 7.1. The period of $V = V_0 \sin(\omega t)$ is $2\pi/\omega$, so the area under the first arch is given by

$$\text{Area} = \int_{0}^{\pi/\omega} V_0 \sin(\omega t)\, dt$$

$$= -\frac{V_0}{\omega}\cos(\omega t)\bigg|_{0}^{\pi/\omega}$$

$$= -\frac{V_0}{\omega}\cos(\pi) + \frac{V_0}{\omega}\cos(0)$$

$$= -\frac{V_0}{\omega}(-1) + \frac{V_0}{\omega}(1) = \frac{2V_0}{\omega}.$$

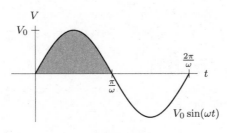

Figure 7.1

85. We substitute $w = 1 - x$ into $I_{m,n}$. Then $dw = -dx$, and $x = 1 - w$.

When $x = 0$, $w = 1$, and when $x = 1$, $w = 0$, so

$$I_{m,n} = \int_0^1 x^m (1-x)^n \, dx = \int_1^0 (1-w)^m w^n (-dw)$$

$$= -\int_1^0 w^n (1-w)^m \, dw = \int_0^1 w^n (1-w)^m \, dw = I_{n,m}.$$

89. Since $v = \dfrac{dh}{dt}$, it follows that $h(t) = \displaystyle\int v(t) \, dt$ and $h(0) = h_0$. Since

$$v(t) = \frac{mg}{k}\left(1 - e^{-\frac{k}{m}t}\right) = \frac{mg}{k} - \frac{mg}{k}e^{-\frac{k}{m}t},$$

we have

$$h(t) = \int v(t) \, dt = \frac{mg}{k}\int dt - \frac{mg}{k}\int e^{-\frac{k}{m}t} \, dt.$$

The first integral is simply $\dfrac{mg}{k}t + C$. To evaluate the second integral, make the substitution $w = -\frac{k}{m}t$. Then

$$dw = -\frac{k}{m}\,dt,$$

so

$$\int e^{-\frac{k}{m}t} \, dt = \int e^w \left(-\frac{m}{k}\right) dw = -\frac{m}{k}e^w + C = -\frac{m}{k}e^{-\frac{k}{m}t} + C.$$

Thus

$$h(t) = \int v \, dt = \frac{mg}{k}t - \frac{mg}{k}\left(-\frac{m}{k}e^{-\frac{k}{m}t}\right) + C$$

$$= \frac{mg}{k}t + \frac{m^2 g}{k^2}e^{-\frac{k}{m}t} + C.$$

Since $h(0) = h_0$,

$$h_0 = \frac{mg}{k}\cdot 0 + \frac{m^2 g}{k^2}e^0 + C;$$

$$C = h_0 - \frac{m^2 g}{k^2}.$$

Thus

$$h(t) = \frac{mg}{k}t + \frac{m^2 g}{k^2}e^{-\frac{k}{m}t} - \frac{m^2 g}{k^2} + h_0$$

$$h(t) = \frac{mg}{k}t - \frac{m^2 g}{k^2}\left(1 - e^{-\frac{k}{m}t}\right) + h_0.$$

Solutions for Section 7.2

Exercises

1. Let $u = \arctan x$, $v' = 1$. Then $v = x$ and $u' = \dfrac{1}{1+x^2}$. Integrating by parts, we get:

$$\int 1 \cdot \arctan x \, dx = x \cdot \arctan x - \int x \cdot \frac{1}{1+x^2} \, dx.$$

To compute the second integral use the substitution, $z = 1 + x^2$.

$$\int \frac{x}{1+x^2} \, dx = \frac{1}{2}\int \frac{dz}{z} = \frac{1}{2}\ln|z| + C = \frac{1}{2}\ln(1+x^2) + C.$$

Thus,

$$\int \arctan x \, dx = x \cdot \arctan x - \frac{1}{2}\ln(1+x^2) + C.$$

5. Let $u = t^2$ and $v' = e^{5t}$, so $u' = 2t$ and $v = \frac{1}{5}e^{5t}$.

Then $\int t^2 e^{5t}\, dt = \frac{1}{5}t^2 e^{5t} - \frac{2}{5}\int te^{5t}\, dt.$

Using Problem 4, we have $\int t^2 e^{5t}\, dt = \frac{1}{5}t^2 e^{5t} - \frac{2}{5}(\frac{1}{5}te^{5t} - \frac{1}{25}e^{5t}) + C$
$= \frac{1}{5}t^2 e^{5t} - \frac{2}{25}te^{5t} + \frac{2}{125}e^{5t} + C.$

9. Let $u = \ln x$ and $v' = x^3$, so $u' = \frac{1}{x}$ and $v = \frac{x^4}{4}$.

Then

$$\int x^3 \ln x\, dx = \frac{x^4}{4}\ln x - \int \frac{x^3}{4}\, dx = \frac{x^4}{4}\ln x - \frac{x^4}{16} + C.$$

13. Let $u = \cos(3\alpha + 1)$ and $v' = \cos(3\alpha + 1)$, so $u' = -3\sin(3\alpha + 1)$, and $v = \frac{1}{3}\sin(3\alpha + 1)$. Then

$$\int \cos^2(3\alpha + 1)\, d\alpha = \int (\cos(3\alpha + 1))\cos(3\alpha + 1)\, d\alpha$$

$$= \frac{1}{3}\cos(3\alpha + 1)\sin(3\alpha + 1) + \int \sin^2(3\alpha + 1)\, d\alpha$$

$$= \frac{1}{3}\cos(3\alpha + 1)\sin(3\alpha + 1) + \int \left(1 - \cos^2(3\alpha + 1)\right)\, d\alpha$$

$$= \frac{1}{3}\cos(3\alpha + 1)\sin(3\alpha + 1) + \alpha - \int \cos^2(3\alpha + 1)\, d\alpha.$$

By adding $\int \cos^2(3\alpha + 1)\, d\alpha$ to both sides of the above equation, we find that

$$2\int \cos^2(3\alpha + 1)\, d\alpha = \frac{1}{3}\cos(3\alpha + 1)\sin(3\alpha + 1) + \alpha + C,$$

which gives

$$\int \cos^2(3\alpha + 1)\, d\alpha = \frac{1}{6}\cos(3\alpha + 1)\sin(3\alpha + 1) + \frac{\alpha}{2} + C.$$

17. Let $u = \theta + 1$ and $v' = \sin(\theta + 1)$, so $u' = 1$ and $v = -\cos(\theta + 1)$.

$$\int (\theta + 1)\sin(\theta + 1)\, d\theta = -(\theta + 1)\cos(\theta + 1) + \int \cos(\theta + 1)\, d\theta$$

$$= -(\theta + 1)\cos(\theta + 1) + \sin(\theta + 1) + C.$$

21. $\int \dfrac{t + 7}{\sqrt{5 - t}}\, dt = \int \dfrac{t}{\sqrt{5 - t}}\, dt + 7\int (5 - t)^{-1/2}\, dt.$

To calculate the first integral, we use integration by parts. Let $u = t$ and $v' = \frac{1}{\sqrt{5-t}}$, so $u' = 1$ and $v = -2(5-t)^{1/2}$.
Then

$$\int \frac{t}{\sqrt{5 - t}}\, dt = -2t(5 - t)^{1/2} + 2\int (5 - t)^{1/2}\, dt = -2t(5 - t)^{1/2} - \frac{4}{3}(5 - t)^{3/2} + C.$$

We can calculate the second integral directly: $7\int (5 - t)^{-1/2} = -14(5 - t)^{1/2} + C_1.$ Thus

$$\int \frac{t + 7}{\sqrt{5 - t}}\, dt = -2t(5 - t)^{1/2} - \frac{4}{3}(5 - t)^{3/2} - 14(5 - t)^{1/2} + C_2.$$

25. This integral can first be simplified by making the substitution $w = x^2$, $dw = 2x\, dx$. Then

$$\int x \arctan x^2\, dx = \frac{1}{2}\int \arctan w\, dw.$$

To evaluate $\int \arctan w\, dw$, we'll use integration by parts. Let $u = \arctan w$ and $v' = 1$, so $u' = \frac{1}{1+w^2}$ and $v = w$.
Then

$$\int \arctan w\, dw = w \arctan w - \int \frac{w}{1 + w^2}\, dw = w \arctan w - \frac{1}{2}\ln|1 + w^2| + C.$$

Since $1 + w^2$ is never negative, we can drop the absolute value signs. Thus, we have

$$\int x \arctan x^2 \, dx = \frac{1}{2} \left(x^2 \arctan x^2 - \frac{1}{2} \ln(1 + (x^2)^2) + C \right)$$
$$= \frac{1}{2} x^2 \arctan x^2 - \frac{1}{4} \ln(1 + x^4) + C.$$

29. Let $u = x - 1, u' = 1$ and $v' = \cosh x, v = \sinh x$. Integrating by parts, we get

$$\int (x - 1) \cosh x \, dx = (x - 1) \sinh x - \int \sinh x \, dx$$
$$= (x - 1) \sinh x - \cosh x + C.$$

33. $\displaystyle\int_1^3 t \ln t \, dt = \left(\frac{1}{2} t^2 \ln t - \frac{1}{4} t \right) \Big|_1^3 = \frac{9}{2} \ln 3 - 2 \approx 2.944.$

37. To simplify the integral, we first make the substitution $z = u^2$, so $dz = 2u \, du$. Then

$$\int_{u=0}^{u=1} u \arcsin u^2 \, du = \frac{1}{2} \int_{z=0}^{z=1} \arcsin z \, dz.$$

From Problem 36, we know that $\int_0^1 \arcsin z \, dz = \frac{\pi}{2} - 1$. Thus,

$$\int_0^1 u \arcsin u^2 \, du = \frac{1}{2} \left(\frac{\pi}{2} - 1 \right) \approx 0.285.$$

Problems

41. Integration by parts: let $u = \cos\theta$ and $v' = \cos\theta$, so $u' = -\sin\theta$ and $v = \sin\theta$.

$$\int \cos^2\theta \, d\theta = \sin\theta \cos\theta - \int (-\sin\theta)(\sin\theta) \, d\theta$$
$$= \sin\theta \cos\theta + \int \sin^2\theta \, d\theta.$$

Now use $\sin^2\theta = 1 - \cos^2\theta$.

$$\int \cos^2\theta \, d\theta = \sin\theta \cos\theta + \int (1 - \cos^2\theta) \, d\theta$$
$$= \sin\theta \cos\theta + \int d\theta - \int \cos^2\theta \, d\theta.$$

Adding $\int \cos^2\theta \, d\theta$ to both sides, we have

$$2 \int \cos^2\theta \, d\theta = \sin\theta \cos\theta + \theta + C$$
$$\int \cos^2\theta \, d\theta = \frac{1}{2} \sin\theta \cos\theta + \frac{1}{2}\theta + C'.$$

Use the identity $\cos^2\theta = \frac{1 + \cos 2\theta}{2}$.

$$\int \cos^2\theta \, d\theta = \int \frac{1 + \cos 2\theta}{2} \, d\theta = \frac{1}{2}\theta + \frac{1}{4} \sin 2\theta + C.$$

The only difference is in the two terms $\frac{1}{2} \sin\theta \cos\theta$ and $\frac{1}{4} \sin 2\theta$, but since $\sin 2\theta = 2\sin\theta \cos\theta$, we have $\frac{1}{4} \sin 2\theta = \frac{1}{4}(2 \sin\theta \cos\theta) = \frac{1}{2} \sin\theta \cos\theta$, so there is no real difference between the formulas.

45. Again we use Problems 42 and 43. Integrate by parts, letting $u = \theta$ and $v' = e^\theta \cos \theta$, so $u' = 1$ and $v = \frac{1}{2}e^\theta(\sin \theta + \cos \theta)$. Then

$$\int \theta e^\theta \cos \theta \, d\theta = \frac{1}{2}\theta e^\theta(\sin \theta + \cos \theta) - \frac{1}{2}\int e^\theta(\sin \theta + \cos \theta) \, d\theta$$

$$= \frac{1}{2}\theta e^\theta(\sin \theta + \cos \theta) - \frac{1}{2}\int e^\theta \sin \theta \, d\theta - \frac{1}{2}\int e^\theta \cos \theta \, d\theta$$

$$= \frac{1}{2}\theta e^\theta(\sin \theta + \cos \theta) - \frac{1}{4}e^\theta(\sin \theta - \cos \theta) - \frac{1}{4}(\sin \theta + \cos \theta) + C$$

$$= \frac{1}{2}\theta e^\theta(\sin \theta + \cos \theta) - \frac{1}{2}e^\theta \sin \theta + C.$$

49. We integrate by parts. Since we know what the answer is supposed to be, it's easier to choose u and v'. Let $u = \cos^{n-1} x$ and $v' = \cos x$, so $u' = (n-1)\cos^{n-2} x(-\sin x)$ and $v = \sin x$.

Then

$$\int \cos^n x \, dx = \cos^{n-1} x \sin x + (n-1)\int \cos^{n-2} x \sin^2 x \, dx$$

$$= \cos^{n-1} x \sin x + (n-1)\int \cos^{n-2} x(1 - \cos^2 x) \, dx$$

$$= \cos^{n-1} x \sin x - (n-1)\int \cos^n x \, dx + (n-1)\int \cos^{n-2} x \, dx.$$

Thus, by adding $(n-1)\int \cos^n x \, dx$ to both sides of the equation, we find

$$n\int \cos^n x \, dx = \cos^{n-1} x \sin x + (n-1)\int \cos^{n-2} x \, dx,$$

so $$\int \cos^n \, dx = \frac{1}{n}\cos^{n-1} x \sin x + \frac{n-1}{n}\int \cos^{n-2} x \, dx.$$

53. (a) We have

$$F(a) = \int_0^a x^2 e^{-x} \, dx$$

$$= -x^2 e^{-x}\Big|_0^a + \int_0^a 2xe^{-x} \, dx$$

$$= (-x^2 e^{-x} - 2xe^{-x})\Big|_0^a + 2\int_0^a e^{-x} \, dx$$

$$= (-x^2 e^{-x} - 2xe^{-x} - 2e^{-x})\Big|_0^a$$

$$= -a^2 e^{-a} - 2ae^{-a} - 2e^{-a} + 2.$$

(b) $F(a)$ is increasing because $x^2 e^{-x}$ is positive, so as a increases, the area under the curve from 0 to a also increases and thus the integral increases.

(c) We have $F'(a) = a^2 e^{-a}$, so
$$F''(a) = 2ae^{-a} - a^2 e^{-a} = a(2-a)e^{-a}.$$

We see that $F''(a) > 0$ for $0 < a < 2$, so F is concave up on this interval.

57. (a) We want to compute C_1, with $C_1 > 0$, such that

$$\int_0^1 (\Psi_1(x))^2 \, dx = \int_0^1 (C_1 \sin(\pi x))^2 \, dx = C_1^2 \int_0^1 \sin^2(\pi x) \, dx = 1.$$

We use integration by parts with $u = v' = \sin(\pi x)$.
So $u' = \pi \cos(\pi x)$ and $v = -\frac{1}{\pi} \cos(\pi x)$. Thus

$$\int_0^1 \sin^2(\pi x)\, dx = -\frac{1}{\pi} \sin(\pi x) \cos(\pi x)\Big|_0^1 + \int_0^1 \cos^2(\pi x)\, dx$$

$$= -\frac{1}{\pi} \sin(\pi x) \cos(\pi x)\Big|_0^1 + \int_0^1 (1 - \sin^2(\pi x))\, dx.$$

Moving $\int_0^1 \sin^2(\pi x)\, dx$ from the right side to the left side of the equation and solving, we get

$$2\int_0^1 \sin^2(\pi x)\, dx = -\frac{1}{\pi} \sin(\pi x) \cos(\pi x)\Big|_0^1 + \int_0^1 1\, dx = 0 + 1 = 1,$$

so

$$\int_0^1 \sin^2(\pi x)\, dx = \frac{1}{2}.$$

Thus, we have

$$\int_0^1 (\Psi_1(x))^2\, dx = C_1^2 \int_0^1 \sin^2(\pi x)\, dx = \frac{C_1^2}{2}.$$

So, to normalize Ψ_1, we take $C_1 > 0$ such that

$$\frac{C_1^2}{2} = 1 \quad \text{so} \quad C_1 = \sqrt{2}.$$

(b) To normalize Ψ_n, we want to compute C_n, with $C_n > 0$, such that

$$\int_0^1 (\Psi_n(x))^2\, dx = C_n^2 \int_0^1 \sin^2(n\pi x)\, dx = 1.$$

The solution to part (a) shows us that

$$\int \sin^2(\pi t)\, dt = -\frac{1}{2\pi} \sin(\pi t) \cos(\pi t) + \frac{1}{2} \int 1\, dt.$$

In the integral for Ψ_n, we make the substitution $t = nx$, so $dx = \frac{1}{n} dt$. Since $t = 0$ when $x = 0$ and $t = n$ when $x = 1$, we have

$$\int_0^1 \sin^2(n\pi x)\, dx = \frac{1}{n} \int_0^n \sin^2(\pi t)\, dt$$

$$= \frac{1}{n} \left(-\frac{1}{2\pi} \sin(\pi t) \cos(\pi t)\Big|_0^n + \frac{1}{2} \int_0^n 1\, dt \right)$$

$$= \frac{1}{n} \left(0 + \frac{n}{2} \right) = \frac{1}{2}.$$

Thus, we have

$$\int_0^1 (\Psi_n(x))^2\, dx = C_n^2 \int_0^1 \sin^2(n\pi x)\, dx = \frac{C_n^2}{2}.$$

So to normalize Ψ_n, we take C_n such that

$$\frac{C_n^2}{2} = 1 \quad \text{so} \quad C_n = \sqrt{2}.$$

Solutions for Section 7.3

Exercises

1. $\frac{1}{10} e^{(-3\theta)} (-3 \cos \theta + \sin \theta) + C.$
(Let $a = -3, b = 1$ in II-9.)

5. Note that you can't use substitution here: letting $w = x^3 + 5$ does not work, since there is no $dw = 3x^2\, dx$ in the integrand. What will work is simply multiplying out the square: $(x^3 + 5)^2 = x^6 + 10x^3 + 25$. Then use I-1:

$$\int (x^3 + 5)^2\, dx = \int x^6\, dx + 10 \int x^3\, dx + 25 \int 1\, dx = \frac{1}{7}x^7 + 10 \cdot \frac{1}{4}x^4 + 25x + C.$$

9. $\left(\dfrac{1}{2}x^3 - \dfrac{3}{4}x^2 + \dfrac{3}{4}x - \dfrac{3}{8}\right)e^{2x} + C.$

(Let $a = 2, p(x) = x^3$ in III-14.)

13. $\dfrac{5}{16}\sin 3\theta \sin 5\theta + \dfrac{3}{16}\cos 3\theta \cos 5\theta + C.$

(Let $a = 3, b = 5$ in II-12.)

17. $\left(\dfrac{1}{3}x^4 - \dfrac{4}{9}x^3 + \dfrac{4}{9}x^2 - \dfrac{8}{27}x + \dfrac{8}{81}\right)e^{3x} + C.$

(Let $a = 3, p(x) = x^4$ in III-14.)

21. Substitute $w = x^2, dw = 2x\,dx.$ Then $\displaystyle\int x^3 \sin x^2\,dx = \frac{1}{2}\int w \sin w\,dw.$ By III-15, we have

$$\int w \sin w\,dw = -\frac{1}{2}w\cos w + \frac{1}{2}\sin w + C = -\frac{1}{2}x^2 \cos x^2 + \frac{1}{2}\sin x^2 + C.$$

25. Use IV-21 twice to get the exponent down to 1:

$$\int \frac{1}{\cos^5 x}\,dx = \frac{1}{4}\frac{\sin x}{\cos^4 x} + \frac{3}{4}\int \frac{1}{\cos^3 x}\,dx$$

$$\int \frac{1}{\cos^3 x}\,dx = \frac{1}{2}\frac{\sin x}{\cos^2 x} + \frac{1}{2}\int \frac{1}{\cos x}\,dx.$$

Now use IV-22 to get

$$\int \frac{1}{\cos x}\,dx = \frac{1}{2}\ln\left|\frac{(\sin x)+1}{(\sin x)-1}\right| + C.$$

Putting this all together gives

$$\int \frac{1}{\cos^5 x}\,dx = \frac{1}{4}\frac{\sin x}{\cos^4 x} + \frac{3}{8}\frac{\sin x}{\cos^2 x} + \frac{3}{16}\ln\left|\frac{(\sin x)+1}{(\sin x)-1}\right| + C.$$

29.

$$\int \frac{1}{x^2 + 4x + 3}\,dx = \int \frac{1}{(x+1)(x+3)}\,dx = \frac{1}{2}(\ln|x+1| - \ln|x+3|) + C.$$

(Let $a = -1$ and $b = -3$ in V-26).

33. $\arctan(z+2) + C.$

(Substitute $w = z + 2$ and use V-24, letting $a = 1$.)

37.

$$\int \sin^3 3\theta \cos^2 3\theta\,d\theta = \int (\sin 3\theta)(\cos^2 3\theta)(1 - \cos^2 3\theta)\,d\theta$$

$$= \int \sin 3\theta(\cos^2 3\theta - \cos^4 3\theta)\,d\theta.$$

Using an extension of the tip given in rule IV-23, we let $w = \cos 3\theta,\ dw = -3\sin 3\theta\,d\theta.$

$$\int \sin 3\theta(\cos^2 3\theta - \cos^4 3\theta)\,d\theta = -\frac{1}{3}\int (w^2 - w^4)\,dw$$

$$= -\frac{1}{3}\left(\frac{w^3}{3} - \frac{w^5}{5}\right) + C$$

$$= -\frac{1}{9}(\cos^3 3\theta) + \frac{1}{15}(\cos^5 3\theta) + C.$$

41. Substitute $w = 3\alpha$, $dw = 3\,d\alpha$. Then $d\alpha = \dfrac{1}{3}\,dw$. We have

$$\int_{\alpha=0}^{\alpha=\frac{\pi}{12}} \sin 3\alpha\,d\alpha = \frac{1}{3}\int_{w=0}^{w=\frac{\pi}{4}} \sin w\,dw$$

$$= -\frac{1}{3}\cos w\,\Big|_{0}^{\frac{\pi}{4}}$$

$$= -\frac{1}{3}\left(\frac{\sqrt{2}}{2} - 1\right) = \frac{1}{3}\left(1 - \frac{\sqrt{2}}{2}\right).$$

Problems

45. Using II-10 in the integral table, if $m \neq \pm n$, then

$$\int_{-\pi}^{\pi} \sin m\theta \sin n\theta\,d\theta \;=\; \frac{1}{n^2 - m^2}[m\cos m\theta \sin n\theta - n\sin m\theta \cos n\theta]\Big|_{-\pi}^{\pi}$$

$$= \frac{1}{n^2 - m^2}[(m\cos m\pi \sin n\pi - n\sin m\pi \cos n\pi) - (m\cos(-m\pi)\sin(-n\pi) - n\sin(-m\pi)\cos(-n\pi))]$$

But $\sin k\pi = 0$ for all integers k, so each term reduces to 0, making the whole integral reduce to 0.

49. We want to calculate

$$\int_{0}^{1} C_n \sin(n\pi x)\cdot C_m \sin(m\pi x)\,dx.$$

We use II-11 from the table of integrals with $a = n\pi$, $b = m\pi$. Since $n \neq m$, we see that

$$\int_{0}^{1} \Psi_n(x)\cdot \Psi_m(x)\,dx = C_n C_m \int_{0}^{1} \sin(n\pi x)\sin(m\pi x)\,dx$$

$$= \frac{C_n C_m}{m^2\pi^2 - n^2\pi^2}\left(n\pi \cos(n\pi x)\sin(m\pi x) - m\pi \sin(n\pi x)\cos(m\pi x)\right)\Big|_{0}^{1}$$

$$= \frac{C_n C_m}{(m^2 - n^2)\pi^2}\Big(n\pi \cos(n\pi)\sin(m\pi) - m\pi \sin(n\pi)\cos(m\pi)$$

$$-n\pi \cos(0)\sin(0) + m\pi \sin(0)\cos(0)\Big)$$

$$= 0$$

since $\sin(0) = \sin(n\pi) = \sin(m\pi) = 0$.

Solutions for Section 7.4

Exercises

1. Since $25 - x^2 = (5 - x)(5 + x)$, we take

$$\frac{20}{25 - x^2} = \frac{A}{5 - x} + \frac{B}{5 + x}.$$

So,

$$20 = A(5 + x) + B(5 - x)$$
$$20 = (A - B)x + 5A + 5B,$$

giving

$$A - B = 0$$
$$5A + 5B = 20.$$

Thus $A = B = 2$ and

$$\frac{20}{25 - x^2} = \frac{2}{5 - x} + \frac{2}{5 + x}.$$

5. Since $s^4 - 1 = (s^2 - 1)(s^2 + 1) = (s - 1)(s + 1)(s^2 + 1)$, we have

$$\frac{2}{s^4 - 1} = \frac{A}{s - 1} + \frac{B}{s + 1} + \frac{Cs + D}{s^2 + 1}.$$

Thus,

$$2 = A(s + 1)(s^2 + 1) + B(s - 1)(s^2 + 1) + (Cs + D)(s - 1)(s + 1)$$
$$2 = (A + B + C)s^3 + (A - B + D)s^2 + (A + B - C)s + (A - B - D),$$

giving

$$A + B + C = 0$$
$$A - B + D = 0$$
$$A + B - C = 0$$
$$A - B - D = 2.$$

From the first and third equations we find $A + B = 0$ and $C = 0$. From the second and fourth we find $A - B = 1$ and $D = -1$. Thus $A = 1/2$ and $B = -1/2$ and

$$\frac{2}{s^4 - 1} = \frac{1}{2(s - 1)} - \frac{1}{2(s + 1)} - \frac{1}{s^2 + 1}.$$

9. Using the result of Problem 2, we have

$$\int \frac{x + 1}{6x + x^2} \, dx = \int \frac{1/6}{x} \, dx + \int \frac{5/6}{6 + x} \, dx = \frac{1}{6} \left(\ln |x| + 5 \ln |6 + x| \right) + C.$$

13. Using the result of Problem 6, we have

$$\int \frac{2y}{y^3 - y^2 + y - 1} \, dy = \int \frac{1}{y - 1} \, dy + \int \frac{1 - y}{y^2 + 1} \, dy = \ln |y - 1| + \arctan y - \frac{1}{2} \ln \left| y^2 + 1 \right| + C.$$

17. We let

$$\frac{10x + 2}{x^3 - 5x^2 + x - 5} = \frac{10x + 2}{(x - 5)(x^2 + 1)} = \frac{A}{x - 5} + \frac{Bx + C}{x^2 + 1}$$

giving

$$10x + 2 = A(x^2 + 1) + (Bx + C)(x - 5)$$
$$10x + 2 = (A + B)x^2 + (C - 5B)x + A - 5C$$

so

$$A + B = 0$$
$$C - 5B = 10$$
$$A - 5C = 2.$$

Thus, $A = 2, B = -2, C = 0$, so

$$\int \frac{10x + 2}{x^3 - 5x^2 + x - 5} \, dx = \int \frac{2}{x - 5} \, dx - \int \frac{2x}{x^2 + 1} \, dx = 2 \ln |x - 5| - \ln \left| x^2 + 1 \right| + K.$$

21. Completing the square gives $x^2 + 4x + 5 = 1 + (x + 2)^2$. Since $x + 2 = \tan t$ and $dx = (1/\cos^2 t)dt$, we have

$$\int \frac{1}{x^2 + 4x + 5} \, dx = \int \frac{1}{1 + \tan^2 t} \cdot \frac{1}{\cos^2 t} \, dt = \int dt = t + C = \arctan(x + 2) + C.$$

Problems

25. Since $x^2 + 6x + 9$ is a perfect square, we write

$$\int \frac{1}{x^2 + 6x + 25} \, dx = \int \frac{1}{(x^2 + 6x + 9) + 16} \, dx = \int \frac{1}{(x + 3)^2 + 16} \, dx.$$

We use the trigonometric substitution $x + 3 = 4 \tan \theta$, so $x = 4 \tan \theta - 3$.

29. Since $2z - z^2 = 1 - (z - 1)^2$, we have

$$\int \frac{z - 1}{\sqrt{2z - z^2}} \, dz = \int \frac{z - 1}{\sqrt{1 - (z - 1)^2}} \, dz.$$

Substitute $w = 1 - (z - 1)^2$, so $dw = -2(z - 1) \, dz$.

33. We write

$$\frac{1}{(x + 2)(x + 3)} = \frac{A}{x + 2} + \frac{B}{x + 3},$$

giving

$$1 = A(x + 3) + B(x + 2)$$
$$1 = (A + B)x + (3A + 2B)$$

so

$$A + B = 0$$
$$3A + 2B = 1.$$

Thus, $A = 1$, $B = -1$, so

$$\int \frac{1}{(x + 2)(x + 3)} \, dx = \int \frac{1}{x + 2} \, dx - \int \frac{1}{x + 3} \, dx = \ln|x + 2| - \ln|x + 3| + C.$$

37. We know $x^2 + 5x + 4 = (x + 1)(x + 4)$, so we can use V-26 of the integral table with $a = -1$ and $b = -4$ to write

$$\int \frac{dx}{x^2 + 5x + 4} = \frac{1}{3}(\ln|x + 1| - \ln|x + 4|) + C.$$

41. Since $x^3 + x = x(x^2 + 1)$ cannot be factored further, we write

$$\frac{x + 1}{x^3 + x} = \frac{A}{x} + \frac{Bx + C}{x^2 + 1}.$$

Multiplying by $x(x^2 + 1)$ gives

$$x + 1 = A(x^2 + 1) + (Bx + C)x$$
$$x + 1 = (A + B)x^2 + Cx + A,$$

so

$$A + B = 0$$
$$C = 1$$
$$A = 1.$$

Thus, $A = C = 1$, $B = -1$, and we have

$$\int \frac{x + 1}{x^3 + x} \, dx = \int \left(\frac{1}{x} + \frac{-x + 1}{x^2 + 1} \right) = \int \frac{dx}{x} - \int \frac{x \, dx}{x^2 + 1} + \int \frac{dx}{x^2 + 1}$$

$$= \ln|x| - \frac{1}{2} \ln \left| x^2 + 1 \right| + \arctan x + K.$$

45. Let $t = \tan \theta$ so $dt = (1/\cos^2 \theta)d\theta$. Since $\sqrt{1 + \tan^2 \theta} = 1/\cos \theta$, we have

$$\int \frac{dt}{t^2 \sqrt{1 + t^2}} = \int \frac{1/\cos^2 \theta}{\tan^2 \theta \sqrt{1 + \tan^2 \theta}} \, d\theta = \int \frac{\cos \theta}{\tan^2 \theta \cos^2 \theta} \, d\theta = \int \frac{\cos \theta}{\sin^2 \theta} \, d\theta.$$

The last integral can be evaluated by guess-and-check or by substituting $w = \sin \theta$. The result is

$$\int \frac{dt}{t^2 \sqrt{1 + t^2}} = \int \frac{\cos \theta}{\sin^2 \theta} \, d\theta = -\frac{1}{\sin \theta} + C.$$

Since $t = \tan \theta$ and $1/\cos^2 \theta = 1 + \tan^2 \theta$, we have

$$\cos \theta = \frac{1}{\sqrt{1 + \tan^2 \theta}} = \frac{1}{\sqrt{1 + t^2}}.$$

In addition, $\tan \theta = \sin \theta / \cos \theta$ so

$$\sin \theta = \tan \theta \cos \theta = \frac{t}{\sqrt{1 + t^2}}.$$

Thus

$$\int \frac{dt}{t^2 \sqrt{1 + t^2}} = -\frac{\sqrt{1 + t^2}}{t} + C.$$

49. Using the substitution $w = e^x$, we get $dw = e^x dx$, so we have

$$\int \frac{e^x}{(e^x - 1)(e^x + 2)} \, dx = \int \frac{dw}{(w - 1)(w + 2)}.$$

But

$$\frac{1}{(w - 1)(w + 2)} = \frac{1}{3} \left(\frac{1}{w - 1} - \frac{1}{w + 2} \right),$$

so

$$\int \frac{e^x}{(e^x - 1)(e^x + 2)} \, dx = \int \frac{1}{3} \left(\frac{1}{w - 1} - \frac{1}{w + 2} \right) dw$$
$$= \frac{1}{3} \left(\ln |w - 1| - \ln |w + 2| \right) + C$$
$$= \frac{1}{3} \left(\ln |e^x - 1| - \ln |e^x + 2| \right) + C.$$

53. We have

$$\text{Area} = \int_0^{\sqrt{2}} \frac{x^3}{\sqrt{4 - x^2}} \, dx.$$

Let $x = 2 \sin \theta$ so $dx = 2 \cos \theta \, d\theta$ and $\sqrt{4 - x^2} = \sqrt{4 - 4 \sin^2 \theta} = 2 \cos \theta$. When $x = 0, \theta = 0$ and when $x = \sqrt{2}, \theta = \pi/4$.

$$\int_0^{\sqrt{2}} \frac{x^3}{\sqrt{4 - x^2}} dx = \int_0^{\pi/4} \frac{(2 \sin \theta)^3}{\sqrt{4 - (2 \sin \theta)^2}} 2 \cos \theta \, d\theta$$
$$= 8 \int_0^{\pi/4} \sin^3 \theta \, d\theta = 8 \int_0^{\pi/4} (\sin \theta - \sin \theta \cos^2 \theta) \, d\theta$$
$$= 8 \left(-\cos \theta + \frac{\cos^3 \theta}{3} \right) \Bigg|_0^{\pi/4} = 8 \left(\frac{2}{3} - \frac{5}{6\sqrt{2}} \right).$$

57. Using partial fractions, we write

$$\frac{2x}{x^2 - 1} = \frac{A}{x + 1} + \frac{B}{x - 1}$$
$$2x = A(x - 1) + B(x + 1) = (A + B)x - A + B.$$

So, $A + B = 2$ and $-A + B = 0$, giving $A = B = 1$. Thus

$$\int \frac{2x}{x^2 - 1} \, dx = \int \left(\frac{1}{x + 1} + \frac{1}{x - 1} \right) \, dx = \ln|x + 1| + \ln|x - 1| + C.$$

Using the substitution $w = x^2 - 1$, we get $dw = 2x \, dx$, so we have

$$\int \frac{2x}{x^2 - 1} \, dx = \int \frac{dw}{w} = \ln|w| + C = \ln\left|x^2 - 1\right| + C.$$

The properties of logarithms show that the two results are the same:

$$\ln|x + 1| + \ln|x - 1| = \ln|(x + 1)(x - 1)| = \ln\left|x^2 - 1\right|.$$

61. (a) If $a \neq b$, we have

$$\int \frac{x}{(x - a)(x - b)} \, dx = \int \frac{1}{a - b} \left(\frac{a}{x - a} - \frac{b}{x - b} \right) \, dx = \frac{1}{a - b}(a \ln|x - a| - b \ln|x - b|) + C.$$

(b) If $a = b$, we have

$$\int \frac{x}{(x - a)^2} \, dx = \int \left(\frac{1}{x - a} + \frac{a}{(x - a)^2} \right) \, dx = \ln|x - a| - \frac{a}{x - a} + C.$$

Solutions for Section 7.5

Exercises

1. (a) The approximation LEFT(2) uses two rectangles, with the height of each rectangle determined by the left-hand endpoint. See Figure 7.2. We see that this approximation is an underestimate.

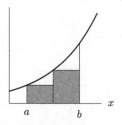

Figure 7.2

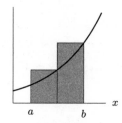

Figure 7.3

(b) The approximation RIGHT(2) uses two rectangles, with the height of each rectangle determined by the right-hand endpoint. See Figure 7.3. We see that this approximation is an overestimate.

(c) The approximation TRAP(2) uses two trapezoids, with the height of each trapezoid given by the secant line connecting the two endpoints. See Figure 7.4. We see that this approximation is an overestimate.

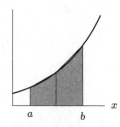

Figure 7.4

(d) The approximation MID(2) uses two rectangles, with the height of each rectangle determined by the height at the midpoint. Alternately, we can view MID(2) as a trapezoid rule where the height is given by the tangent line at the midpoint. Both interpretations are shown in Figure 7.5. We see from the tangent line interpretation that this approximation is an underestimate

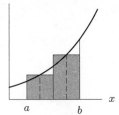 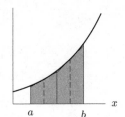

Figure 7.5

5. (a) The approximation LEFT(2) uses two rectangles, with the height of each rectangle determined by the left-hand endpoint. See Figure 7.6. We see that this approximation is an underestimate (that is, it is more negative).

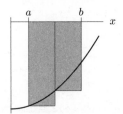 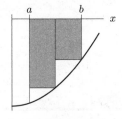

Figure 7.6   **Figure 7.7**

(b) The approximation RIGHT(2) uses two rectangles, with the height of each rectangle determined by the right-hand endpoint. See Figure 7.7. We see that this approximation is an overestimate (that is, it is less negative).

(c) The approximation TRAP(2) uses two trapezoids, with the height of each trapezoid given by the secant line connecting the two endpoints. See Figure 7.8. We see that this approximation is an overestimate (that is, it is less negative).

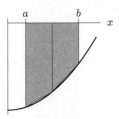

Figure 7.8

(d) The approximation MID(2) uses two rectangles, with the height of each rectangle determined by the height at the midpoint. Alternately, we can view MID(2) as a trapezoid rule where the height is given by the tangent line at the midpoint. Both interpretations are shown in Figure 7.9. We see from the tangent line interpretation that this approximation is an underestimate (that is, it is more negative).

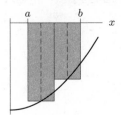 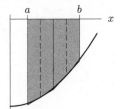

Figure 7.9

9. (a)

$$MID(2) = 2 \cdot f(1) + 2 \cdot f(3)$$
$$= 2 \cdot 2 + 2 \cdot 10$$
$$= 24$$
$$TRAP(2) = \frac{LEFT(2) + RIGHT(2)}{2}$$
$$= \frac{12 + 44}{2} \quad \text{(see Problem 8)}$$
$$= 28$$

(b)

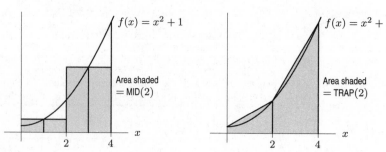

MID(2) is an underestimate, since $f(x) = x^2 + 1$ is concave up and a tangent line will be below the curve. TRAP(2) is an overestimate, since a secant line lies above the curve.

Problems

13. Since the function is decreasing, LEFT is an overestimate and RIGHT is an underestimate. Since the graph is concave down, secant lines lie below the graph so TRAP is an underestimate and tangent lines lie above the graph so MID is an overestimate. We can see that MID and TRAP are closer to the exact value than LEFT and RIGHT. In order smallest to largest, we have:

$$RIGHT(n) < TRAP(n) < \text{Exact value} < MID(n) < LEFT(n).$$

17. $f(x)$ is decreasing and concave up, so LEFT and TRAP give overestimates and RIGHT and MID give underestimates.

21. (a) $\displaystyle\int_0^{2\pi} \sin\theta \, d\theta = -\cos\theta \Big|_0^{2\pi} = 0.$

   (b) MID(1) is 0 since the midpoint of 0 and 2π is π, and $\sin\pi = 0$. Thus MID(1) $= 2\pi(\sin\pi) = 0$. The midpoints we use for MID(2) are $\pi/2$ and $3\pi/2$, and $\sin(\pi/2) = -\sin(3\pi/2)$. Thus MID(2) $= \pi\sin(\pi/2) + \pi\sin(3\pi/2) = 0$.

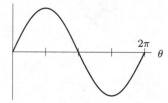

   (c) MID(3) = 0.

   In general, MID(n) = 0 for all n, even though your calculator (because of round-off error) might not return it as such. The reason is that $\sin(x) = -\sin(2\pi - x)$. If we use MID(n), we will always take sums where we are adding pairs of the form $\sin(x)$ and $\sin(2\pi - x)$, so the sum will cancel to 0. (If n is odd, we will get a $\sin\pi$ in the sum which does not pair up with anything — but $\sin\pi$ is already 0.)

25.

$$TRAP(n) = \frac{LEFT(n) + RIGHT(n)}{2}$$
$$= \frac{LEFT(n) + LEFT(n) + f(b)\Delta x - f(a)\Delta x}{2}$$
$$= LEFT(n) + \frac{1}{2}(f(b) - f(a))\Delta x$$

Solutions for Section 7.6

Exercises

1. We saw in Problem 7 in Section 7.5 that, for this definite integral, we have LEFT(2) = 27, RIGHT(2) = 135, TRAP(2) = 81, and MID(2) = 67.5. Thus,

$$\text{SIMP(2)} = \frac{2\text{MID(2)} + \text{TRAP(2)}}{3} = \frac{2(67.5) + 81}{3} = 72.$$

Notice that

$$\int_0^6 x^2 dx = \left.\frac{x^3}{3}\right|_0^6 = \frac{6^3}{3} - \frac{0^3}{3} = 72,$$

and so SIMP(2) gives the exact value of the integral in this case.

Problems

5. (a) $\displaystyle\int_0^4 e^x \, dx = \left. e^x \right|_0^4 = e^4 - e^0 \approx 53.598\ldots.$

(b) Computing the sums directly, since $\Delta x = 2$, we have
LEFT(2)= $2 \cdot e^0 + 2 \cdot e^2 \approx 2(1) + 2(7.389) = 16.778$;   error $= 36.820$.
RIGHT(2)= $2 \cdot e^2 + 2 \cdot e^4 \approx 2(7.389) + 2(54.598) = 123.974$;   error $= -70.376$.
TRAP(2)= $\dfrac{16.778 + 123.974}{2} = 70.376$;   error $= 16.778$.
MID(2)= $2 \cdot e^1 + 2 \cdot e^3 \approx 2(2.718) + 2(20.086) = 45.608$;   error $= 7.990$.
SIMP(2)= $\dfrac{2(45.608) + 70.376}{3} = 53.864$;   error $= -0.266$.

(c) Similarly, since $\Delta x = 1$, we have LEFT(4)= 31.193;   error $= 22.405$
RIGHT(4)= 84.791;   error $= -31.193$
TRAP(4)= 57.992;   error $= -4.394$
MID(4)= 51.428;   error $= 2.170$
SIMP(4)= 53.616;   error $= -0.018$

(d) For LEFT and RIGHT, we expect the error to go down by $1/2$, and this is very roughly what we see. For MID and TRAP, we expect the error to go down by $1/4$, and this is approximately what we see. For SIMP, we expect the error to go down by $1/2^4 = 1/16$, and this is approximately what we see.

9. (a) Suppose $q_i(x)$ is the quadratic function approximating $f(x)$ on the subinterval $[x_i, x_{i+1}]$, and m_i is the midpoint of the interval, $m_i = (x_i + x_{i+1})/2$. Then, using the equation in Problem 8, with $a = x_i$ and $b = x_{i+1}$ and $h = \Delta x = x_{i+1} - x_i$:

$$\int_{x_i}^{x_{i+1}} f(x)dx \approx \int_{x_i}^{x_{i+1}} q_i(x)dx = \frac{\Delta x}{3}\left(\frac{q_i(x_i)}{2} + 2q_i(m_i) + \frac{q_i(x_{i+1})}{2}\right).$$

(b) Summing over all subintervals gives

$$\int_a^b f(x)dx \approx \sum_{i=0}^{n-1} \int_{x_i}^{x_{i+1}} q_i(x)dx = \sum_{i=0}^{n-1} \frac{\Delta x}{3}\left(\frac{q_i(x_i)}{2} + 2q_i(m_i) + \frac{q_i(x_{i+1})}{2}\right).$$

Splitting the sum into two parts:

$$= \frac{2}{3}\sum_{i=0}^{n-1} q_i(m_i)\Delta x + \frac{1}{3}\sum_{i=0}^{n-1} \frac{q_i(x_i) + q_i(x_{i+1})}{2}\Delta x$$

$$= \frac{2}{3}\,\text{MID}(n) + \frac{1}{3}\,\text{TRAP}(n)$$

$$= \text{SIMP}(n).$$

Solutions for Section 7.7

Exercises

1. (a) See Figure 7.10. The area extends out infinitely far along the positive x-axis.

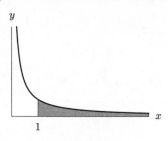

Figure 7.10

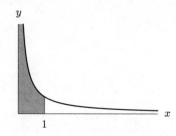

Figure 7.11

(b) See Figure 7.11. The area extends up infinitely far along the positive y-axis.

5. We have

$$\int_1^\infty \frac{1}{5x+2}\,dx = \lim_{b\to\infty}\int_1^b \frac{1}{5x+2}\,dx = \lim_{b\to\infty}\left(\frac{1}{5}\ln(5x+2)\right)\Big|_1^b = \lim_{b\to\infty}\left(\frac{1}{5}\ln(5b+2) - \frac{1}{5}\ln(7)\right).$$

As $b \leftarrow \infty$, we know that $\ln(5b+2) \to \infty$, and so this integral diverges.

9. Using integration by parts with $u = x$ and $v' = e^{-x}$, we find that

$$\int xe^{-x}\,dx = -xe^{-x} - \int -e^{-x}\,dx = -(1+x)e^{-x}$$

so

$$\int_0^\infty \frac{x}{e^x}\,dx = \lim_{b\to\infty}\int_0^b \frac{x}{e^x}\,dx$$

$$= \lim_{b\to\infty} -1(1+x)e^{-x}\Big|_0^b$$

$$= \lim_{b\to\infty}\left[1 - (1+b)e^{-b}\right]$$

$$= 1.$$

13. This is an improper integral because $\sqrt{16-x^2} = 0$ at $x = 4$. So

$$\int_0^4 \frac{dx}{\sqrt{16-x^2}} = \lim_{b\to4^-}\int_0^b \frac{dx}{\sqrt{16-x^2}}$$

$$= \lim_{b\to4^-} (\arcsin x/4)\Big|_0^b$$

$$= \lim_{b\to4^-} [\arcsin(b/4) - \arcsin(0)] = \pi/2 - 0 = \pi/2.$$

17.

$$\int_1^\infty \frac{1}{x^2+1}\,dx = \lim_{b\to\infty}\int_1^b \frac{1}{x^2+1}\,dx$$

$$= \lim_{b\to\infty} \arctan(x)\Big|_1^b$$

$$= \lim_{b\to\infty} [\arctan(b) - \arctan(1)]$$

$$= \pi/2 - \pi/4 = \pi/4.$$

21. With the substitution $w = \ln x$, $dw = \frac{1}{x}dx$,

$$\int \frac{dx}{x \ln x} = \int \frac{1}{w} dw = \ln |w| + C = \ln |\ln x| + C$$

so

$$\int_2^\infty \frac{dx}{x \ln x} = \lim_{b \to \infty} \int_2^b \frac{dx}{x \ln x}$$

$$= \lim_{b \to \infty} \ln |\ln x| \Big|_2^b$$

$$= \lim_{b \to \infty} [\ln |\ln b| - \ln |\ln 2|].$$

As $b \to \infty$, the limit goes to ∞ and hence the integral diverges.

25. Using the substitution $w = -x^{\frac{1}{2}}$, $-2dw = x^{-\frac{1}{2}} dx$,

$$\int e^{-x^{\frac{1}{2}}} x^{-\frac{1}{2}} dx = -2 \int e^w dw = -2e^{-x^{\frac{1}{2}}} + C.$$

So

$$\int_0^\pi \frac{1}{\sqrt{x}} e^{-\sqrt{x}} dx = \lim_{b \to 0^+} \int_b^\pi \frac{1}{\sqrt{x}} e^{-\sqrt{x}} dx$$

$$= \lim_{b \to 0^+} -2e^{-\sqrt{x}} \Big|_b^\pi$$

$$= 2 - 2e^{-\sqrt{\pi}}.$$

29. $\int \frac{dx}{x^2 - 1} = \int \frac{dx}{(x-1)(x+1)} = \frac{1}{2}(\ln |x-1| - \ln |x+1|) + C = \frac{1}{2}\left(\ln \frac{|x-1|}{|x+1|}\right) + C$, so

$$\int_4^\infty \frac{dx}{x^2 - 1} = \lim_{b \to \infty} \int_4^b \frac{dx}{x^2 - 1}$$

$$= \lim_{b \to \infty} \frac{1}{2}\left(\ln \frac{|x-1|}{|x+1|}\right)\Big|_4^b$$

$$= \lim_{b \to \infty} \left[\frac{1}{2}\ln\left(\frac{b-1}{b+1}\right) - \frac{1}{2}\ln \frac{3}{5}\right]$$

$$= -\frac{1}{2}\ln \frac{3}{5} = \frac{1}{2}\ln \frac{5}{3}.$$

Problems

33. Since the graph is above the x-axis for $x \geq 0$, we have

$$\text{Area} = \int_0^\infty xe^{-x} dx = \lim_{b \to \infty} \int_0^b xe^{-x} dx$$

$$= \lim_{b \to \infty} \left(-xe^{-x}\Big|_0^b + \int_0^b e^{-x} dx\right)$$

$$= \lim_{b \to \infty} \left(-be^{-b} - e^{-x}\Big|_0^b\right)$$

$$= \lim_{b \to \infty} (-be^{-b} - e^{-b} + e^0) = 1.$$

37. (a) We have
$$\int_0^\infty \frac{e^{-y/\alpha}}{\alpha}\,dy = \lim_{b\to\infty} -e^{-y/\alpha}\Big|_0^b = \lim_{b\to\infty}\left(1 - e^{-b/\alpha}\right) = 1.$$

(b) Using integration by parts with $u = y$ and $v' = (1/\alpha)e^{-y/\alpha}$, so $u' = 1$, $v = -e^{-y/\alpha}$, we have
$$\int_0^\infty \frac{ye^{-y/\alpha}}{\alpha}\,dy = \lim_{b\to\infty} \int_0^b \frac{ye^{-y/\alpha}}{\alpha}\,dy$$
$$= \lim_{b\to\infty}\left(-ye^{-y/\alpha}\Big|_0^b + \int_0^b e^{-y/\alpha}\,dy\right)$$
$$= \lim_{b\to\infty}\left(-be^{-b/\alpha} - \alpha e^{-y/\alpha}\Big|_0^b\right)$$
$$= \lim_{b\to\infty}\left(-be^{-b/\alpha} - \alpha e^{-b/\alpha} + \alpha\right)$$

Since $\lim_{b\to\infty} -be^{-b/\alpha} = \lim_{b\to\infty} e^{-b/\alpha} = 0$, we have
$$\int \frac{ye^{-y/\alpha}}{\alpha}\,dy = \alpha.$$

(c) Using integration by parts, this time with $u = y^2$, $v' = (1/\alpha)e^{-y/\alpha}$, so $u' = 2y$, $v = -e^{-y/\alpha}$, we have
$$\int_0^\infty \frac{y^2 e^{-y/\alpha}}{\alpha}\,dy = \lim_{b\to\infty} \int_0^b \frac{y^2 e^{-y/\alpha}}{\alpha}\,dy$$
$$= \lim_{b\to\infty}\left(-y^2 e^{-y/\alpha}\Big|_0^b + 2\int_0^b ye^{-y/\alpha}\,dy\right)$$
$$= \lim_{b\to\infty} -b^2 e^{-b/\alpha} + 2\int_0^\infty ye^{-y/\alpha}\,dy$$

Now $\lim_{b\to\infty} -b^2 e^{-b/\alpha} = 0$ and in part (b) we found
$$\int_0^\infty \frac{ye^{-y/\alpha}}{\alpha}\,dy = \alpha,$$
so
$$\int_0^\infty ye^{-y/\alpha}\,dy = \alpha^2.$$
Thus,
$$\int_0^\infty \frac{y^2 e^{-y/\alpha}}{\alpha}\,dy = 2\int_0^\infty ye^{-y/\alpha}\,dy = 2\alpha^2.$$

41. We calculate
$$m_3 = \frac{1}{\sqrt{2\pi}} \int_{-\infty}^\infty x^3 e^{-x^2/2}\,dx.$$

Since the integrand is odd, for any b, the integral
$$\int_{-b}^b x^3 e^{-x^2/2}\,dx = 0.$$

Thus,
$$m_3 = \frac{1}{\sqrt{2\pi}} \int_{-\infty}^\infty x^3 e^{-x^2/2}\,dx = \frac{1}{\sqrt{2\pi}} \lim_{b\to\infty} \int_{-b}^b x^3 e^{-x^2/2}\,dx = 0.$$

45. The energy required is
$$E = \int_1^\infty \frac{kq_1 q_2}{r^2}\,dr = kq_1 q_2 \lim_{b\to\infty} -\frac{1}{r}\Big|_1^b$$
$$= (9\times 10^9)(1)(1)(1) = 9\times 10^9 \text{ joules}$$

Solutions for Section 7.8

Exercises

1. For large x, the integrand behaves like $1/x^2$ because

$$\frac{x^2}{x^4+1} \approx \frac{x^2}{x^4} = \frac{1}{x^2}.$$

Since $\int_1^\infty \frac{dx}{x^2}$ converges, we expect our integral to converge. More precisely, since $x^4 + 1 > x^4$, we have

$$\frac{x^2}{x^4+1} < \frac{x^2}{x^4} = \frac{1}{x^2}.$$

Since $\int_1^\infty \frac{dx}{x^2}$ is convergent, the comparison test tells us that $\int_1^\infty \frac{x^2}{x^4+1}\,dx$ converges also.

5. The integrand is continuous for all $x \geq 1$, so whether the integral converges or diverges depends only on the behavior of the function as $x \to \infty$. As $x \to \infty$, polynomials behave like the highest powered term. Thus, as $x \to \infty$, the integrand $\frac{x}{x^2+2x+4}$ behaves like $\frac{x}{x^2}$ or $\frac{1}{x}$. Since $\int_1^\infty \frac{1}{x}\,dx$ diverges, we predict that the given integral will diverge.

9. The integrand is continuous for all $x \geq 1$, so whether the integral converges or diverges depends only on the behavior of the function as $x \to \infty$. As $x \to \infty$, polynomials behave like the highest powered term. Thus, as $x \to \infty$, the integrand $\frac{x^2+4}{x^4+3x^2+11}$ behaves like $\frac{x^2}{x^4}$ or $\frac{1}{x^2}$. Since $\int_1^\infty \frac{1}{x^2}\,dx$ converges, we predict that the given integral will converge.

13. The integrand is unbounded as $t \to 5$. We substitute $w = t - 5$, so $dw = dt$. When $t = 5$, $w = 0$ and when $t = 8$, $w = 3$.

$$\int_5^8 \frac{6}{\sqrt{t-5}}\,dt = \int_0^3 \frac{6}{\sqrt{w}}\,dw.$$

Since

$$\int_0^3 \frac{6}{\sqrt{w}}\,dw = \lim_{a \to 0^+} 6\int_a^3 \frac{1}{\sqrt{w}}\,dw = 6\lim_{a \to 0^+} 2w^{1/2}\Big|_a^3 = 12\lim_{a \to 0^+}(\sqrt{3}-\sqrt{a}) = 12\sqrt{3},$$

our integral converges.

17. Since $\frac{1}{u+u^2} < \frac{1}{u^2}$ for $u \geq 1$, and since $\int_1^\infty \frac{du}{u^2}$ converges, $\int_1^\infty \frac{du}{u+u^2}$ converges.

21. Since $\frac{1}{1+e^y} \leq \frac{1}{e^y} = e^{-y}$ and $\int_0^\infty e^{-y}\,dy$ converges, the integral $\int_0^\infty \frac{dy}{1+e^y}$ converges.

25. Since $\frac{3+\sin\alpha}{\alpha} \geq \frac{2}{\alpha}$ for $\alpha \geq 4$, and since $\int_4^\infty \frac{2}{\alpha}\,d\alpha$ diverges, then $\int_4^\infty \frac{3+\sin\alpha}{\alpha}\,d\alpha$ diverges.

Problems

29. The convergence or divergence of an improper integral depends on the long-term behavior of the integrand, not on its short-term behavior. Figure 7.12 suggests that $g(x) \leq f(x)$ for all values of x beyond $x = k$. Since $\int_k^\infty f(x)\,dx$ converges, we expect $\int_k^\infty g(x)\,dx$ converges also.

However we are interested in $\int_a^\infty g(x)\,dx$. Breaking the integral into two parts enables us to use the fact that $\int_k^\infty g(x)\,dx$ is finite:

$$\int_a^\infty g(x)\,dx = \int_a^k g(x)\,dx + \int_k^\infty g(x)\,dx.$$

The first integral is also finite because the interval from a to k is finite. Therefore, we expect $\int_a^\infty g(x)\,dx$ converges.

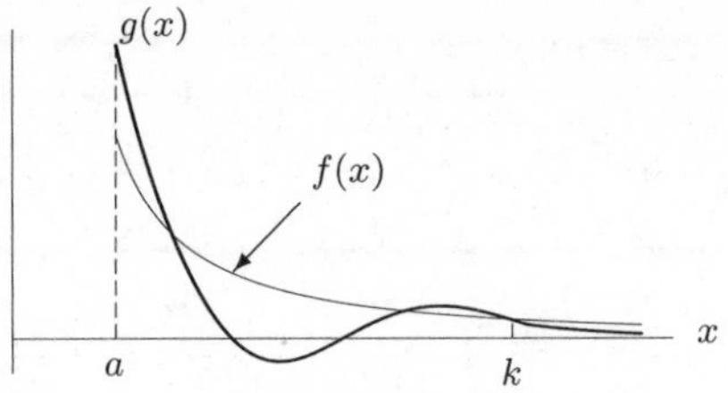

Figure 7.12

33. (a) If we substitute $w = x - k$ and $dw = dx$, we find

$$\int_{-\infty}^{\infty} ae^{-\frac{(x-k)^2}{2}} \, dx = \int_{-\infty}^{\infty} ae^{-\frac{w^2}{2}} \, dw.$$

This integral is the same as the integral in Problem 32, so the value of a will be the same, namely 0.399.

(b) The answer is the same because $g(x)$ is the same as $f(x)$ in Problem 32 except that it is shifted by k to the right. Since we are integrating from $-\infty$ to ∞, however, this shift does not mean anything for the integral.

Solutions for Chapter 7 Review

Exercises

1. Since $\dfrac{d}{dt} \cos t = -\sin t$, we have

$$\int \sin t \, dt = -\cos t + C, \quad \text{where } C \text{ is a constant.}$$

5. Since $\int \sin w \, d\theta = -\cos w + C$, the substitution $w = 2\theta$, $dw = 2 \, d\theta$ gives $\int \sin 2\theta \, d\theta = -\dfrac{1}{2} \cos 2\theta + C$.

9. Either expand $(r + 1)^3$ or use the substitution $w = r + 1$. If $w = r + 1$, then $dw = dr$ and

$$\int (r + 1)^3 \, dr = \int w^3 \, dw = \frac{1}{4}w^4 + C = \frac{1}{4}(r + 1)^4 + C.$$

13. Substitute $w = t^2$, so $dw = 2t \, dt$.

$$\int te^{t^2} \, dt = \frac{1}{2} \int e^{t^2} 2t \, dt = \frac{1}{2} \int e^w \, dw = \frac{1}{2}e^w + C = \frac{1}{2}e^{t^2} + C.$$

Check:

$$\frac{d}{dt}\left(\frac{1}{2}e^{t^2} + C\right) = 2t\left(\frac{1}{2}e^{t^2}\right) = te^{t^2}.$$

17. Integration by parts with $u = \ln x$, $v' = x$ gives

$$\int x \ln x \, dx = \frac{x^2}{2} \ln x - \int \frac{1}{2}x \, dx = \frac{1}{2}x^2 \ln x - \frac{1}{4}x^2 + C.$$

Or use the integral table, III-13, with $n = 1$.

21. Using the exponent rules and the chain rule, we have

$$\int e^{0.5 - 0.3t} \, dt = e^{0.5} \int e^{-0.3t} \, dt = -\frac{e^{0.5}}{0.3}e^{-0.3t} + C = -\frac{e^{0.5 - 0.3t}}{0.3} + C.$$

25. Substitute $w = \sqrt{y}$, $dw = 1/(2\sqrt{y})\,dy$. Then

$$\int \frac{\cos\sqrt{y}}{\sqrt{y}}\,dy = 2\int \cos w\,dw = 2\sin w + C = 2\sin\sqrt{y} + C.$$

Check:

$$\frac{d}{dy}2\sin\sqrt{y} + C = \frac{2\cos\sqrt{y}}{2\sqrt{y}} = \frac{\cos\sqrt{y}}{\sqrt{y}}.$$

29. Substitute $w = 2x - 6$. Then $dw = 2\,dx$ and

$$\int \tan(2x - 6)\,dx = \frac{1}{2}\int \tan w\,dw = \frac{1}{2}\int \frac{\sin w}{\cos w}\,dw$$

$$= -\frac{1}{2}\ln|\cos w| + C \text{ by substitution or by I-7 of the integral table.}$$

$$= -\frac{1}{2}\ln|\cos(2x - 6)| + C.$$

33. Dividing and then integrating, we obtain

$$\int \frac{t + 1}{t^2}\,dt = \int \frac{1}{t}\,dt + \int \frac{1}{t^2}\,dt = \ln|t| - \frac{1}{t} + C, \text{ where } C \text{ is a constant.}$$

37. Using substitution,

$$\int \frac{x}{x^2 + 1}\,dx = \int \frac{1/2}{w}\,dw \qquad (x^2 + 1 = w, 2x\,dx = dw, x\,dx = \frac{1}{2}\,dw)$$

$$= \frac{1}{2}\int \frac{1}{w}\,dw = \frac{1}{2}\ln|w| + C = \frac{1}{2}\ln|x^2 + 1| + C,$$

where C is a constant.

41. Let $\cos 5\theta = w$, then $-5\sin 5\theta\,d\theta = dw$, $\sin 5\theta\,d\theta = -\frac{1}{5}dw$. So

$$\int \sin 5\theta \cos^3 5\theta\,d\theta = \int w^3 \cdot (-\frac{1}{5})\,dw = -\frac{1}{5}\int w^3\,dw = -\frac{1}{20}w^4 + C$$

$$= -\frac{1}{20}\cos^4 5\theta + C,$$

where C is a constant.

45.

$$\int xe^x\,dx = xe^x - \int e^x\,dx \qquad (\text{let } x = u, e^x = v', e^x = v)$$

$$= xe^x - e^x + C,$$

where C is a constant.

49. Rewrite $9 + u^2$ as $9[1 + (u/3)^2]$ and let $w = u/3$, then $dw = du/3$ so that

$$\int \frac{du}{9 + u^2} = \frac{1}{3}\int \frac{dw}{1 + w^2} = \frac{1}{3}\arctan w + C = \frac{1}{3}\arctan\left(\frac{u}{3}\right) + C.$$

53. Let $u = 2x$, then $du = 2\,dx$ so that

$$\int \frac{dx}{\sqrt{1 - 4x^2}} = \frac{1}{2}\int \frac{du}{\sqrt{1 - u^2}} = \frac{1}{2}\arcsin u + C = \frac{1}{2}\arcsin(2x) + C.$$

57. Let $w = \ln x$. Then $dw = (1/x)dx$ which gives

$$\int \frac{dx}{x \ln x} = \int \frac{dw}{w} = \ln |w| + C = \ln |\ln x| + C.$$

61. Using integration by parts, let $r = u$ and $dt = e^{ku}du$, so $dr = du$ and $t = (1/k)e^{ku}$. Thus

$$\int ue^{ku}\,du = \frac{u}{k}e^{ku} - \frac{1}{k}\int e^{ku}\,du = \frac{u}{k}e^{ku} - \frac{1}{k^2}e^{ku} + C.$$

65. $\displaystyle\int (e^x + x)^2 dx = \int (e^{2x} + 2xe^x + x^2)dy$. Separating into three integrals, we have

$$\int e^{2x}\,dx = \frac{1}{2}\int e^{2x}2\,dx = \frac{1}{2}e^{2x} + C_1,$$

$$\int 2xe^x\,dx = 2\int xe^x\,dx = 2xe^x - 2e^x + C_2$$

from Formula II-13 of the integral table or integration by parts, and

$$\int x^2\,dx = \frac{x^3}{3} + C_3.$$

Combining the results and writing $C = C_1 + C_2 + C_3$, we get

$$\frac{1}{2}e^{2x} + 2xe^x - 2e^x + \frac{x^3}{3} + C.$$

69. We can factor $r^2 - 100 = (r - 10)(r + 10)$ so we can use Table V-26 (with $a = 10$ and $b = -10$) to get

$$\int \frac{dr}{r^2 - 100} = \frac{1}{20}\left[\ln |r - 10| + \ln |r + 10|\right] + C.$$

73. Since $\displaystyle\int (x^{\sqrt{k}} + (\sqrt{k})^x)dx = \int x^{\sqrt{k}}dx + \int (\sqrt{k})^x\,dx$, for the first integral, use Formula I-1 with $n = \sqrt{k}$. For the second integral, use Formula I-3 with $a = \sqrt{k}$. The result is

$$\int (x^{\sqrt{x}} + (\sqrt{k})^x)\,dx = \frac{x^{(\sqrt{k})+1}}{(\sqrt{k}) + 1} + \frac{(\sqrt{k})^x}{\ln \sqrt{k}} + C.$$

77. First divide $x^2 + 1$ by $x^2 - 3x + 2$ to obtain

$$\frac{x^2 + 1}{x^2 - 3x + 2} = 1 + \frac{3x - 1}{x^2 - 3x + 2}.$$

Factoring $x^2 - 3x + 2 = (x - 2)(x - 1)$ we can use V-27 (with $c = 3$, $d = -1$, $a = 2$ and $b = 1$) to write

$$\int \frac{3x - 1}{x^2 - 3x + 2}\,dx = 5\ln |x - 2| - 2\ln |x - 1| + C.$$

Remembering to include the extra term of $+1$ we got when dividing, we get

$$\int \frac{x^2 + 1}{x^2 - 3x + 2}\,dx = \int \left(1 + \frac{3x - 1}{x^2 - 3x + 2}\right)\,dx = x + 5\ln |x - 2| - 2\ln |x - 1| + C.$$

81. If $u = 2^t + 1$, $du = 2^t(\ln 2)\,dt$, so

$$\int \frac{2^t}{2^t + 1}\,dt = \frac{1}{\ln 2}\int \frac{2^t \ln 2}{2^t + 1}\,dt = \frac{1}{\ln 2}\int \frac{1}{u} = \frac{1}{\ln 2}\ln |u| + C = \frac{1}{\ln 2}\ln |2^t + 1| + C.$$

85. By VI-30 in the table of integrals, we have

$$\int \sqrt{4 - x^2} \, dx = \frac{x\sqrt{4 - x^2}}{2} + 2 \int \frac{1}{\sqrt{4 - x^2}} \, dx.$$

The same table informs us in formula VI-28 that

$$\int \frac{1}{\sqrt{4 - x^2}} \, dx = \arcsin \frac{x}{2} + C.$$

Thus

$$\int \sqrt{4 - x^2} \, dx = \frac{x\sqrt{4 - x^2}}{2} + 2 \arcsin \frac{x}{2} + C.$$

89. $\int \frac{x}{\cos^2 x} dx = \int x \frac{1}{\cos^2 x} dx$. Using integration by parts with $u = x$, $du = dx$ and $dv = \frac{1}{\cos^2 x} dx$, $v = \tan x$, we have

$$\int x \left(\frac{1}{\cos^2 x} dx \right) = x \tan x - \int \tan x dx.$$

Formula I-7 gives the final result of $x \tan x - (-\ln|\cos x|) + C = x \tan x + \ln|\cos x| + C$.

93. If $u = e^{2y} + 1$, then $du = e^{2y} 2 \, dy$, so

$$\int \frac{e^{2y}}{e^{2y} + 1} \, dy = \frac{1}{2} \int \frac{2e^{2y}}{e^{2y} + 1} \, dy = \frac{1}{2} \int \frac{1}{u} \, du = \frac{1}{2} \ln|u| + C = \frac{1}{2} \ln|e^{2y} + 1| + C.$$

97. $\int \frac{(2x - 1)e^{x^2}}{e^x} dx = \int e^{x^2 - x}(2x - 1)dx$. If $u = x^2 - x$, $du = (2x - 1)dx$, so

$$\int e^{x^2 - x}(2x - 1)dx = \int e^u \, du$$
$$= e^u + c$$
$$= e^{x^2 - x} + C.$$

101. Let $x = 2\theta$, then $dx = 2d\theta$. Thus

$$\int \sin^2(2\theta) \cos^3(2\theta) \, d\theta = \frac{1}{2} \int \sin^2 x \cos^3 x \, dx.$$

We let $w = \sin x$ and $dw = \cos x \, dx$. Then

$$\frac{1}{2} \int \sin^2 x \cos^3 x \, dx = \frac{1}{2} \int \sin^2 x \cos^2 x \cos x \, dx$$
$$= \frac{1}{2} \int \sin^2 x(1 - \sin^2 x) \cos x \, dx$$
$$= \frac{1}{2} \int w^2(1 - w^2) \, dw = \frac{1}{2} \int (w^2 - w^4) \, dw$$
$$= \frac{1}{2} \left(\frac{w^3}{3} - \frac{w^5}{5} \right) + C = \frac{1}{6} \sin^3 x - \frac{1}{10} \sin^5 x + C$$
$$= \frac{1}{6} \sin^3(2\theta) - \frac{1}{10} \sin^5(2\theta) + C.$$

105. Use the substitution $w = \sinh x$ and $dw = \cosh x dx$ so

$$\int \sinh^2 x \cosh x \, dx = \int w^2 \, dw = \frac{w^3}{3} + C = \frac{1}{3} \sinh^3 x + C.$$

Check this answer by taking the derivative: $\frac{d}{dx} \left[\frac{1}{3} \sinh^3 x + C \right] = \sinh^2 x \cosh x$.

109. We substitute $w = \cos\theta + 5$, $dw = -\sin\theta\,d\theta$. Then

$$\int_{\theta=0}^{\theta=\pi} \sin\theta\,d\theta(\cos\theta + 5)^7 = -\int_{w=6}^{w=4} w^7\,dw = \int_{w=4}^{w=6} w^7\,dw = \frac{w^8}{8}\bigg|_4^6 = 201{,}760.$$

113. In Problem 19, we found that

$$\int (\ln x)^2\,dx = x(\ln x)^2 - 2x\ln x + 2x + C.$$

Thus

$$\int_1^e (\ln x)^2\,dx = \left[x(\ln x)^2 - 2x\ln x + 2x\right]\bigg|_1^e = e - 2 \approx 0.71828.$$

This matches the approximation given by Simpson's rule with 10 intervals.

117. Let $\sqrt{x} = w$, $\frac{1}{2}x^{-\frac{1}{2}}\,dx = dw$, $\frac{dx}{\sqrt{x}} = 2\,dw$. If $x = 1$ then $w = 1$, and if $x = 4$ so $w = 2$. So we have

$$\int_1^4 \frac{e^{\sqrt{x}}}{\sqrt{x}}\,dx = \int_1^2 e^w \cdot 2\,dw = 2e^w\bigg|_1^2 = 2(e^2 - e) \approx 9.34.$$

121. Splitting the integrand into partial fractions with denominators $(x-2)$ and $(x+2)$, we have

$$\frac{1}{(x-2)(x+2)} = \frac{A}{x-2} + \frac{B}{x+2}.$$

Multiplying by $(x-2)(x+2)$ gives the identity

$$1 = A(x+2) + B(x-2)$$

so

$$1 = (A+B)x + 2A - 2B.$$

Since this equation holds for all x, the constant terms on both sides must be equal. Similarly, the coefficient of x on both sides must be equal. So

$$2A - 2B = 1$$
$$A + B = 0.$$

Solving these equations gives $A = 1/4$, $B = -1/4$ and the integral becomes

$$\int \frac{1}{(x-2)(x+2)}\,dx = \frac{1}{4}\int \frac{1}{x-2}\,dx - \frac{1}{4}\int \frac{1}{x+2}\,dx = \frac{1}{4}\left(\ln|x-2| - \ln|x+2|\right) + C.$$

125. Splitting the integrand into partial fractions with denominators x, $(x+2)$ and $(x-1)$, we have

$$\frac{2x+3}{x(x+2)(x-1)} = \frac{A}{x} + \frac{B}{x+2} + \frac{C}{x-1}.$$

Multiplying by $x(x+2)(x-1)$ gives the identity

$$2x + 3 = A(x+2)(x-1) + Bx(x-1) + Cx(x+2)$$

so

$$2x + 3 = (A+B+C)x^2 + (A - B + 2C)x - 2A.$$

Since this equation holds for all x, the constant terms on both sides must be equal. Similarly, the coefficient of x on both sides must be equal. So

$$-2A = 3$$
$$A - B + 2C = 2$$
$$A + B + C = 0.$$

Solving these equations gives $A = -3/2$, $B = -1/6$ and $C = 5/3$. The integral becomes

$$\int \frac{2x+3}{x(x+2)(x-1)} dx = -\frac{3}{2} \int \frac{1}{x} dx - \frac{1}{6} \int \frac{1}{x+2} + \frac{5}{3} \int \frac{1}{x-1} dx$$
$$= -\frac{3}{2} \ln|x| - \frac{1}{6} \ln|x+2| + \frac{5}{3} \ln|x-1| + C.$$

129. Completing the square in the denominator gives

$$\int \frac{dx}{x^2 + 4x + 5} = \int \frac{dx}{(x+2)^2 + 1}.$$

We make the substitution $\tan\theta = x + 2$. Then $dx = \frac{1}{\cos^2\theta} d\theta$.

$$\int \frac{dx}{(x+2)^2 + 1} = \int \frac{d\theta}{\cos^2\theta(\tan^2\theta + 1)}$$
$$= \int \frac{d\theta}{\cos^2\theta(\frac{\sin^2\theta}{\cos^2\theta} + 1)}$$
$$= \int \frac{d\theta}{\sin^2\theta + \cos^2\theta}$$
$$= \int d\theta = \theta + C$$

But since $\tan\theta = x + 2$, $\theta = \arctan(x+2)$, and so $\theta + C = \arctan(x+2) + C$.

133. $\int_4^\infty \frac{dt}{t^{3/2}}$ should converge, since $\int_1^\infty \frac{dt}{t^n}$ converges for $n > 1$.
We calculate its value.
$$\int_4^\infty \frac{dt}{t^{3/2}} = \lim_{b\to\infty} \int_4^b t^{-3/2} dt = \lim_{b\to\infty} -2t^{-1/2} \Big|_4^b = \lim_{b\to\infty} \left(1 - \frac{2}{\sqrt{b}}\right) = 1.$$

137. Since the value of $\tan\theta$ is between -1 and 1 on the interval $-\pi/4 \le \theta \le \pi/4$, our integral is not improper and so converges. Moreover, since $\tan\theta$ is an odd function, we have

$$\int_{-\frac{\pi}{4}}^{\frac{\pi}{4}} \tan\theta \, d\theta = \int_{-\frac{\pi}{4}}^0 \tan\theta \, d\theta + \int_0^{\frac{\pi}{4}} \tan\theta \, d\theta$$
$$= -\int_{-\frac{\pi}{4}}^0 \tan(-\theta) \, d\theta + \int_0^{\frac{\pi}{4}} \tan\theta \, d\theta$$
$$= -\int_0^{\frac{\pi}{4}} \tan\theta \, d\theta + \int_0^{\frac{\pi}{4}} \tan\theta \, d\theta = 0.$$

141. Since $\sin\phi < \phi$ for $\phi > 0$,

$$\int_0^{\frac{\pi}{2}} \frac{1}{\sin\phi} \, d\phi > \int_0^{\frac{\pi}{2}} \frac{1}{\phi} \, d\phi,$$

The integral on the right diverges, so the integral on the left must also. Alternatively, we use IV-20 in the integral table to get

$$\int_0^{\frac{\pi}{2}} \frac{1}{\sin\phi} \, d\phi = \lim_{b\to 0^+} \int_b^{\frac{\pi}{2}} \frac{1}{\sin\phi} \, d\phi$$
$$= \lim_{b\to 0^+} \frac{1}{2} \ln\left|\frac{\cos\phi - 1}{\cos\phi + 1}\right| \Big|_b^{\frac{\pi}{2}}$$
$$= -\frac{1}{2} \lim_{b\to 0^+} \ln\left|\frac{\cos b - 1}{\cos b + 1}\right|.$$

As $b \to 0^+$, $\cos b - 1 \to 0$ and $\cos b + 1 \to 2$, so $\ln|\frac{\cos b-1}{\cos b+1}| \to -\infty$. Thus the integral diverges.

145. $\int_0^\pi \tan^2\theta\, d\theta = \tan\theta - \theta + C$, by formula IV-23. The integrand blows up at $\theta = \frac{\pi}{2}$, so

$$\int_0^\pi \tan^2\theta\, d\theta = \int_0^{\frac{\pi}{2}} \tan^2\theta\, d\theta + \int_{\frac{\pi}{2}}^\pi \tan^2\theta\, d\theta = \lim_{b\to\frac{\pi}{2}} [\tan\theta - \theta]_0^b + \lim_{a\to\frac{\pi}{2}} [\tan\theta - \theta]_a^\pi$$

which is undefined.

Problems

149. (a) Since $h(z)$ is even, we know that $\int_0^1 h(z)\, dz = \int_{-1}^0 h(z)\, dz$. Since $\int_{-1}^1 h(z)\, dz = \int_{-1}^0 h(z)\, dz + \int_0^1 h(z)\, dz$, we
see that $\int_{-1}^1 h(z)\, dz = 2\int_0^1 h(z)\, dz = 7$. Thus $\int_0^1 h(z)\, dz = 3.5$
(b) If $w = z + 3$, then $dw = dz$. When $z = -4$, $w = -1$; when $z = -2$, $w = 1$. Thus,

$$\int_{-4}^{-2} 5h(z+3)\, dt = 5\int_{-1}^1 h(w)\,(dw) = 5 \cdot 7 = 35.$$

153. (a) i. 0   ii. $\frac{2}{\pi}$ iii. $\frac{1}{2}$
(b) Average value of $f(t) <$ Average value of $k(t) <$ Average value of $g(t)$
   We can look at the three functions in the range $-\frac{\pi}{2} \le x \le \frac{3\pi}{2}$, since they all have periods of 2π ($|\cos t|$
and $(\cos t)^2$ also have a period of π, but that does not hurt our calculation). It is clear from the graphs of the three
functions below that the average value for $\cos t$ is 0 (since the area above the x-axis is equal to the area below it),
while the average values for the other two are positive (since they are everywhere positive, except where they are 0).

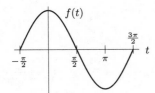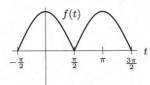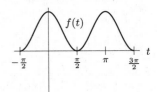

   It is also fairly clear from the graphs that the average value of $g(t)$ is greater than the average value of $k(t)$; it is
also possible to see this algebraically, since

$$(\cos t)^2 = |\cos t|^2 \le |\cos t|$$

because $|\cos t| \le 1$ (and both of these $\le$'s are $<$'s at all the points where the functions are not 0 or 1).

157. Use integration by parts, with $u = x$ and $dv = xe^{-x^2}$. Then $v = -(1/2)e^{-x^2}$, and

$$\int_0^b x^2 e^{-x^2}\, dx = -\frac{1}{2}xe^{-x^2}\Big|_0^b + \int_0^b \frac{1}{2}e^{-x^2}\, dx$$

$$= -\frac{1}{2}be^{-b^2} + \frac{1}{2}\int_0^b e^{-x^2}\, dx.$$

Since the exponential grows faster than any power,

$$\lim_{b\to\infty} be^{-b^2} = \lim_{b\to\infty} \frac{b}{e^{b^2}} = 0.$$

So

$$\int_0^b x^2 e^{-x^2}\, dx = 0 + \frac{1}{2}\int_0^\infty e^{-x^2}\, dx = \frac{1}{2}\cdot\frac{\sqrt{\pi}}{2} = \frac{\sqrt{\pi}}{4}.$$

161. (a) Since the rate is given by $r(t) = 2te^{-2t}$ ml/sec, by the Fundamental Theorem of Calculus, the total quantity is given by the definite integral:

$$\text{Total quantity} \approx \int_0^\infty 2te^{-2t}\, dt = 2 \lim_{b \to \infty} \int_0^b te^{-2t}\, dt.$$

Integration by parts with $u = t$, $v' = e^{-2t}$ gives

$$\text{Total quantity} \approx 2 \lim_{b \to \infty} \left(-\frac{t}{2}e^{-2t} - \frac{1}{4}e^{-2t} \right)\Big|_0^b$$

$$= 2 \lim_{b \to \infty} \left(\frac{1}{4} - \left(\frac{b}{2} + \frac{1}{4} \right) e^{-2b} \right) = 2 \cdot \frac{1}{4} = 0.5 \text{ ml.}$$

(b) At the end of 5 seconds,

$$\text{Quantity received} = \int_0^5 2te^{-2t}\, dt \approx 0.49975 \text{ ml.}$$

Since $0.49975/0.5 = 0.9995 = 99.95\%$, the patient has received 99.95% of the dose in the first 5 seconds.

CAS Challenge Problems

165. (a) A CAS gives

$$\int \ln x\, dx = -x + x \ln x$$

$$\int (\ln x)^2\, dx = 2x - 2x \ln x + x(\ln x)^2$$

$$\int (\ln x)^3\, dx = -6x + 6x \ln x - 3x(\ln x)^2 + x(\ln x)^3$$

$$\int (\ln x)^4\, dx = 24x - 24x \ln x + 12x(\ln x)^2 - 4x(\ln x)^3 + x(\ln x)^4$$

(b) In each of the cases in part (a), the expression for the integral $\int (\ln x)^n\, dx$ has two parts. The first part is simply a multiple of the expression for $\int (\ln x)^{n-1}\, dx$. For example, $\int (\ln x)^2\, dx$ starts out with $2x - 2x \ln x = -2 \int \ln x\, dx$. Similarly, $\int (\ln x)^3\, dx$ starts out with $-6x + 6x \ln x - 3(\ln x)^2 = -3 \int (\ln x)^2\, dx$, and $\int (\ln x)^4\, dx$ starts out with $-4 \int (\ln x)^3\, dx$. The remaining part of each antiderivative is a single term: it's $x(\ln x)^2$ in the case $n = 2$, it's $x(\ln x)^3$ for $n = 3$, and it's $x(\ln x)^4$ for $n = 4$. The general pattern is

$$\int (\ln x)^n\, dx = -n \int (\ln x)^{n-1}\, dx + x(\ln x)^n.$$

To check this formula, we use integration by parts. Let $u = (\ln x)^n$ so $u' = n(\ln x)^{n-1}/x$ and $v' = 1$ so $v = x$. Then

$$\int (\ln x)^n\, dx = x(\ln x)^n - \int n\frac{(\ln x)^{n-1}}{x} \cdot x\, dx$$

$$\int (\ln x)^n\, dx = x(\ln x)^n - n \int (\ln x)^{n-1}\, dx.$$

This is the result we obtained before.

Alternatively, we can check our result by differentiation:

$$\frac{d}{dx}\left(-n \int (\ln x)^{n-1}\, dx + x(\ln x)^n \right) = -n(\ln x)^{n-1} + \frac{d}{dx}(x(\ln x)^n)$$

$$= -n(\ln x)^{n-1} + (\ln x)^n + x \cdot n(\ln x)^{n-1}\frac{1}{x}$$

$$= -n(\ln x)^{n-1} + (\ln x)^n + n(\ln x)^{n-1} = (\ln x)^n.$$

Therefore,

$$\int (\ln x)^n\, dx = -n \int (\ln x)^{n-1}\, dx + x(\ln x)^n.$$

CHECK YOUR UNDERSTANDING

1. False. The subdivision size $\Delta x = (1/10)(6 - 2) = 4/10$.

5. True. We have

$$\text{LEFT}(n) - \text{RIGHT}(n) = (f(x_0) + f(x_1) + \cdots + f(x_{n-1}))\Delta x - (f(x_1) + f(x_2) + \cdots + f(x_n))\Delta x.$$

On the right side of the equation, all terms cancel except the first and last, so:

$$\text{LEFT}(n) - \text{RIGHT}(n) = (f(x_0) - f(x_n))\Delta x = (f(2) - f(6))\Delta x.$$

This is also discussed in Section 5.1.

9. False. This is true if f is an increasing function or if f is a decreasing function, but it is not true in general. For example, suppose that $f(2) = f(6)$. Then $\text{LEFT}(n) = \text{RIGHT}(n)$ for all n, which means that if $\int_2^6 f(x)dx$ lies between $\text{LEFT}(n)$ and $\text{RIGHT}(n)$, then it must equal $\text{LEFT}(n)$, which is not always the case.

   For example, if $f(x) = (x - 4)^2$ and $n = 1$, then $f(2) = f(6) = 4$, so

$$\text{LEFT}(1) = \text{RIGHT}(1) = 4 \cdot (6 - 2) = 16.$$

However

$$\int_2^6 (x - 4)^2 dx = \left.\frac{(x - 4)^3}{3}\right|_2^6 = \frac{2^3}{3} - \left(-\frac{2^3}{3}\right) = \frac{16}{3}.$$

In this example, since $\text{LEFT}(n) = \text{RIGHT}(n)$, we have $\text{TRAP}(n) = \text{LEFT}(n)$. However trapezoids overestimate the area, since the graph of f is concave up. This is also discussed in Section 7.5.

13. True. Rewrite $\sin^7 \theta = \sin \theta \sin^6 \theta = \sin \theta(1 - \cos^2 \theta)^3$. Expanding, substituting $w = \cos \theta, dw = -\sin \theta\, d\theta$, and integrating gives a polynomial in w, which is a polynomial in $\cos \theta$.

17. True. Let $u = t, v' = \sin(5 - t)$, so $u' = 1, v = \cos(5 - t)$. Then the integral $\int 1 \cdot \cos(5 - t)\, dt$ can be done by guess-and-check or by substituting $w = 5 - t$.

21. False. Let $f(x) = x + 1$. Then

$$\int_0^\infty \frac{1}{x + 1} dx = \lim_{b \to \infty} \ln|x + 1|\Big|_0^b = \lim_{b \to \infty} \ln(b + 1),$$

but $\lim_{b \to \infty} \ln(b + 1)$ does not exist.

25. True. Make the substitution $w = ax$. Then $dw = a\, dx$, so

$$\int_0^b f(ax)\, dx = \frac{1}{a} \int_0^c f(w)\, dw,$$

where $c = ab$. As b approaches infinity, so does c, since a is constant. Thus the limit of the left side of the equation as b approaches infinity is finite exactly when the limit of the right side of the equation as c approaches infinity is finite. That is, $\int_0^\infty f(ax)\, dx$ converges exactly when $\int_0^\infty f(x)\, dx$ converges.

CHAPTER EIGHT

Solutions for Section 8.1

Exercises

1. Each strip is a rectangle of length 3 and width Δx, so

$$\text{Area of strip } = 3\Delta x, \quad \text{so}$$

$$\text{Area of region } = \int_0^5 3\,dx = 3x\Big|_0^5 = 15.$$

Check: This area can also be computed using Length $\times$ Width $= 5 \cdot 3 = 15$.

5. The strip has width Δy, so the variable of integration is y. The length of the strip is x. Since $x^2 + y^2 = 10$ and the region is in the first quadrant, solving for x gives $x = \sqrt{10 - y^2}$. Thus

$$\text{Area of strip } \approx x\Delta y = \sqrt{10 - y^2}\,dy.$$

The region stretches from $y = 0$ to $y = \sqrt{10}$, so

$$\text{Area of region } = \int_0^{\sqrt{10}} \sqrt{10 - y^2}\,dy.$$

Evaluating using VI-30 from the Table of Integrals, we have

$$\text{Area } = \frac{1}{2}\left(y\sqrt{10 - y^2} + 10\arcsin\left(\frac{y}{\sqrt{10}}\right)\right)\Bigg|_0^{\sqrt{10}} = 5(\arcsin 1 - \arcsin 0) = \frac{5}{2}\pi.$$

Check: This area can also be computed using the formula $\frac{1}{4}\pi r^2 = \frac{1}{4}\pi(\sqrt{10})^2 = \frac{5}{2}\pi$.

9. Each slice is a circular disk with radius $r = 2$ cm.

$$\text{Volume of disk } = \pi r^2 \Delta x = 4\pi\Delta x \text{ cm}^3.$$

Summing over all disks, we have

$$\text{Total volume } \approx \sum 4\pi\Delta x \text{ cm}^3.$$

Taking a limit as $\Delta x \to 0$, we get

$$\text{Total volume } = \lim_{\Delta x \to 0} \sum 4\pi\Delta x = \int_0^9 4\pi\,dx \text{ cm}^3.$$

Evaluating gives

$$\text{Total volume } = 4\pi x\Big|_0^9 = 36\pi \text{ cm}^3.$$

Check: The volume of the cylinder can also be calculated using the formula $V = \pi r^2 h = \pi 2^2 \cdot 9 = 36\pi$ cm^3.

13. Each slice is a circular disk. See Figure 8.1. The radius of the sphere is 5 mm, and the radius r at height y is given by the Pythagorean Theorem

$$y^2 + r^2 = 5^2.$$

Solving gives $r = \sqrt{5^2 - y^2}$ mm. Thus,

$$\text{Volume of disk } \approx \pi r^2 \Delta y = \pi(5^2 - y^2)\Delta y \text{ mm}^3.$$

Summing over all disks, we have

$$\text{Total volume} \approx \sum \pi(5^2 - y^2)\Delta y \text{ mm}^3.$$

Taking the limit as $\Delta y \to 0$, we get

$$\text{Total volume} = \lim_{\Delta y \to 0} \sum \pi(5^2 - y^2)\Delta y = \int_0^5 \pi(5^2 - y^2)\, dy \text{ mm}^3.$$

Evaluating gives

$$\text{Total volume} = \pi\left(25y - \frac{y^3}{3}\right)\Big|_0^5 = \frac{250}{3}\pi \text{ mm}^3.$$

Check: The volume of a hemisphere can be calculated using the formula $V = \frac{2}{3}\pi r^3 = \frac{2}{3}\pi 5^3 = \frac{250}{3}\pi \text{ mm}^3$.

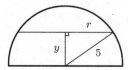

Figure 8.1

Problems

17. Quarter circle of radius $r = \sqrt{15}$. See Figure 8.2.

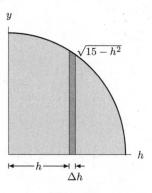

Figure 8.2

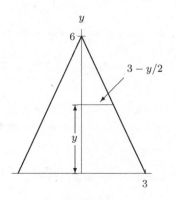

Figure 8.3

21. Cone with height 6 and radius 3. See Figure 8.3.

25.

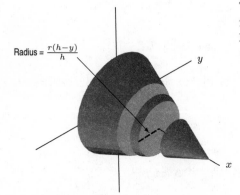

This cone is what you get when you rotate the line $x = r(h - y)/h$ about the y–axis. So slicing perpendicular to the y–axis yields

$$V = \int_{y=0}^{y=h} \pi x^2\, dy = \pi \int_0^h \left(\frac{(h-y)r}{h}\right)^2 dy$$

$$= \pi \frac{r^2}{h^2} \int_0^h (h^2 - 2hy + y^2)\, dy$$

$$= \frac{\pi r^2}{h^2}\left[h^2 y - hy^2 + \frac{y^3}{3}\right]\Big|_0^h = \frac{\pi r^2 h}{3}.$$

Solutions for Section 8.2

Exercises

1. The volume is given by

$$V = \int_0^1 \pi y^2 \, dx = \int_0^1 \pi x^4 \, dx = \pi \frac{x^5}{5}\Big|_0^1 = \frac{\pi}{5}.$$

5. The volume is given by

$$V = \int_{-1}^1 \pi y^2 \, dx = \int_{-1}^1 \pi (e^x)^2 \, dx = \int_{-1}^1 \pi e^{2x} \, dx = \frac{\pi}{2} e^{2x}\Big|_{-1}^1 = \frac{\pi}{2}(e^2 - e^{-2}).$$

9. Since the graph of $y = x^2$ is below the graph of $y = x$ for $0 \le x \le 1$, the volume is given by

$$V = \int_0^1 \pi x^2 \, dx - \int_0^1 \pi (x^2)^2 \, dx = \pi \int_0^1 (x^2 - x^4) \, dx = \pi \left(\frac{x^3}{3} - \frac{x^5}{5}\right)\Big|_0^1 = \frac{2\pi}{15}.$$

13. Since $f'(x) = \sinh x$, the arc length is given by

$$L = \int_0^2 \sqrt{1 + \sinh^2 x} \, dx = \int_0^2 \sqrt{\cosh^2 x} \, dx = \int_0^2 \cosh x \, dx = \sinh x \Big|_0^2 = \sinh 2.$$

Problems

17. (a) Slicing the region perpendicular to the x-axis gives disks of radius y. See Figure 8.4.

$$\text{Volume of slice} \approx \pi y^2 \Delta x = \pi (x^2 - 1)\Delta x.$$

Thus,

$$\text{Total volume} = \lim_{\Delta x \to 0} \sum \pi (x^2 - 1)\Delta x = \int_2^3 \pi (x^2 - 1) \, dx = \pi \left(\frac{x^3}{3} - x\right)\Big|_2^3$$

$$= \pi \left(9 - 3 - \left(\frac{8}{3} - 2\right)\right) = \frac{16\pi}{3}.$$

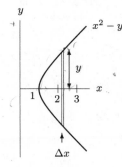

Figure 8.4

(b) The arc length, L, of the curve $y = f(x)$ is given by $L = \int_a^b \sqrt{1 + (f'(x))^2} \, dx$. In this problem y is an implicit function of x. Solving for y gives $y = \sqrt{x^2 - 1}$ as the equation of the top half of the hyperbola. Differentiating gives

$$\frac{dy}{dx} = \frac{1}{2}(x^2 - 1)^{-1/2}(2x) = \frac{x}{\sqrt{x^2 - 1}}.$$

Thus

$$\text{Arc length} = \int_2^3 \sqrt{1 + \left(\frac{x}{\sqrt{x^2 - 1}}\right)^2} \, dx = \int_2^3 \sqrt{1 + \frac{x^2}{x^2 - 1}} \, dx = \int_2^3 \sqrt{\frac{2x^2 - 1}{x^2 - 1}} \, dx = 1.48.$$

21.

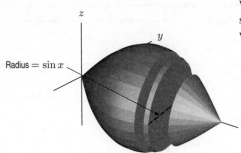

We take slices perpendicular to the x–axis. The Riemann sum for approximating the volume is $\sum \pi \sin^2 x \Delta x$. The volume is the integral corresponding to that sum, namely

$$V = \int_0^\pi \pi \sin^2 x \, dx$$

$$= \pi \left[-\frac{1}{2} \sin x \cos x + \frac{1}{2} x \right] \Big|_0^\pi = \frac{\pi^2}{2} \approx 4.935.$$

25. Slicing perpendicularly to the x-axis gives squares whose thickness is Δx and whose side is $1 - y = 1 - x^2$. See Figure 8.5. Thus

$$\text{Volume of square slice} \approx (1 - x^2)^2 \Delta x = (1 - 2x^2 + x^4) \, \Delta x.$$

$$\text{Volume of solid} = \int_0^1 (1 - 2x^2 + x^4) \, dx = x - \frac{2}{3} x^3 + \frac{x^5}{5} \Big|_0^1 = \frac{8}{15}.$$

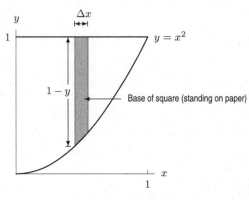

Figure 8.5: Base of solid

29.

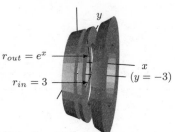

We slice the volume with planes perpendicular to the line $y = -3$. This divides the curve into thin washers, as in Example 3 on page 376 of the text, whose volumes are

$$\pi r_{\text{out}}^2 dx - \pi r_{\text{in}}^2 dx = \pi (3 + y)^2 dx - \pi 3^2 dx.$$

So the integral we get from adding all these washers up is

$$V = \int_{x=0}^{x=1} \left[\pi (3 + y)^2 - \pi 3^2 \right] dx$$

$$= \pi \int_0^1 \left[(3 + e^x)^2 - 9 \right] dx$$

$$= \pi \int_0^1 [e^{2x} + 6e^x] \, dx = \pi \left[\frac{e^{2x}}{2} + 6e^x \right] \Big|_0^1$$

$$= \pi [(e^2/2 + 6e) - (1/2 + 6)] \approx 42.42.$$

33. (a) We can begin by slicing the pie into horizontal slabs of thickness Δh located at height h. To find the radius of each slice, we note that radius increases linearly with height. Since $r = 4.5$ when $h = 3$ and $r = 3.5$ when $h = 0$, we should have $r = 3.5 + h/3$. Then the volume of each slab will be $\pi r^2 \, \Delta h = \pi (3.5 + h/3)^2 \, \Delta h$. To find the total volume of the pie, we integrate this from $h = 0$ to $h = 3$:

$$V = \pi \int_0^3 \left(3.5 + \frac{h}{3} \right)^2 dh$$

$$= \pi \left[\frac{h^3}{27} + \frac{7h^2}{6} + \frac{49h}{4} \right] \Bigg|_0^3$$

$$= \pi \left[\frac{3^3}{27} + \frac{7(3^2)}{6} + \frac{49(3)}{4} \right] \approx 152 \text{ in}^3.$$

(b) We use 1.5 in as a rough estimate of the radius of an apple. This gives us a volume of $(4/3)\pi(1.5)^3 \approx 10 \text{ in}^3$. Since $152/10 \approx 15$, we would need about 15 apples to make a pie.

37. (a) The volume, V, contained in the bowl when the surface has height h is

$$V = \int_0^h \pi x^2 \, dy.$$

However, since $y = x^4$, we have $x^2 = \sqrt{y}$ so that

$$V = \int_0^h \pi \sqrt{y} \, dy = \frac{2}{3} \pi h^{3/2}.$$

Differentiating gives $dV/dh = \pi h^{1/2} = \pi \sqrt{h}$. We are given that $dV/dt = -6\sqrt{h}$, where the negative sign reflects the fact that V is decreasing. Using the chain rule we have

$$\frac{dh}{dt} = \frac{dh}{dV} \cdot \frac{dV}{dt} = \frac{1}{dV/dh} \cdot \frac{dV}{dt} = \frac{1}{\pi \sqrt{h}} \cdot (-6\sqrt{h}) = -\frac{6}{\pi}.$$

Thus, $dh/dt = -6/\pi$, a constant.

(b) Since $dh/dt = -6/\pi$ we know that $h = -6t/\pi + C$. However, when $t = 0$, $h = 1$, therefore $h = 1 - 6t/\pi$. The bowl is empty when $h = 0$, that is when $t = \pi/6$ units.

41. Since $y = (e^x + e^{-x})/2$, $y' = (e^x - e^{-x})/2$. The length of the catenary is

$$\int_{-1}^1 \sqrt{1 + (y')^2} \, dx = \int_{-1}^1 \sqrt{1 + \left[\frac{e^x - e^{-x}}{2} \right]^2} \, dx = \int_{-1}^1 \sqrt{1 + \frac{e^{2x}}{4} - \frac{1}{2} + \frac{e^{-2x}}{4}} \, dx$$

$$= \int_{-1}^1 \sqrt{\left[\frac{e^x + e^{-x}}{2} \right]^2} \, dx = \int_{-1}^1 \frac{e^x + e^{-x}}{2} \, dx$$

$$= \left[\frac{e^x - e^{-x}}{2} \right] \Bigg|_{-1}^1 = e - e^{-1}.$$

Solutions for Section 8.3

Exercises

1. With $r = 1$ and $\theta = 2\pi/3$, we find $x = r \cos \theta = 1 \cdot \cos(2\pi/3) = -1/2$ and $y = r \sin \theta = 1 \cdot \sin(2\pi/3) = \sqrt{3}/2$. The rectangular coordinates are $(-1/2, \sqrt{3}/2)$.

5. With $x = 1$ and $y = 1$, find r from $r = \sqrt{x^2 + y^2} = \sqrt{1^2 + 1^2} = \sqrt{2}$. Find θ from $\tan \theta = y/x = 1/1 = 1$. Thus, $\theta = \tan^{-1}(1) = \pi/4$. Since $(1, 1)$ is in the first quadrant this is a correct θ. The polar coordinates are $(\sqrt{2}, \pi/4)$.

9. (a) Table 8.1 contains values of $r = 1 - \sin \theta$, both exact and rounded to one decimal.

Table 8.1

θ	0	$\pi/3$	$\pi/2$	$2\pi/3$	π	$4\pi/3$	$3\pi/2$	$5\pi/3$	2π	$7\pi/3$	$5\pi/2$	$8\pi/3$
r	1	$1 - \sqrt{3}/2$	0	$1 - \sqrt{3}/2$	1	$1 + \sqrt{3}/2$	2	$1 + \sqrt{3}/2$	1	$1 - \sqrt{3}/2$	0	$1 - \sqrt{3}/2$
r	1	0.134	0	0.134	1	1.866	2	1.866	1	0.134	0	0.134

(b) See Figure 8.6.

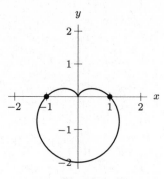

Figure 8.6

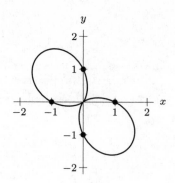

Figure 8.7

(c) The circle has equation $r = 1/2$. The cardioid is $r = 1 - \sin\theta$. Solving these two simultaneously gives

$$1/2 = 1 - \sin\theta,$$

or

$$\sin\theta = 1/2.$$

Thus, $\theta = \pi/6$ or $5\pi/6$. This gives the points $(x, y) = ((1/2)\cos\pi/6, (1/2)\sin\pi/6) = (\sqrt{3}/4, 1/4)$ and $(x, y) = ((1/2)\cos 5\pi/6, (1/2)\sin 5\pi/6) = (-\sqrt{3}/4, 1/4)$ as the location of intersection.

(d) The curve $r = 1 - \sin 2\theta$, pictured in Figure 8.7, has two regions instead of the one region that $r = 1 - \sin\theta$ has. This is because $1 - \sin 2\theta$ will be 0 twice for every 2π cycle in θ, as opposed to once for every 2π cycle in θ for $1 - \sin\theta$.

13. See Figures 8.8 and 8.9. The first curve will be similar to the second curve, except the cardioid (heart) will be rotated clockwise by $90°$ ($\pi/2$ radians). This makes sense because of the identity $\sin\theta = \cos(\theta - \pi/2)$.

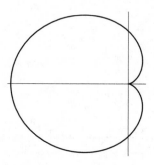

Figure 8.8: $r = 1 - \cos\theta$

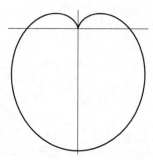

Figure 8.9: $r = 1 - \sin\theta$

17. The region is given by $\sqrt{8} \leq r \leq \sqrt{18}$ and $\pi/4 \leq \theta \leq \pi/2$.

Problems

21. The spiral is shown in Figure 8.10.

$$\text{Area} = \frac{1}{2}\int_0^{2\pi} \theta^2 \, d\theta = \frac{1}{6}\theta^3 \Big|_0^{2\pi} = \frac{8\pi^3}{6}.$$

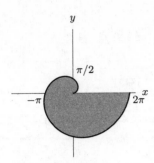

Figure 8.10: Spiral $r = \theta$

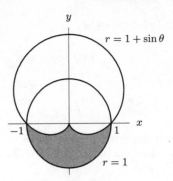

Figure 8.11

25. See Figure 8.11. Notice that the curves intersect at $(1, 0)$, where $\theta = 0, 2\pi$, and at $(-1, 0)$, where $\theta = \pi$, so

$$\text{Area} = \frac{1}{2} \int_{\pi}^{2\pi} (1^2 - (1 + \sin \theta)^2) \, d\theta = \frac{1}{2} \int_{\pi}^{2\pi} (-2 \sin \theta - \sin^2 \theta) \, d\theta.$$

Using a calculator, integration by parts, or formula IV-17 in the integral table, we have

$$\text{Area} = \frac{1}{2} \left(2 \cos \theta + \frac{1}{2} \sin \theta \cos \theta - \frac{1}{2} \theta \right) \Big|_{\pi}^{2\pi} = \frac{1}{2} \left(2 \cdot 2 + 0 - \frac{1}{2} \pi \right) = 2 - \frac{\pi}{4}.$$

29. The area is

$$A = \frac{1}{2} \int_0^a r^2 \, d\theta = \frac{1}{2} \int_0^a \theta^2 \, d\theta = 1$$
$$\frac{1}{2} \left(\frac{\theta^3}{3} \right) \Big|_0^a = 1$$
$$\frac{a^3}{6} = 1$$
$$a^3 = 6$$
$$a = \sqrt[3]{6}.$$

33. We can express x and y in terms of θ as a parameter. Since $r = \theta$, we have

$$x = r \cos \theta = \theta \cos \theta \quad \text{and} \quad y = r \sin \theta = \theta \sin \theta.$$

Calculating the slope using the parametric formula,

$$\frac{dy}{dx} = \frac{dy/d\theta}{dx/d\theta},$$

we have

$$\frac{dy}{dx} = \frac{\sin \theta + \theta \cos \theta}{\cos \theta - \theta \sin \theta}.$$

Horizontal tangents occur where $dy/dx = 0$, so

$$\sin \theta + \theta \cos \theta = 0$$
$$\theta = -\tan \theta.$$

Solving this equation numerically gives

$$\theta = 0, 2.029, 4.913.$$

Vertical tangents occur where dy/dx is undefined, so

$$\cos \theta - \theta \sin \theta = 0$$
$$\theta = \frac{1}{\tan \theta} = \cot \theta.$$

Solving this equation numerically gives

$$\theta = 0.860, 3.426.$$

37. Since $x = \cos\theta/\theta$ and $y = \sin\theta/\theta$, we have

$$\text{Arc length} = \int_{\pi}^{2\pi} \sqrt{\left(\frac{-\theta\sin\theta - \cos\theta}{\theta^2}\right)^2 + \left(\frac{\theta\cos\theta - \sin\theta}{\theta^2}\right)^2}\, d\theta$$

$$= \int_{\pi}^{2\pi} \sqrt{\frac{\theta^2 + 1}{\theta^4}}\, d\theta = 0.712.$$

Solutions for Section 8.4

Exercises

1. Since density is e^{-x} gm/cm,

$$\text{Mass} = \int_0^{10} e^{-x}\, dx = -e^{-x}\Big|_0^{10} = 1 - e^{-10}\text{ gm.}$$

5. (a) Figure 8.12 shows a graph of the density function.

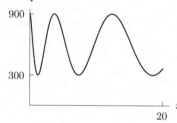

900

300

x

20

Figure 8.12

(b) Suppose we choose an x, $0 \le x \le 20$. We approximate the density of the number of the cars between x and $x + \Delta x$ miles as $\delta(x)$ cars per mile. Therefore, the number of cars between x and $x + \Delta x$ is approximately $\delta(x)\Delta x$. If we slice the 20 mile strip into N slices, we get that the total number of cars is

$$C \approx \sum_{i=1}^{N} \delta(x_i)\Delta x = \sum_{i=1}^{N} \left[600 + 300\sin(4\sqrt{x_i + 0.15})\right]\Delta x,$$

where $\Delta x = 20/N$. (This is a right-hand approximation; the corresponding left-hand approximation is $\sum_{i=0}^{N-1} \delta(x_i)\Delta x$.)

(c) As $N \to \infty$, the Riemann sum above approaches the integral

$$C = \int_0^{20} (600 + 300\sin 4\sqrt{x + 0.15})\, dx.$$

If we calculate the integral numerically, we find $C \approx 11513$. We can also find the integral exactly as follows:

$$C = \int_0^{20} (600 + 300\sin 4\sqrt{x + 0.15})\, dx$$

$$= \int_0^{20} 600\, dx + \int_0^{20} 300\sin 4\sqrt{x + 0.15}\, dx$$

$$= 12000 + 300 \int_0^{20} \sin 4\sqrt{x + 0.15}\, dx.$$

Let $w = \sqrt{x + 0.15}$, so $x = w^2 - 0.15$ and $dx = 2w\, dw$. Then

$$\int_{x=0}^{x=20} \sin 4\sqrt{x + 0.15}\, dx = 2\int_{w=\sqrt{0.15}}^{w=\sqrt{20.15}} w\sin 4w\, dw, \text{ (using integral table III-15)}$$

$$= 2\left[-\frac{1}{4}w\cos 4w + \frac{1}{16}\sin 4w\right]\Big|_{\sqrt{0.15}}^{\sqrt{20.15}}$$

$$\approx -1.624.$$

Using this, we have $C \approx 12000 + 300(-1.624) \approx 11513$, which matches our numerical approximation.

Problems

9. Since the density varies with x, the region must be sliced perpendicular to the x-axis. This has the effect of making the density approximately constant on each strip. See Figure 8.13. Since a strip is of height y, its area is approximately $y\Delta x$. The density on the strip is $\delta(x) = 1 + x$ gm/cm^2. Thus

$$\text{Mass of strip} \approx \text{Density} \cdot \text{Area} \approx (1 + x)y\Delta x \text{ gm.}$$

Because the tops of the strips end on two different lines, one for $x \geq 0$ and the other for $x < 0$, the mass is calculated as the sum of two integrals. See Figure 8.13. For the left part of the region, $y = x + 1$, so

$$\text{Mass of left part} = \lim_{\Delta x \to 0} \sum (1 + x)y\Delta x = \int_{-1}^{0} (1 + x)(x + 1)\, dx$$

$$= \int_{-1}^{0} (1 + x)^2\, dx = \left.\frac{(x + 1)^3}{3}\right|_{-1}^{0} = \frac{1}{3} \text{ gm.}$$

From Figure 8.13, we see that for the right part of the region, $y = -x + 1$, so

$$\text{Mass of right part} = \lim_{\Delta x \to 0} \sum (1 + x)y\Delta x = \int_{0}^{1} (1 + x)(-x + 1)\, dx$$

$$= \int_{0}^{1} (1 - x^2)\, dx = \left. x - \frac{x^3}{3}\right|_{0}^{1} = \frac{2}{3} \text{ gm.}$$

$$\text{Total mass} = \frac{1}{3} + \frac{2}{3} = 1 \text{ gm.}$$

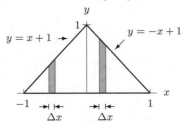

Figure 8.13

13. (a) Use the formula for the volume of a cylinder:

$$\text{Volume} = \pi r^2 l.$$

Since it is only a half cylinder

$$\text{Volume of shed} = \frac{1}{2}\pi r^2 l.$$

(b) Set up the axes as shown in Figure 8.14. The density can be defined as

$$\text{Density} = ky.$$

Now slice the sawdust horizontally into slabs of thickness Δy as shown in Figure 8.15, and calculate

$$\text{Volume of slab} \approx 2xl\Delta y = 2l(\sqrt{r^2 - y^2})\Delta y.$$

$$\text{Mass of slab} = \text{Density} \cdot \text{Volume} \approx 2kly\sqrt{r^2 - y^2}\Delta y.$$

Finally, we compute the total mass of sawdust:

$$\text{Total mass of sawdust} = \int_{0}^{r} 2kly\sqrt{r^2 - y^2}\, dy = \left.-\frac{2}{3}kl(r^2 - y^2)^{3/2}\right|_{0}^{r} = \frac{2klr^3}{3}.$$

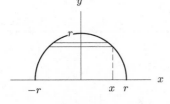

Figure 8.14

Figure 8.15

17. We need the numerator of $\bar{x}$, to be zero, i.e. $\sum x_i m_i = 0$. Since all of the masses are the same, we can factor them out and write $4 \sum x_i = 0$. Thus the fourth mass needs to be placed so that all of the positions sum to zero. The first three positions sum to $(-6 + 1 + 3) = -2$, so the fourth mass needs to be placed at $x = 2$.

21. (a) The density is minimum at $x = -1$ and increases as x increases, so more of the mass of the rod is in the right half of the rod. We thus expect the balancing point to be to the right of the origin.

(b) We need to compute

$$\int_{-1}^{1} x(3 - e^{-x}) \, dx = \left(\frac{3}{2} x^2 + x e^{-x} + e^{-x} \right) \Big|_{-1}^{1} \quad \text{(using integration by parts)}$$

$$= \frac{3}{2} + e^{-1} + e^{-1} - \left(\frac{3}{2} - e^1 + e^1 \right) = \frac{2}{e}.$$

We must divide this result by the total mass, which is given by

$$\int_{-1}^{1} (3 - e^{-x}) \, dx = (3x + e^{-x}) \Big|_{-1}^{1} = 6 - e + \frac{1}{e}.$$

We therefore have

$$\bar{x} = \frac{2/e}{6 - e + (1/e)} = \frac{2}{1 + 6e - e^2} \approx 0.2.$$

25. The triangle is symmetric about the x axis, so $\bar{y} = 0$.

To find $\bar{x}$, we first calculate the density. The area of the triangle is $ab/2$, so it has density $2m/(ab)$ where m is the total mass of the triangle. We need to find the mass of a small strip of width Δx located at x_i (see Figure 8.16).

$$\text{Area of the small strip} \approx A_x(x) \Delta x = 2 \cdot \frac{b(a - x)}{2a} \Delta x.$$

Multiplying by the density $2m/(ab)$ gives

$$\text{Mass of the strip} \approx 2m \frac{(a - x)}{a^2} \Delta x.$$

We then sum the product of these masses with x_i, and take the limit as $\Delta x \to 0$ to get

$$\text{Moment} = \int_0^a \frac{2mx(a - x)}{a^2} \, dx = \frac{2m}{a^2} \left(\frac{ax^2}{2} - \frac{x^3}{3} \right) \Big|_0^a = \frac{2m}{a^2} \left(\frac{a^3}{2} - \frac{a^3}{3} \right) = \frac{ma}{3}.$$

Finally, we divide by the total mass m to get the desired result $\bar{x} = a/3$, which is independent of the length of the base b.

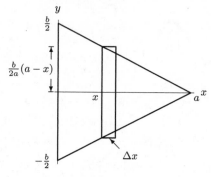

Figure 8.16

Solutions for Section 8.5

Exercises

1. The work done is given by

$$W = \int_1^2 3x \, dx = \frac{3}{2} x^2 \Big|_1^2 = \frac{9}{2} \text{ joules.}$$

5. The force exerted on the satellite by the earth (and vice versa!) is GMm/r^2, where r is the distance from the center of the earth to the center of the satellite, m is the mass of the satellite, M is the mass of the earth, and G is the gravitational constant. So the total work done is

$$\int_{6.4\cdot10^6}^{8.4\cdot10^6} F \, dr = \int_{6.4\cdot10^6}^{8.4\cdot10^6} \frac{GMm}{r^2} \, dr = \left(\frac{-GMm}{r}\right)\Bigg|_{6.4\cdot10^6}^{8.4\cdot10^6} \approx 1.489 \cdot 10^{10} \text{ joules.}$$

Problems

9. The bucket moves upward at $40/10 = 4$ meters/minute. If time is in minutes, at time t the bucket is at a height of $x = 4t$ meters above the ground. See Figure 8.17.

    The water drips out at a rate of $5/10 = 0.5$ kg/minute. Initially there is 20 kg of water in the bucket, so at time t minutes, the mass of water remaining is

$$m = 20 - 0.5t \text{ kg.}$$

Consider the time interval between t and $t + \Delta t$. During this time the bucket moves a distance $\Delta x = 4\Delta t$ meters. So, during this interval,

$$\text{Work done} \approx mg\Delta x = (20 - 0.5t)g4\Delta t \text{ joules.}$$

$$\text{Total work done} = \lim_{\Delta t \to 0} \sum (20 - 0.5t)g4\Delta t = 4g \int_0^{10} (20 - 0.5t) \, dt$$

$$= 4g(20t - 0.25t^2)\Big|_0^{10} = 700g = 700(9.8) = 6860 \text{ joules.}$$

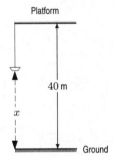

Platform

40 m

x

Ground

Figure 8.17

13. Let x be the distance from the bottom of the tank. See Figure 8.18. To pump a layer of water of thickness Δx at x feet from the bottom to 10 feet above the tank, the work done is $(62.4)\pi6^2(30 - x)\Delta x$. Thus the total work is

$$\int_0^{20} 36 \cdot (62.4)\pi(30 - x)dx$$

$$= 36 \cdot (62.4)\pi \left(30x - \frac{1}{2}x^2\right)\Bigg|_0^{20}$$

$$= 36 \cdot (62.4)\pi(30(20) - \frac{1}{2}20^2)$$

$$\approx 2{,}822{,}909.50 \text{ ft-lb.}$$

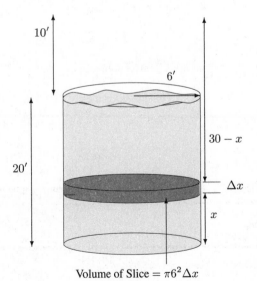

10′

6′

30 − x

Δx

20′

x

Volume of Slice $= \pi6^2\Delta x$

Figure 8.18

17. Let h represent distance below the surface in feet. We slice the tank up into horizontal slabs of thickness Δh. From looking at Figure 8.19, we can see that the slabs will be rectangular. The length of any slab is 12 feet. The width w of a slab h units below the ground will equal $2x$, where $(14 - h)^2 + x^2 = 16$, so $w = 2\sqrt{4^2 - (14 - h)^2}$. The volume of such a slab is therefore $12w\,\Delta h = 24\sqrt{16 - (14 - h)^2}\,\Delta h$ cubic feet; the slab weighs $42 \cdot 24\sqrt{16 - (14 - h)^2}\,\Delta h = 1008\sqrt{16 - (14 - h)^2}\,\Delta h$ pounds. So the total work done in pumping out all the gasoline is

$$\int_{10}^{18} 1008 h \sqrt{16 - (14 - h)^2}\,dh = 1008\int_{10}^{18} h\sqrt{16 - (14 - h)^2}\,dh.$$

Substitute $s = 14 - h$, $ds = -dh$. We get

$$1008\int_{10}^{18} h\sqrt{16 - (14 - h)^2}\,dh = -1008\int_{4}^{-4} (14 - s)\sqrt{16 - s^2}\,ds$$

$$= 1008 \cdot 14 \int_{-4}^{4} \sqrt{16 - s^2}\,ds - 1008\int_{-4}^{4} s\sqrt{16 - s^2}\,ds.$$

The first integral represents the area of a semicircle of radius 4, which is 8π. The second is the integral of an odd function, over the interval $-4 \le s \le 4$, and is therefore 0. Hence, the total work is $1008 \cdot 14 \cdot 8\pi \approx 354{,}673$ foot-pounds.

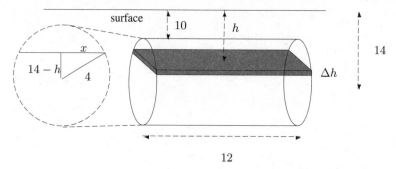

Figure 8.19

21. See Figure 8.20.

For the bottom: The bottom of the tank is at constant depth 15 feet, and therefore is under constant pressure, $15 \cdot 62.4 = 936$ lb/ft^2. The area of the base is 200 ft^2, so

$$\text{Total force on bottom } = 200 \text{ ft}^2 \cdot 936 \text{ lb/ft}^2 = 187200 \text{ lb}.$$

For the 15×10 side: The area of a horizontal strip of width dh is $10\,dh$ square feet, and the pressure at height h is $62.4h$ pounds per square foot. Therefore, the force on such a strip is $62.4h(10\,dh)$ pounds. Hence,

$$\text{Total force on the } 15 \times 10 \text{ side } = \int_{0}^{15} (62.4h)(10)\,dh = 624\frac{h^2}{2}\bigg|_{0}^{15} = 70200 \text{ lbs}.$$

For the 15×20 side: Similarly,

$$\text{Total force on the } 15 \times 20 \text{ side } = \int_{0}^{15} (62.4h)(20)\,dh = 1248\frac{h^2}{2}\bigg|_{0}^{15} = 140400 \text{ lbs}.$$

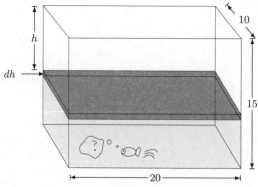

Figure 8.20

25. (a) At a depth of 350 feet,

$$\text{Pressure} = 62.4 \cdot 350 = 21{,}840 \text{ lb/ft}^2.$$

To imagine this pressure, we convert to pounds per square inch, giving a pressure of $21{,}840/144 = 151.7 \text{ lb/in}^2$.

(b)   (i) When the square is held horizontally, the pressure is constant at $21{,}840 \text{ lbs/ft}^2$, so

$$\text{Force} = \text{Pressure} \cdot \text{Area} = 21{,}840 \cdot 5^2 = 546{,}000 \text{ pounds.}$$

(ii) When the square is held vertically, only the bottom is at 350 feet. Dividing into horizontal strips, as in Figure 8.21, we have

$$\text{Area of strip} = 5\Delta h \text{ ft}^2.$$

Since the pressure on a strip at a depth of h feet is $62.4h \text{ lb/ft}^2$,

$$\text{Force on strip} \approx 62.4h \cdot 5\Delta h = 312h\Delta h \text{ pounds.}$$

Summing over all strips and taking the limit as $\Delta h \to 0$ gives a definite integral. The strips vary between a depth of 350 feet and 345 feet, so

$$\text{Total force} = \lim_{\Delta h \to 0} \sum 312h\Delta h = \int_{345}^{350} 312h\, dh \text{ pounds.}$$

Evaluating gives

$$\text{Total force} = 312\frac{h^2}{2}\bigg|_{345}^{350} = 156(350^2 - 345^2) = 542{,}100 \text{ pounds.}$$

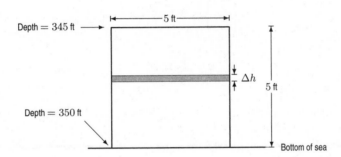

Figure 8.21

29. The density of the rod is $10 \text{ kg}/6 \text{ m} = \frac{5}{3}\frac{\text{kg}}{\text{m}}$. A little piece, dx m, of the rod thus has mass $5/3\, dx$ kg. If this piece has an angular velocity of 2 rad/sec, then its actual velocity is $2|x|$ m/sec. This is because a radian angle sweeps out an arc length equal to the radius of the circle, and in this case the little piece moves in circles about the origin of radius $|x|$. See Figure 8.22. The kinetic energy of the little piece is $mv^2/2 = (5/3\, dx)(2|x|)^2/2 = \frac{10}{3}x^2\, dx$.

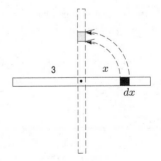

Figure 8.22

Therefore,

$$\text{Total Kinetic Energy} = \int_{-3}^{3} \frac{10x^2}{3}\, dx = \frac{20}{3}\left[\frac{x^3}{3}\right]\bigg|_{0}^{3} = 60 \text{ kg} \cdot \text{m}^2/\text{sec}^2 = 60 \text{ joules.}$$

33. In Figure 8.23, consider a small piece of the ring of length Δl and mass

$$\Delta M = \frac{\Delta l M}{2\pi a}.$$

The gravitational force exerted by the small piece of the ring is along the line QP. As we sum over all pieces of the ring, the components perpendicular to the line OP cancel. The components of the force toward the point O are all in the same direction, so the net force is in this direction. The small piece of length Δl and mass $\Delta l M/2\pi a$ is at a distance of $\sqrt{a^2 + y^2}$ from P, so

$$\text{Gravitational force from small piece} = \Delta F = \frac{G\frac{\Delta l M}{2\pi a}m}{(\sqrt{a^2 + y^2})^2} = \frac{GMm\Delta l}{2\pi a(a^2 + y^2)}.$$

Thus the force toward O exerted by the small piece is given by

$$\Delta F \cos\theta = \Delta F \frac{y}{\sqrt{a^2 + y^2}} = \frac{GMm\Delta l}{2\pi a(a^2 + y^2)} \frac{y}{\sqrt{a^2 + y^2}} = \frac{GMmy\Delta l}{2\pi a(a^2 + y^2)^{3/2}}.$$

The total force toward O is given by $F \approx \sum \Delta F \cos\theta$, so

$$F = \frac{GMmy \cdot \text{Total length}}{2\pi a(a^2 + y^2)^{3/2}} = \frac{GMmy2\pi a}{2\pi a(a^2 + y^2)^{3/2}} = \frac{GMmy}{(a^2 + y^2)^{3/2}}.$$

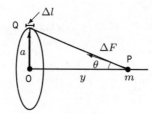

Figure 8.23

Solutions for Section 8.6

Exercises

1. At any time t, in a time interval Δt, an amount of $1000\Delta t$ is deposited into the account. This amount earns interest for $(10 - t)$ years giving a future value of $1000e^{(0.08)(10-t)}$. Summing all such deposits, we have

$$\text{Future value} = \int_0^{10} 1000e^{0.08(10-t)} \, dt = \$15{,}319.30.$$

Problems

5.

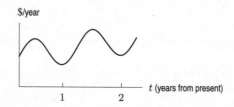

The graph reaches a peak each summer, and a trough each winter. The graph shows sunscreen sales increasing from cycle to cycle. This gradual increase may be due in part to inflation and to population growth.

9. You should choose the payment which gives you the highest present value. The immediate lump-sum payment of $2800 obviously has a present value of exactly $2800, since you are getting it now. We can calculate the present value of the installment plan as:

$$PV = 1000e^{-0.06(0)} + 1000e^{-0.06(1)} + 1000e^{-0.06(2)}$$
$$\approx \$2828.68.$$

Since the installment payments offer a (slightly) higher present value, you should accept this option.

13. (a) Suppose the oil extracted over the time period $[0, M]$ is S. (See Figure 8.24.) Since $q(t)$ is the rate of oil extraction, we have:

$$S = \int_0^M q(t)dt = \int_0^M (a - bt)dt = \int_0^M (10 - 0.1t)\,dt.$$

To calculate the time at which the oil is exhausted, set $S = 100$ and try different values of M. We find $M = 10.6$ gives

$$\int_0^{10.6} (10 - 0.1t)\,dt = 100,$$

so the oil is exhausted in 10.6 years.

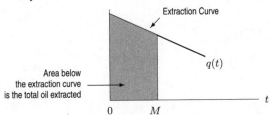

Figure 8.24

(b) Suppose p is the oil price, C is the extraction cost per barrel, and r is the interest rate. We have the present value of the profit as

$$\text{Present value of profit} = \int_0^M (p - C)q(t)e^{-rt}dt$$
$$= \int_0^{10.6} (20 - 10)(10 - 0.1t)e^{-0.1t}\,dt$$
$$= 624.9 \text{ million dollars.}$$

17.

$$\int_0^{q^*} (p^* - S(q))\,dq = \int_0^{q^*} p^*\,dq - \int_0^{q^*} S(q)\,dq$$
$$= p^*q^* - \int_0^{q^*} S(q)\,dq.$$

Using Problem 16, this integral is the extra amount consumers pay (i.e., suppliers earn over and above the minimum they would be willing to accept for supplying the good). It results from charging the equilibrium price.

Solutions for Section 8.7

Exercises

1.

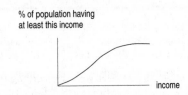

Figure 8.25: Density function   **Figure 8.26**: Cumulative distribution function

5. Since the function is decreasing, it cannot be a cdf (whose values never decrease). Thus, the function is a pdf.
The area under a pdf is 1, so, using the formula for the area of a triangle, we have

$$\frac{1}{2}4c = 1, \quad \text{giving} \quad c = \frac{1}{2}.$$

The pdf is

$$p(x) = \frac{1}{2} - \frac{1}{8}x \quad \text{for} \quad 0 \le x \le 4,$$

so the cdf is given in Figure 8.27 by

$$P(x) = \begin{cases} 0 & \text{for} \quad x < 0 \\ \dfrac{x}{2} - \dfrac{x^2}{16} & \text{for} \quad 0 \le x \le 4 \\ 1 & \text{for} \quad x > 4. \end{cases}$$

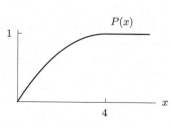

Figure 8.27

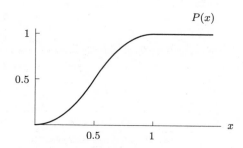

Figure 8.28

9. This function does not level off to 1, and it is not always increasing. Thus, the function is a pdf. Since the area under the curve must be 1, using the formula for the area of a triangle,

$$\frac{1}{2} \cdot c \cdot 1 = 1 \quad \text{so} \quad c = 2.$$

Thus, the pdf is given by

$$p(x) = \begin{cases} 0 & \text{for} \quad x < 0 \\ 4x & \text{for} \quad 0 \le x \le 0.5 \\ 2 - 4(x - 0.5) = 4 - 4x & \text{for} \quad 0.5 < x \le 1 \\ 0 & \text{for} \quad x > 0. \end{cases}$$

To find the cdf, we integrate each part of the function separately, making sure that the constants of integration are arranged so that the cdf is continuous.

Since $\int 4x\,dx = 2x^2 + C$ and $P(0) = 0$, we have $2(0)^2 + C = 0$ so $C = 0$. Thus $P(x) = 2x^2$ on $0 \le x \le 0.5$. At $x = 0.5$, the cdf has value $P(0.5) = 2(0.5)^2 = 0.5$. Thus, we arrange that the integral of $4 - 4x$ goes through the point $(0.5, 0.5)$. Since $\int (4 - 4x)\,dx = 4x - 2x^2 + C$, we have

$$4(0.5) - 2(0.5)^2 + C = 0.5 \quad \text{giving} \quad C = -1.$$

Thus

$$P(x) = \begin{cases} 0 & \text{for} \quad x < 0 \\ 2x^2 & \text{for} \quad 0 \le x \le 0.5 \\ 4x - 2x^2 - 1 & \text{for} \quad 0.5 < x \le 1 \\ 1 & \text{for} \quad x > 1. \end{cases}$$

See Figure 8.28.

Problems

13. For a small interval Δx around 68, the fraction of the population of American men with heights in this interval is about $(0.2)\Delta x$. For example, taking $\Delta x = 0.1$, we can say that approximately $(0.2)(0.1) = 0.02 = 2\%$ of American men have heights between 68 and 68.1 inches.

17. (a) The percentage of calls lasting from 1 to 2 minutes is given by the integral

$$\int_1^2 p(x)\,dx = \int_1^2 0.4e^{-0.4x}\,dx = e^{-0.4} - e^{-0.8} \approx 22.1\%.$$

(b) A similar calculation (changing the limits of integration) gives the percentage of calls lasting 1 minute or less as

$$\int_0^1 p(x)\,dx = \int_0^1 0.4e^{-0.4x}\,dx = 1 - e^{-0.4} \approx 33.0\%.$$

(c) The percentage of calls lasting 3 minutes or more is given by the improper integral

$$\int_3^\infty p(x)\,dx = \lim_{b\to\infty}\int_3^b 0.4e^{-0.4x}\,dx = \lim_{b\to\infty}\left(e^{-1.2} - e^{-0.4b}\right) = e^{-1.2} \approx 30.1\%.$$

(d) The cumulative distribution function is the integral of the probability density; thus,

$$C(h) = \int_0^h p(x)\,dx = \int_0^h 0.4e^{-0.4x}\,dx = 1 - e^{-0.4h}.$$

Solutions for Section 8.8

Exercises

1.

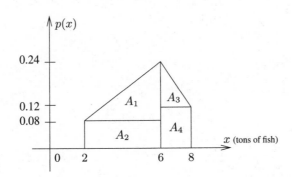

Splitting the figure into four pieces, we see that

$$\begin{aligned}
\text{Area under the curve} &= A_1 + A_2 + A_3 + A_4 \\
&= \frac{1}{2}(0.16)4 + 4(0.08) + \frac{1}{2}(0.12)2 + 2(0.12) \\
&= 1.
\end{aligned}$$

We expect the area to be 1, since $\displaystyle\int_{-\infty}^\infty p(x)\,dx = 1$ for any probability density function, and $p(x)$ is 0 except when $2 \le x \le 8$.

Problems

5. (a) We can find the proportion of students by integrating the density $p(x)$ between $x = 1.5$ and $x = 2$:

$$\begin{aligned}
P(2) - P(1.5) &= \int_{1.5}^2 \frac{x^3}{4}\,dx \\
&= \left.\frac{x^4}{16}\right|_{1.5}^2 \\
&= \frac{(2)^4}{16} - \frac{(1.5)^4}{16} = 0.684,
\end{aligned}$$

so that the proportion is $0.684 : 1$ or 68.4%.

(b) We find the mean by integrating x times the density over the relevant range:

$$\text{Mean} = \int_0^2 x \left(\frac{x^3}{4}\right)\, dx$$

$$= \int_0^2 \frac{x^4}{4}\, dx$$

$$= \left.\frac{x^5}{20}\right|_0^2$$

$$= \frac{2^5}{20} = 1.6 \text{ hours.}$$

(c) The median will be the time T such that exactly half of the students are finished by time T, or in other words

$$\frac{1}{2} = \int_0^T \frac{x^3}{4}\, dx$$

$$\frac{1}{2} = \left.\frac{x^4}{16}\right|_0^T$$

$$\frac{1}{2} = \frac{T^4}{16}$$

$$T = \sqrt[4]{8} = 1.682 \text{ hours.}$$

9. (a) Since $\mu = 100$ and $\sigma = 15$:

$$p(x) = \frac{1}{15\sqrt{2\pi}} e^{-\frac{1}{2}\left(\frac{x-100}{15}\right)^2}.$$

(b) The fraction of the population with IQ scores between 115 and 120 is (integrating numerically)

$$\int_{115}^{120} p(x)\, dx = \int_{115}^{120} \frac{1}{15\sqrt{2\pi}} e^{-\frac{(x-100)^2}{450}}\, dx$$

$$= \frac{1}{15\sqrt{2\pi}} \int_{115}^{120} e^{-\frac{(x-100)^2}{450}}\, dx$$

$$\approx 0.067 = 6.7\% \text{ of the population.}$$

13. It is not (a) since a probability density must be a non-negative function; not (c) since the total integral of a probability density must be 1; (b) and (d) are probability density functions, but (d) is not a good model. According to (d), the probability that the next customer comes after 4 minutes is 0. In real life there should be a positive probability of not having a customer in the next 4 minutes. So (b) is the best answer.

Solutions for Chapter 8 Review

Exercises

1. Vertical slices are circular. Horizontal slices would be similar to ellipses in cross-section, or at least ovals (a word derived from *ovum*, the Latin word for egg).

Figure 8.29

5. We slice the region vertically. Each rotated slice is approximately a cylinder with radius $y = x^2 + 1$ and thickness Δx. See Figure 8.30. The volume of a typical slice is $\pi(x^2 + 1)^2 \Delta x$. The volume, V, of the object is the sum of the volumes of the slices:

$$V \approx \sum \pi(x^2 + 1)^2 \Delta x.$$

As $\Delta x \to 0$ we obtain an integral.

$$V = \int_0^4 \pi(x^2 + 1)^2 dx = \pi \int_0^4 (x^4 + 2x^2 + 1)dx = \pi \left(\frac{x^5}{5} + \frac{2x^3}{3} + x \right) \Big|_0^4 = \frac{3772\pi}{15} = 790.006.$$

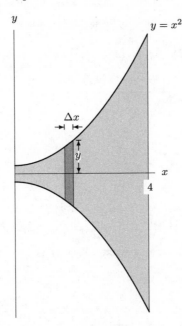

Figure 8.30

9. We divide the region into vertical strips of thickness Δx. As a slice is rotated about the x-axis, it creates a disk of radius r_{out} from which has been removed a smaller circular disk of inside radius r_{in}. We see in Figure 8.31 that $r_{\text{out}} = 2x$ and $r_{\text{in}} = x$. Thus,

$$\text{Volume of a slice} \approx \pi(r_{\text{out}})^2 \Delta x - \pi(r_{\text{in}})^2 \Delta x = \pi(2x)^2 \Delta x - \pi(x)^2 \Delta x.$$

To find the total volume, V, we integrate this quantity between $x = 0$ and $x = 3$:

$$V = \int_0^3 (\pi(2x)^2 - \pi(x)^2)dx = \pi \int_0^3 (4x^2 - x^2) \, dx = \pi \int_0^3 3x^2 dx = \pi x^3 \Big|_0^3 = 27\pi = 84.823.$$

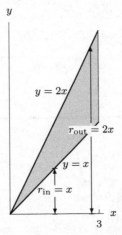

Figure 8.31

13. We slice the tank horizontally. There is an outside radius r_{out} and an inside radius r_{in} and, at height h,

$$\text{Volume of a slice} \approx \pi(r_{\text{out}})^2 \Delta h - \pi(r_{\text{in}})^2 \Delta h.$$

See Figure 8.32. We see that $r_{\text{out}} = 3$ for every slice. We use similar triangles to find r_{in} in terms of the height h:

$$\frac{r_{\text{in}}}{h} = \frac{3}{6} \quad \text{so} \quad r_{\text{in}} = \frac{1}{2}h.$$

At height h,

$$\text{Volume of slice} \approx \pi(3)^2 \Delta h - \pi\left(\frac{1}{2}h\right)^2 \Delta h.$$

To find the total volume, we integrate this quantity from $h = 0$ to $h = 6$.

$$V = \int_0^6 \left(\pi(3)^2 - \pi\left(\frac{1}{2}h\right)^2\right) dh = \pi \int_0^6 \left(9 - \frac{1}{4}h^2\right) dh = \pi\left(9h - \frac{h^3}{12}\right)\Bigg|_0^6 = 36\pi = 113.097 \text{ m}^3.$$

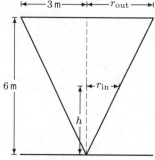

Figure 8.32

17. Since $f'(x) = 10x$, we have

$$L = \int_0^3 \sqrt{1 + (f'(x))^2}\, dx = \int_0^3 \sqrt{1 + (10x)^2}\, dx = \int_0^3 \sqrt{1 + 100x^2}\, dx = 45.230.$$

We see in Figure 8.33 that the length of the curve is definitely longer than 45 and slightly longer than $\sqrt{45^2 + 3^2} = 45.10$, so the answer of 45.230 is reasonable.

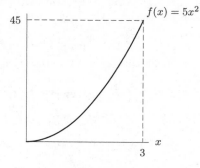

Figure 8.33

Problems

21. (a) Since $y = ax^2$ is non-negative, we integrate to find the area:

$$\text{Area} = \int_0^2 (ax^2)dx = a\frac{x^3}{3}\Bigg|_0^2 = \frac{8a}{3}.$$

(b) Each slice of the object is approximately a cylinder with radius ax^2 and thickness Δx. We have

$$\text{Volume} = \int_0^2 \pi(ax^2)^2 dx = \pi a^2 \frac{x^5}{5}\Bigg|_0^2 = \frac{32}{5}a^2\pi.$$

25. (a) We divide the region into vertical strips of thickness Δx. As a slice is rotated about the x-axis, it creates a disk of radius r_{out} from which has been removed a disk of radius r_{in}. We see in Figure 8.34 that $r_{\text{out}} = 2 + x^2$ and $r_{\text{in}} = 2$. Thus,

$$\text{Volume of a slice} \approx \pi(r_{\text{out}})^2 \Delta x - \pi(r_{\text{in}})^2 \Delta x = \pi(2 + x^2)^2 \Delta x - \pi(2)^2 \Delta x.$$

To find the total volume, V, we integrate this quantity between $x = 0$ and $x = 3$:

$$V = \int_0^3 (\pi(2 + x^2)^2 - \pi(2)^2) dx = \pi \int_0^3 ((2 + x^2)^2 - 4) \, dx = \frac{\pi}{15}(3x^5 + 20x^3)\Big|_0^3 = \frac{423\pi}{5} = 265.778.$$

(b) We see in Figure 8.35 that $r_{\text{out}} = 10$ and $r_{\text{in}} = 10 - x^2$. Thus,

$$\text{Volume of a slice} \approx \pi(r_{\text{out}})^2 \Delta x - \pi(r_{\text{in}})^2 \Delta x = \pi(10)^2 \Delta x - \pi(10 - x^2)^2 \Delta x.$$

To find the total volume, V, we integrate this quantity between $x = 0$ and $x = 3$:

$$V = \int_0^3 (\pi(10)^2 - \pi(10 - x^2)^2) dx = \pi \int_0^3 (100 - (10 - x^2)^2) \, dx = \frac{\pi}{15}(100x^3 - 3x^5)\Big|_0^3 = \frac{657\pi}{5} = 412.805.$$

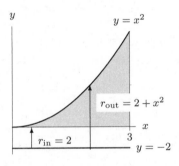

Figure 8.34

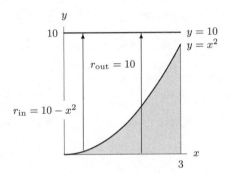

Figure 8.35

29. Slicing perpendicularly to the x-axis gives squares whose thickness is Δx and whose side is $y = \sqrt{1 - x^2}$. See Figure 8.36. Thus,

$$\text{Volume of square slice} \approx (\sqrt{1 - x^2})^2 \Delta x = (1 - x^2) \, \Delta x.$$

$$\text{Volume of solid} = \int_0^1 (1 - x^2) \, dx = x - \frac{x^3}{3}\Big|_0^1 = \frac{2}{3}.$$

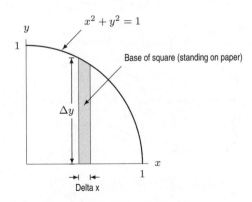

Figure 8.36: Base of solid

33. (a) The line $y = ax$ must pass through (l, b). Hence $b = al$, so $a = b/l$.

(b) Cut the cone into N slices, slicing perpendicular to the x–axis. Each piece is almost a cylinder. The radius of the ith cylinder is $r(x_i) = \dfrac{bx_i}{l}$, so the volume

$$V \approx \sum_{i=1}^{N} \pi \left(\frac{bx_i}{l} \right)^2 \Delta x.$$

Therefore, as $N \to \infty$, we get

$$V = \int_0^l \pi b^2 l^{-2} x^2 dx$$

$$= \pi \frac{b^2}{l^2} \left[\frac{x^3}{3} \right]_0^l = \left(\pi \frac{b^2}{l^2} \right) \left(\frac{l^3}{3} \right) = \frac{1}{3}\pi b^2 l.$$

37. The area is given by

$$\int_{-\pi/2}^{\pi/2} \frac{1}{2} r^2 \, d\theta = \int_{-\pi/2}^{\pi/2} \frac{1}{2} (2a \cos \theta)^2 \, d\theta = 2a^2 \int_{-\pi/2}^{\pi/2} \cos^2 \theta \, d\theta = 2a^2 \left(\frac{1}{2} \cos \theta \sin \theta + \frac{\theta}{2} \right) \Bigg|_{-\pi/2}^{\pi/2} = \pi a^2.$$

(We have used formula IV-18 from the integral table. The integral can also be done using a calculator or integration by parts.)

41. The total mass is 12 gm, so the center of mass is located at $\overline{x} = \frac{1}{12}(-5 \cdot 3 - 3 \cdot 3 + 2 \cdot 3 + 7 \cdot 3) = \frac{1}{4}$.

45.

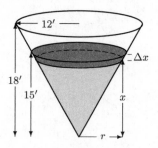

Figure 8.37

Let x be the depth of the water measured from the bottom of the tank. See Figure 8.37. It follows that $0 \leq x \leq 15$. Let r be the radius of the section of the cone with height x. By similar triangles, $\frac{r}{x} = \frac{12}{18}$, so $r = \frac{2}{3}x$. Then the work required to pump a layer of water with thickness of Δx at depth x over the top of the tank is $62.4\pi \left(\frac{2}{3}x \right)^2 \Delta x(18 - x)$. So the total work done by pumping the water over the top of the tank is

$$W = \int_0^{15} 62.4\pi \left(\frac{2}{3}x \right)^2 (18 - x)dx$$

$$= \frac{4}{9} 62.4\pi \int_0^{15} x^2 (18 - x)dx$$

$$= \frac{4}{9} 62.4\pi \left(6x^3 - \frac{1}{4}x^4 \right) \Bigg|_0^{15}$$

$$= \frac{4}{9} 62.4\pi(7593.75) \approx 661{,}619.41 \text{ ft-lb.}$$

49. (a)

$$\text{Future Value} = \int_0^{20} 100e^{0.10(20-t)} dt$$

$$= 100 \int_0^{20} e^2 e^{-0.10t} dt$$

$$= \frac{100e^2}{-0.10}e^{-0.10t}\bigg|_0^{20}$$

$$= \frac{100e^2}{0.10}(1 - e^{-0.10(20)}) \approx \$6389.06.$$

The present value of the income stream is

$$\int_0^{20} 100e^{-0.10t}\,dt = 100\left(\frac{1}{-0.10}\right)e^{-0.10t}\bigg|_0^{20}$$

$$= 1000\left(1 - e^{-2}\right) = \$864.66.$$

Note that this is also the present value of the sum \$6389.06.

(b) Let T be the number of years for the balance to reach \$5000. Then

$$5000 = \int_0^T 100e^{0.10(T-t)}\,dt$$

$$50 = e^{0.10T}\int_0^T e^{-0.10t}\,dt$$

$$= \frac{e^{0.10T}}{-0.10}e^{-0.10t}\bigg|_0^T$$

$$= 10e^{0.10T}(1 - e^{-0.10T}) = 10e^{0.10T} - 10.$$

So, $60 = 10e^{0.10T}$, and $T = 10\ln 6 \approx 17.92$ years.

53. Look at the disc-shaped slab of water at height y and of thickness Δy. The rate at which water is flowing out when it is at depth y is $k\sqrt{y}$ (Torricelli's Law, with k constant). Then, if $x = g(y)$, we have

$$\Delta t = \left(\begin{array}{c}\text{Time for water to}\\ \text{drop by this amount}\end{array}\right) = \frac{\text{Volume}}{\text{Rate}} = \frac{\pi(g(y))^2\Delta y}{k\sqrt{y}}.$$

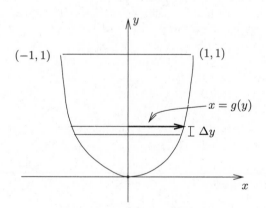

If the rate at which the depth of the water is dropping is constant, then dy/dt is constant, so we want

$$\frac{\pi(g(y))^2}{k\sqrt{y}} = \text{constant},$$

so $g(y) = c\sqrt[4]{y}$, for some constant c. Since $x = 1$ when $y = 1$, we have $c = 1$ and so $x = \sqrt[4]{y}$, or $y = x^4$.

57. First we find the volume of the body up to the horizontal line through Q.

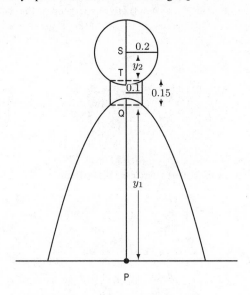

We put the origin at P, the x-axis horizontal and the y-axis pointing upward, and compute the volume obtained by rotating the curve $y = 1 - 4x^2$ around the y-axis up to Q. At Q, we have $x = 0.1$, so

$$y_1 = 1 - 4(0.1^2) = 0.96.$$

Slicing the body horizontally into disks of radius x, thickness Δy, we have

$$\text{Volume of disk in body} \approx \pi x^2 \Delta y = \frac{\pi}{4}(1 - y)\Delta y.$$

Thus,

$$\text{Volume of body up to Q} = \int_0^{0.96} \frac{\pi}{4}(1 - y)dy = \frac{\pi}{4}\left(y - \frac{y^2}{2}\right)\bigg|_0^{0.96} = 0.3921.$$

To find the volume of the head, it is easiest to consider the origin at S, the x-axis horizontal, and the y-axis pointed upward. Then think of the head as the volume obtained by rotating the circle $x^2 + y^2 = (0.2)^2$ about the y-axis. We compute the volume of the head down to the horizontal line through T, at which point $x = 0.1$. Thus

$$(0.1)^2 + y_2{}^2 = (0.2)^2.$$

So

$$y_2 = -\sqrt{0.03} = -0.1732.$$

Slicing the head into circular disks, we have

$$\text{Volume of disk in head} \approx \pi x^2 \Delta y = \pi(0.2^2 - y^2)\Delta y.$$

Thus,

$$\text{Volume of head down to T} = \int_{-0.1732}^{0.2} \pi(0.2^2 - y^2)dy = \pi(0.2^2 y - \frac{y^3}{3})\bigg|_{-0.1732}^{0.2}$$
$$= 0.0331.$$

The neck is exactly cylindrical, with

$$\text{Volume of neck} = \pi(0.1^2)0.15 = 0.0047.$$

Thus,

$$\text{Total volume} = \text{Vol body} + \text{Vol head} + \text{Vol neck}$$
$$= 0.3921 + 0.0331 + 0.0047$$
$$= 0.4299 \approx 0.43\text{m}^3.$$

61. (a) The volume of water in the centrifuge is $\pi(1^2) \cdot 1 = \pi$ cubic meters. The centrifuge has total volume 2π cubic meters, so the volume of the air in the centrifuge is π cubic meters. Now suppose the equation of the parabola is $y = h + bx^2$. We know that the volume of air in the centrifuge is the volume of the top part (a cylinder) plus the volume of the middle part (shaped like a bowl). See Figure 8.38.

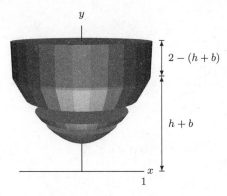

$2 - (h + b)$

$h + b$

1

Figure 8.38: The Volume of Air

To find the volume of the cylinder of air, we find the maximum water depth. If $x = 1$, then $y = h + b$. Therefore the height of the water at the edge of the bowl, 1 meter away from the center, is $h + b$. The volume of the cylinder of air is therefore $[2 - (h + b)] \cdot \pi \cdot (1)^2 = [2 - h - b]\pi$.

To find the volume of the bowl of air, we note that the bowl is a volume of rotation with radius x at height y, where $y = h + bx^2$. Solving for x^2 gives $x^2 = (y - h)/b$. Hence, slicing horizontally as shown in the picture:

$$\text{Bowl Volume} = \int_h^{h+b} \pi x^2 \, dy = \int_h^{h+b} \pi \frac{y-h}{b} \, dy = \left. \frac{\pi(y-h)^2}{2b} \right|_h^{h+b} = \frac{b\pi}{2}.$$

So the volume of both pieces together is $[2 - h - b]\pi + b\pi/2 = (2 - h - b/2)\pi$. But we know the volume of air should be π, so $(2 - h - b/2)\pi = \pi$, hence $h + b/2 = 1$ and $b = 2 - 2h$. Therefore, the equation of the parabolic cross-section is $y = h + (2 - 2h)x^2$.

(b) The water spills out the top when $h + b = h + (2 - 2h) = 2$, or when $h = 0$. The bottom is exposed when $h = 0$. Therefore, the two events happen simultaneously.

CAS Challenge Problems

65. (a) The expression for arc length in terms of a definite integral gives

$$A(t) = \int_0^t \sqrt{1 + 4x^2} \, dx = \frac{2t\sqrt{1 + 4t^2} + \text{arcsinh}(2t)}{4}.$$

The integral was evaluated using a computer algebra system; different systems may give the answer in different forms. Here arcsinh is the inverse function of the hyperbolic sine function.

(b) Figure 8.39 shows that the graphs of $A(t)$ and t^2 look very similar. This suggests that $A(t) \approx t^2$.

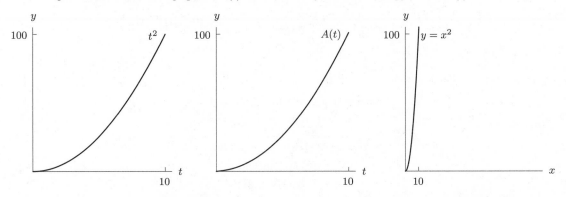

Figure 8.39

Figure 8.40

(c) The graph in Figure 8.40 is approximately vertical and close to the y axis. Thus, if we measure the arc length up to a certain y-value, the answer is approximately the same as if we had measured the length straight up the y-axis. Hence

$$A(t) \approx y = f(t) = t^2.$$

So

$$A(t) \approx t^2.$$

CHECK YOUR UNDERSTANDING

1. True. Since $y = \pm\sqrt{9 - x^2}$ represent the top and bottom halves of the sphere, slicing disks perpendicular to the x-axis gives

$$\text{Volume of slice} \approx \pi y^2 \Delta x = \pi(9 - x^2)\Delta x$$
$$\text{Volume} = \int_{-3}^{3} \pi(9 - x^2)\, dx.$$

5. False. Volume is always positive, like area.

9. True. One way to look at it is that the center of mass should not change if you change the units by which you measure the masses. If you double the masses, that is no different than using as a new unit of mass half the old unit. Alternatively, let the masses be $m_1, m_2,$ and m_3 located at $x_1, x_2,$ and x_3. Then the center of mass is given by:

$$\bar{x} = \frac{x_1 m_1 + x_2 m_2 + x_3 m_3}{m_1 + m_2 + m_3}.$$

Doubling the masses does not change the center of mass, since it doubles both the numerator and the denominator.

13. False. Work is the product of force and distance moved, so the work done in either case is 200 ft-lb.

17. False. The pressure is positive and when integrated gives a positive force.

21. False. It is true that $p(x) \geq 0$ for all x, but we also need $\int_{-\infty}^{\infty} p(x)dx = 1$. Since $p(x) = 0$ for $x \leq 0$, we need only check the integral from 0 to ∞. We have

$$\int_{0}^{\infty} xe^{-x^2}\, dx = \lim_{b \to \infty} \left(-\frac{1}{2}e^{-x^2}\right)\bigg|_{0}^{b} = \frac{1}{2}.$$

25. False. Since f is concave down, this means that $f'(x)$ is decreasing, so $f'(x) \leq f'(0) = 3/4$ on the interval $[0, 4]$. However, it could be that $f'(x)$ becomes negative so that $(f'(x))^2$ becomes large, making the integral for the arc length large also. For example, $f(x) = (3/4)x - x^2$ is concave down and $f'(0) = 3/4$, but $f(0) = 0$ and $f(4) = -13$, so the graph of f on the interval $[0, 4]$ has arc length at least 13.

29. False. Note that p is the density function for the population, not the cumulative density function. Thus $p(10) = p(20)$ means that x values near 10 are as likely as x values near 20.

CHAPTER NINE

Solutions for Section 9.1

Exercises

1. The first term is $2^1 + 1 = 3$. The second term is $2^2 + 1 = 5$. The third term is $2^3 + 1 = 9$, the fourth is $2^4 + 1 = 17$, and the fifth is $2^5 + 1 = 33$. The first five terms are $3, 5, 9, 17, 33$.

5. The first term is $(-1)^2(1/2)^0 = 1$. The second term is $(-1)^3(1/2)^1 = -1/2$. The first five terms are

$$1, -1/2, 1/4, -1/8, 1/16.$$

9. We observe that if we subtract 1 from each term of the sequence, we get $1, 4, 9, 16, 25, \ldots$, namely the squares $1^2, 2^2, 3^2, 4^2, 5^2, \ldots$. Thus $s_n = n^2 + 1$.

13. We have $s_2 = s_1 + 2 = 3$ and $s_3 = s_2 + 3 = 6$. Continuing, we get

$$1, 3, 6, 10, 15, 21.$$

Problems

17. (a) matches (IV), since the sequence increases toward 1.
(b) matches (III), since the odd terms increase toward 1 and the even terms decrease toward 1.
(c) matches (II), since the sequence decreases toward 0.
(d) matches (I), since the sequence decreases toward 1.

21. Since 2^n increases without bound as n increases, the sequence diverges.

25. We have:

$$\lim_{n \to \infty} \left(\frac{n}{10} + \frac{10}{n} \right) = \lim_{n \to \infty} \frac{n}{10} + \lim_{n \to \infty} 10n.$$

Since $n/10$ gets arbitrarily large and $10/n$ approaches 0 as $n \to \infty$, the sequence diverges.

29. Since $\lim_{n \to \infty} 1/n = 0$ and $-1 \le \sin n \le 1$, the terms approach zero and the sequence converges to 0.

33. The first 6 terms of the sequence for the sampling is

$$(-0.5)^2, \ (0.0)^2, \ (0.5)^2, \ (1.0)^2, \ (1.5)^2, \ (2.0)^2,$$

$$= 0.25, \ 0.00, \ 0.25, \ 1.00, \ 2.25, \ 4.00.$$

37. The first smoothing gives

$$1.5, \ 2, \ 3, \ 4, \ 5, \ 6, \ 7 \ldots$$

The second smoothing gives

$$1.75, \ 2.17, \ 3, \ 4, \ 5, \ 6 \ldots$$

Terms which are already the same as their average with their neighbors are not changed.

41. In year 1, the payment is

$$p_1 = 10{,}000 + 0.05(100{,}000) = 15{,}000.$$

The balance in year 2 is $100{,}000 - 10{,}000 = 90{,}000$, so

$$p_2 = 10{,}000 + 0.05(90{,}000) = 14{,}500.$$

The balance in year 3 is $80{,}000$, so

$$p_3 = 10{,}000 + 0.05(80{,}000) = 14{,}000.$$

Thus,

$$p_n = 10{,}000 + 0.05(100{,}000 - (n-1) \cdot 10{,}000)$$
$$= 15{,}500 - 500n.$$

45. We use Theorem 9.1, so we must show that s_n is bounded. Since t_n converges, it is bounded so there is a number M, such that $t_n \leq M$ for all n. Therefore $s_n \leq t_n \leq M$ for all n. Since s_n is increasing, $s_1 \leq s_n$ for all n. Thus if we let $K = s_1$, we have $K \leq s_n \leq M$ for all n, so s_n is bounded. Therefore, s_n converges.

49. The differences between consecutive terms are $4, 9, 16, 25$, so, for example, $s_2 = s_1 + 4$ and $s_3 = s_2 + 9$. Thus, a possible recursive definition is $s_n = s_{n-1} + n^2$ for $n > 1$ and $s_1 = 1$.

53. For $n > 1$, if $s_n = n(n+1)/2$, then $s_{n-1} = (n-1)(n-1+1)/2 = n(n-1)/2$. Since

$$s_n = \frac{1}{2}(n^2 + n) = \frac{n^2}{2} + \frac{n}{2} \quad \text{and} \quad s_{n-1} = \frac{1}{2}(n^2 - n) = \frac{n^2}{2} - \frac{n}{2},$$

we have

$$s_n - s_{n-1} = \frac{n}{2} + \frac{n}{2} = n,$$

so

$$s_n = s_{n-1} + n.$$

In addition, $s_1 = 1(2)/2 = 1$.

57. The sequence seems to converge. By the 25th term it stabilizes to four decimal places at $L = 0.7391$.

Solutions for Section 9.2

Exercises

1. Yes, $a = 2$, ratio $= 1/2$.

5. No. Ratio between successive terms is not constant: $\dfrac{2x^2}{x} = 2x$, while $\dfrac{3x^3}{2x^2} = \dfrac{3}{2}x$.

9. Yes, $a = 1$, ratio $= -x$.

13. Sum $= \dfrac{1}{1-(-x)} = \dfrac{1}{1+x}, |x| < 1$

17. The series has 9 terms. The first term is $a = 0.00002$ and the constant ratio is $x = 0.1$, so

$$\text{Sum} = \frac{0.00002(1 - x^9)}{(1-x)} = \frac{0.00002(1 - (0.1)^9)}{0.9} = 0.0000222.$$

21. Using the formula for the sum of a finite geometric series,

$$\sum_{n=4}^{20} \left(\frac{1}{3}\right)^n = \left(\frac{1}{3}\right)^4 + \left(\frac{1}{3}\right)^5 + \cdots + \left(\frac{1}{3}\right)^{20} = \left(\frac{1}{3}\right)^4 \left(1 + \frac{1}{3} + \left(\frac{1}{3}\right)^2 + \cdots \left(\frac{1}{3}\right)^{16}\right) = \frac{(1/3)^4(1 - (1/3)^{17})}{1 - (1/3)} = \frac{3^{17} - 1}{2 \cdot 3^{20}}.$$

Problems

25. (a) The amount of atenolol in the blood is given by $Q(t) = Q_0 e^{-kt}$, where $Q_0 = Q(0)$ and k is a constant. Since the half-life is 6.3 hours,

$$\frac{1}{2} = e^{-6.3k}, \quad k = -\frac{1}{6.3} \ln \frac{1}{2} \approx 0.11.$$

After 24 hours

$$Q = Q_0 e^{-k(24)} \approx Q_0 e^{-0.11(24)} \approx Q_0(0.07).$$

Thus, the percentage of the atenolol that remains after 24 hours $\approx 7\%$.

(b)

$$Q_0 = 50$$
$$Q_1 = 50 + 50(0.07)$$
$$Q_2 = 50 + 50(0.07) + 50(0.07)^2$$
$$Q_3 = 50 + 50(0.07) + 50(0.07)^2 + 50(0.07)^3$$

$$\vdots$$

$$Q_n = 50 + 50(0.07) + 50(0.07)^2 + \cdots + 50(0.07)^n = \frac{50(1 - (0.07)^{n+1})}{1 - 0.07}$$

(c)

$$P_1 = 50(0.07)$$
$$P_2 = 50(0.07) + 50(0.07)^2$$
$$P_3 = 50(0.07) + 50(0.07)^2 + 50(0.07)^3$$
$$P_4 = 50(0.07) + 50(0.07)^2 + 50(0.07)^3 + 50(0.07)^4$$

$$\vdots$$

$$P_n = 50(0.07) + 50(0.07)^2 + 50(0.07)^3 + \cdots + 50(0.07)^n$$
$$= 50(0.07)\left(1 + (0.07) + (0.07)^2 + \cdots + (0.07)^{n-1}\right) = \frac{0.07(50)(1 - (0.07)^n)}{1 - 0.07}$$

29. (a) The acceleration of gravity is 32 ft/sec^2 so acceleration = 32 and velocity $v = 32t + C$. Since the ball is dropped, its initial velocity is 0 so $v = 32t$. Thus the position is $s = 16t^2 + C$. Calling the initial position $s = 0$, we have $s = 6t$. The distance traveled is h so $h = 16t$. Solving for t we get $t = \frac{1}{4}\sqrt{h}$.

(b) The first drop from 10 feet takes $\frac{1}{4}\sqrt{10}$ seconds. The first full bounce (to $10 \cdot \left(\frac{3}{4}\right)$ feet) takes $\frac{1}{4}\sqrt{10 \cdot \left(\frac{3}{4}\right)}$ seconds to rise, therefore the same time to come down. Thus, the full bounce, up and down, takes $2\left(\frac{1}{4}\right)\sqrt{10 \cdot \left(\frac{3}{4}\right)}$ seconds. The next full bounce takes $2\left(\frac{1}{4}\right)10 \cdot \left(\frac{3}{4}\right)^2 = 2\left(\frac{1}{4}\right)\sqrt{10}\left(\sqrt{\frac{3}{4}}\right)^2$ seconds. The nth bounce takes $2\left(\frac{1}{4}\right)\sqrt{10}\left(\sqrt{\frac{3}{4}}\right)^n$ seconds. Therefore the

Total amount of time

$$= \frac{1}{4}\sqrt{10} + \underbrace{\frac{2}{4}\sqrt{10}\sqrt{\frac{3}{4}} + \frac{2}{4}\sqrt{10}\left(\sqrt{\frac{3}{4}}\right)^2 + \frac{2}{4}\sqrt{10}\left(\sqrt{\frac{3}{4}}\right)^3}_{\text{Geometric series with } a = \frac{2}{4}\sqrt{10}\sqrt{\frac{3}{4}} = \frac{1}{2}\sqrt{10}\sqrt{\frac{3}{4}} \text{ and } x = \sqrt{\frac{3}{4}}} + \cdots$$

$$= \frac{1}{4}\sqrt{10} + \frac{1}{2}\sqrt{10}\sqrt{\frac{3}{4}}\left(\frac{1}{1 - \sqrt{3/4}}\right) \text{ seconds.}$$

Solutions for Section 9.3

Exercises

1. We use the integral test with $f(x) = 1/x^3$ to determine whether this series converges or diverges. We determine whether the corresponding improper integral $\int_1^\infty \frac{1}{x^3}\,dx$ converges or diverges:

$$\int_1^\infty \frac{1}{x^3}\,dx = \lim_{b \to \infty} \int_1^b \frac{1}{x^3}\,dx = \lim_{b \to \infty} \left.\frac{-1}{2x^2}\right|_1^b = \lim_{b \to \infty}\left(\frac{-1}{2b^2} + \frac{1}{2}\right) = \frac{1}{2}.$$

Since the integral $\int_1^\infty \frac{1}{x^3}\,dx$ converges, we conclude from the integral test that the series $\sum_{n=1}^\infty \frac{1}{n^3}$ converges.

5. The improper integral $\int_1^\infty x^{-3}\, dx$ converges to $\dfrac{1}{2}$, since

$$\int_1^b x^{-3}\, dx = \frac{x^{-2}}{-2}\Big|_1^b = \frac{b^{-2}}{-2} - \frac{1^{-2}}{-2} = \frac{1}{-2b^2} + \frac{1}{2}$$

and

$$\lim_{b\to\infty}\left(\frac{1}{-2b^2} + \frac{1}{2}\right) = \frac{1}{2}.$$

The terms of the series $\displaystyle\sum_{n=2}^\infty n^{-3}$ form a right hand sum for the improper integral; each term represents the area of a rectangle of width 1 fitting completely under the graph of the function x^{-3}. (See Figure 9.1.) Thus the sequence of partial sums is bounded above by $1/2$. Since the partial sums are increasing (every new term added is positive) the series is guaranteed to converge to some number less than or equal to $1/2$ by Theorem 9.1.

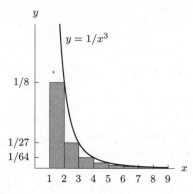

Figure 9.1

9. The integral test requires that $f(x) = e^{-x}\sin x$, which is not positive, nor is it decreasing.

Problems

13. Writing $a_n = n/(n+1)$, we have $\lim_{n\to\infty} a_n = 1$ so the series diverges by Property 3 of Theorem 9.2.

17. The series can be written as

$$\sum_{n=1}^\infty \frac{n + 2^n}{n2^n} = \sum_{n=1}^\infty \left(\frac{1}{2^n} + \frac{1}{n}\right).$$

If this series converges, then $\displaystyle\sum_{n=1}^\infty \left(\frac{1}{2^n} + \frac{1}{n}\right) - \sum_{n=1}^\infty \frac{1}{2^n} = \sum_{n=1}^\infty \frac{1}{n}$ would converge by Theorem 9.2. Since this is the harmonic series, which diverges, then the series $\displaystyle\sum_{n=1}^\infty \frac{n + 2^n}{n}$ diverges.

21. Using $\ln(2^n) = n\ln 2$, we see that

$$\sum_{n=1}^\infty \frac{1}{(\ln(2^n))^2} = \sum_{n=1}^\infty \frac{1}{(\ln 2)^2 n^2}.$$

Since $\sum 1/n^2$ converges, $\sum 1/((\ln(2))^2 n^2)$ converges by property 1 of Theorem 9.2.

25. (a) The partial sum

$$S_4 = \ln\left(\frac{1\cdot 3}{2\cdot 2}\right) + \ln\left(\frac{2\cdot 4}{3\cdot 3}\right) + \ln\left(\frac{3\cdot 5}{4\cdot 4}\right).$$

Using the property $\ln(A) + \ln(B) = \ln(AB)$, we get

$$S_4 = \ln\left(\frac{1\cdot 3\cdot 2\cdot 4\cdot 3\cdot 5}{2\cdot 2\cdot 3\cdot 3\cdot 4\cdot 4}\right).$$

The intermediate factors cancel out, leaving only $\ln\left(\dfrac{1\cdot 5}{2\cdot 4}\right)$, so $S_4 = \ln\left(\dfrac{5}{8}\right).$

(b) For the partial sum S_n, similar steps yield

$$S_n = \ln\left(\frac{1 \cdot 3 \cdot 2 \cdot 4 \cdot 3 \cdot 5 \cdots (n-1)(n+1)}{2 \cdot 2 \cdot 3 \cdot 3 \cdot 4 \cdot 4 \cdots n \cdot n}\right).$$

As before, most of the factors cancel, leaving $S_n = \ln\left(\dfrac{n+1}{2n}\right)$.

(c) The limit of $S_n = \ln\left(\dfrac{n+1}{2n}\right)$ as $n \to \infty$ is $\lim\limits_{n\to\infty} \ln\left(\dfrac{n+1}{2n}\right) = \ln\left(\dfrac{1}{2}\right)$. Thus the series $\sum\limits_{k=2}^{\infty} \ln\left(\dfrac{(k-1)(k+1)}{k^2}\right)$

converges to $\ln\left(\dfrac{1}{2}\right)$.

29. From Property 1 in Theorem 9.2, we know that if $\sum a_n$ converges, then so does $\sum k a_n$.

Now suppose that $\sum a_n$ diverges and $\sum k a_n$ converges for $k \neq 0$. Thus using Property 1 and replacing $\sum a_n$ by $\sum k a_n$, we know that the following series converges:

$$\sum \frac{1}{k}(k a_n) = \sum a_n.$$

Thus, we have arrived at a contradiction, which means our original assumption, that $\sum\limits_{n=1}^{\infty} k a_n$ converged, must be wrong.

33. (a) Show that the sum of each group of fractions is more than $1/2$.

(b) Explain why this shows that the harmonic series does not converge.

(a) Notice that

$$\frac{1}{3} + \frac{1}{4} > \frac{1}{4} + \frac{1}{4} = \frac{2}{4} = \frac{1}{2}$$
$$\frac{1}{5} + \frac{1}{6} + \frac{1}{7} + \frac{1}{8} > \frac{1}{8} + \frac{1}{8} + \frac{1}{8} + \frac{1}{8} = \frac{4}{8} = \frac{1}{2}$$
$$\frac{1}{9} + \frac{1}{10} + \cdots + \frac{1}{16} > \frac{1}{16} + \frac{1}{16} + \cdots + \frac{1}{16} = \frac{8}{16} = \frac{1}{2}.$$

In the same way, we can see that the sum of the fractions in each grouping is greater than $1/2$.

(b) Since the sum of the first n groups is greater than $n/2$, it follows that the partial sums of the harmonic series are not bounded. Thus, the harmonic series diverges.

37. (a) We have $e > 1 + 1 + 1/2 + 1/6 + 1/24 = 65/24 = 2.708$.

(b) We have

$$\frac{1}{n!} = \frac{1}{1 \cdot 2 \cdot 3 \cdot 4 \cdots n} \leq \frac{1}{1 \cdot 2 \cdot 2 \cdot 2 \cdots 2} = \frac{1}{2^{n-1}}.$$

(c) The inequality in part (b) can be used to replace the given series with a geometric series that we can sum.

$$e = \sum_{n=0}^{\infty} \frac{1}{n!} = 1 + \sum_{n=1}^{\infty} \frac{1}{n!} < 1 + \sum_{n=1}^{\infty} \frac{1}{2^{n-1}} = 1 + \frac{1}{1 - 1/2} = 3.$$

Solutions for Section 9.4

Exercises

1. Let $a_n = 1/(n-3)$, for $n \geq 4$. Since $n - 3 < n$, we have $1/(n-3) > 1/n$, so

$$a_n > \frac{1}{n}.$$

The harmonic series $\sum\limits_{n=4}^{\infty} \frac{1}{n}$ diverges, so the comparison test tells us that the series $\sum\limits_{n=4}^{\infty} \frac{1}{n-3}$ also diverges.

5. Let $a_n = 1/(n^4 + e^n)$. Since $n^4 + e^n > n^4$, we have

$$\frac{1}{n^4 + e^n} < \frac{1}{n^4},$$

so

$$0 < a_n < \frac{1}{n^4}.$$

Since the p-series $\sum_{n=1}^{\infty} \frac{1}{n^4}$ converges, the comparison test tells us that the series $\sum_{n=1}^{\infty} \frac{1}{n^4 + e^n}$ also converges.

9. Let $a_n = (2^n + 1)/(n2^n - 1)$. Since $n2^n - 1 < n2^n + n = n(2^n + 1)$, we have

$$\frac{2^n + 1}{n2^n - 1} > \frac{2^n + 1}{n(2^n + 1)} = \frac{1}{n}.$$

Therefore, we can compare the series $\sum_{n=1}^{\infty} \frac{2^n + 1}{n2^n - 1}$ with the divergent harmonic series $\sum_{n=1}^{\infty} \frac{1}{n}$. The comparison test tells us that $\sum_{n=1}^{\infty} \frac{2^n + 1}{n2^n - 1}$ also diverges.

13. Since $a_n = 1/(r^n n!)$, replacing n by $n + 1$ gives $a_{n+1} = 1/(r^{n+1}(n+1)!)$. Thus

$$\frac{|a_{n+1}|}{|a_n|} = \frac{\dfrac{1}{r^{n+1}(n+1)!}}{\dfrac{1}{r^n n!}} = \frac{r^n n!}{r^{n+1}(n+1)!} = \frac{1}{r(n+1)},$$

so

$$L = \lim_{n \to \infty} \frac{|a_{n+1}|}{|a_n|} = \frac{1}{r} \lim_{n \to \infty} \frac{1}{n+1} = 0.$$

Since $L = 0$, the ratio test tells us that $\sum_{n=1}^{\infty} \frac{1}{r^n n!}$ converges for all $r > 0$.

17. Since $\cos(n\pi) = (-1)^n$, this is an alternating series.

21. Let $a_n = 1/(2n + 1)$. Then replacing n by $n + 1$ gives $a_{n+1} = 1/(2n + 3)$. Since $2n + 3 > 2n + 1$, we have

$$0 < a_{n+1} = \frac{1}{2n+3} < \frac{1}{2n+1} = a_n.$$

We also have $\lim_{n \to \infty} a_n = 0$. Therefore, the alternating series test tells us that the series $\sum_{n=1}^{\infty} \frac{(-1)^{n-1}}{2n+1}$ converges.

25. We have

$$\frac{a_n}{b_n} = \frac{((1+n)/(3n))^n}{(1/3)^n} = \left(\frac{n+1}{n}\right)^n = \left(1 + \frac{1}{n}\right)^n,$$

so

$$\lim_{n \to \infty} \frac{a_n}{b_n} = \lim_{n \to \infty} \left(1 + \frac{1}{n}\right)^n = e = c \neq 0.$$

Since $\sum_{n=1}^{\infty} \left(\frac{1}{3}\right)^n$ is a convergent geometric series, the original series converges.

29. The nth term $a_n = 2^n/(3^n - 1)$ behaves like $2^n/3^n$ for large n, so we take $b_n = 2^n/3^n$. We have

$$\lim_{n \to \infty} \frac{a_n}{b_n} = \lim_{n \to \infty} \frac{2^n/(3^n - 1)}{2^n/3^n} = \lim_{n \to \infty} \frac{3^n}{3^n - 1} = \lim_{n \to \infty} \frac{1}{1 - 3^{-n}} = 1.$$

The limit comparison test applies with $c = 1$. The geometric series $\sum 2^n/3^n = \sum (2/3)^n$ converges. Therefore $\sum 2^n/(3^n - 1)$ also converges.

Problems

33. The comparison test requires that $a_n = \sin n$ be positive for all n. It is not.

37. The alternating series test requires $a_n = \sin n$ be positive, which it is not. This is not an alternating series.

41. The partial sums look like: $S_1 = 1$, $S_2 = 0$, $S_3 = 0.5$, $S_4 = 0.3333$, $S_5 = 0.375$, $S_{10} = 0.3679$, $S_{20} = 0.3679$, and higher partial sums agree with these first 4 decimal places. The series appears to be converging to about 0.3679.

Since $a_n = 1/n!$ is positive and decreasing and $\lim_{n\to\infty} 1/n! = 0$, the alternating series test confirms the convergence of this series.

45. The first few terms of the series may be written

$$e + e^2 + e^3 + \cdots = e + e \cdot e + e \cdot e^2 + \cdots;$$

this is a geometric series with $a = e$ and $x = e$. Since $|x| > 1$, this geometric series diverges.

49. Note that $\cos(n\pi)/n = (-1)^n/n$, so this is an alternating series. Therefore, since $1/(n+1) < 1/n$ and $\lim_{n\to\infty} 1/n = 0$, we see that $\sum (\cos(n\pi)/n)$ converges by the alternating series test.

53. Let $a_n = n(n+1)/\sqrt{n^3 + 2n^2}$. Since $n^3 + 2n^2 = n^2(n+2)$, we have

$$a_n = \frac{n(n+1)}{n\sqrt{n+2}} = \frac{n+1}{\sqrt{n+2}}$$

so a_n grows without bound as $n \to \infty$, therefore the series $\displaystyle\sum_{n=1}^{\infty} \frac{n(n+1)}{\sqrt{n^3 + 2n^2}}$ diverges.

57. Both $\displaystyle\sum \frac{(-1)^n}{2^n} = \sum \left(\frac{-1}{2}\right)^n$ and $\displaystyle\sum \frac{1}{2^n} = \sum \left(\frac{1}{2}\right)^n$ are convergent geometric series. Thus $\displaystyle\sum \frac{(-1)^n}{2^n}$ is absolutely convergent.

61. Since $0 \le c_n \le 2^{-n}$ for all n, and since $\sum 2^{-n}$ is a convergent geometric series, $\sum c_n$ converges by the Comparison Test. Similarly, since $2^n \le a_n$, and since $\sum 2^n$ is a divergent geometric series, $\sum a_n$ diverges by the Comparison Test. We do not have enough information to determine whether or not $\sum b_n$ and $\sum d_n$ converge.

65. Each term in $\sum b_n$ is greater than or equal to a_1 times a term in the harmonic series:

$$b_1 = a_1 \cdot 1$$
$$b_2 = \frac{a_1 + a_2}{2} > a_1 \cdot \frac{1}{2}$$
$$b_3 = \frac{a_1 + a_2 + a_3}{3} > a_1 \cdot \frac{1}{3}$$
$$\vdots$$
$$b_n = \frac{a_1 + a_2 + \cdots + a_n}{n} > a_1 \cdot \frac{1}{n}$$

Adding these inequalities gives

$$\sum b_n > a_1 \sum \frac{1}{n}.$$

Since the harmonic series $\sum 1/n$ diverges, a_1 times the harmonic series also diverges. Then, by the comparison test, the series $\sum b_n$ diverges.

69. The limit

$$\lim_{n\to\infty} \sqrt[n]{a_n} = \lim_{n\to\infty} \frac{5n+1}{3n^2} = 0 < 1,$$

so the series converges.

Solutions for Section 9.5

Exercises

1. Yes.

5. The general term can be written as $\dfrac{1 \cdot 3 \cdot 5 \cdots (2n-1)}{2^n \cdot n!} x^n$ for $n \geq 1$. Other answers are possible.

9. The general term can be written as $\dfrac{(x-a)^n}{2^{n-1} \cdot n!}$ for $n \geq 1$. Other answers are possible.

13. Since $C_n = (n+1)/(2^n + n)$, replacing n by $n+1$ gives $C_{n+1} = (n+2)/(2^{n+1} + n + 1)$. Using the ratio test, we have

$$\frac{|a_{n+1}|}{|a_n|} = |x|\frac{|C_{n+1}|}{|C_n|} = |x|\frac{(n+2)/(2^{n+1}+n+1)}{(n+1)/(2^n+n)} = |x|\frac{n+2}{2^{n+1}+n+1} \cdot \frac{2^n+n}{n+1} = |x|\frac{n+2}{n+1} \cdot \frac{2^n+n}{2^{n+1}+n+1}.$$

Since

$$\lim_{n \to \infty} \frac{n+2}{n+1} = 1$$

and

$$\lim_{n \to \infty} \left(\frac{2^n+n}{2^{n+1}+n+1} \right) = \frac{1}{2}\lim_{n \to \infty}\left(\frac{2^n+n}{2^n + (n+1)/2} \right) = \frac{1}{2},$$

because 2^n dominates n as $n \to \infty$, we have

$$\lim_{n \to \infty} \frac{|a_{n+1}|}{|a_n|} = \frac{1}{2}|x|.$$

Thus the radius of convergence is $R = 2$.

17. Here the coefficient of the nth term is $C_n = (2^n/n!)$. Now we have

$$\left| \frac{a_{n+1}}{a_n} \right| = \left| \frac{(2^{n+1}/(n+1)!)x^{n+1}}{(2^n/n!)x^n} \right| = \frac{2|x|}{n+1} \to 0 \text{ as } n \to \infty.$$

Thus, the radius of convergence is $R = \infty$, and the series converges for all x.

21. We write the series as

$$x - \frac{x^3}{3} + \frac{x^5}{5} - \frac{x^7}{7} + \cdots + (-1)^{n-1}\frac{x^{2n-1}}{2n-1} + \cdots,$$

so

$$a_n = (-1)^{n-1}\frac{x^{2n-1}}{2n-1}.$$

Replacing n by $n+1$, we have

$$a_{n+1} = (-1)^{n+1-1}\frac{x^{2(n+1)-1}}{2(n+1)-1} = (-1)^n\frac{x^{2n+1}}{2n+1}.$$

Thus

$$\frac{|a_{n+1}|}{|a_n|} = \left| \frac{(-1)^n x^{2n+1}}{2n+1} \right| \cdot \left| \frac{2n-1}{(-1)^{n-1}x^{2n-1}} \right| = \frac{2n-1}{2n+1}x^2,$$

so

$$L = \lim_{n \to \infty} \frac{|a_{n+1}|}{|a_n|} = \lim_{n \to \infty} \frac{2n-1}{2n+1}x^2 = x^2.$$

By the ratio test, this series converges if $L < 1$, that is, if $x^2 < 1$, so $R = 1$.

Problems

25. We use the ratio test:

$$\left|\frac{a_{n+1}}{a_n}\right| = \left|\frac{(x-3)^{n+1}}{n+1} \cdot \frac{n}{(x-3)^n}\right| = \frac{n}{n+1} \cdot |x-3|.$$

Since $n/(n+1) \to 1$ as $n \to \infty$, we have

$$\lim_{n\to\infty} \left|\frac{a_{n+1}}{a_n}\right| = |x-3|.$$

The series converges for $|x-3| < 1$. The radius of convergence is 1 and the series converges for $2 < x < 4$.

We check the endpoints. For $x = 2$, we have

$$\sum_{n=2}^{\infty} \frac{(x-3)^n}{n} = \sum_{n=2}^{\infty} \frac{(2-3)^n}{n} = \sum_{n=2}^{\infty} \frac{(-1)^n}{n}.$$

This is the alternating harmonic series and converges. For $x = 4$, we have

$$\sum_{n=2}^{\infty} \frac{(x-3)^n}{n} = \sum_{n=2}^{\infty} \frac{(4-3)^n}{n} = \sum_{n=2}^{\infty} \frac{1}{n}.$$

This is the harmonic series and diverges. The series converges at $x = 2$ and diverges at $x = 4$. Therefore, the interval of convergence is $2 \le x < 4$.

29. The kth coefficient in the series $\sum kC_k x^k$ is $D_k = k \cdot C_k$. We are given that the series $\sum C_k x^k$ has radius of convergence R by the ratio test, so

$$|x| \lim_{k\to\infty} \frac{|C_{k+1}|}{|C_k|} = \frac{|x|}{R}.$$

Thus, applying the ratio test to the new series, we have

$$\lim_{k\to\infty} \left|\frac{D_{k+1}x^{k+1}}{D_k x^k}\right| = \lim_{k\to\infty} \left|\frac{(k+1)C_{k+1}}{kC_k}\right| |x| = \frac{|x|}{R}.$$

Hence the new series has radius of convergence R.

33. The series is centered at $x = 3$. Since the series converges at $x = 7$, which is a distance of 4 from $x = 3$, we know $R \ge 4$. Since the series diverges at $x = 10$, which is a distance of 7 from $x = 3$, we know $R \le 7$. That is, $4 \le R \le 7$.

Since $x = 11$ is a distance of 8 from $x = 3$, the series diverges at $x = 11$.

Since $x = 5$ is a distance of 2 from $x = 3$, the series converges there.

Since $x = 0$ is a distance of 3 from $x = 3$, the series converges at $x = 3$.

37. (a) We have

$$(p(x))^2 = \left(1 - \frac{x^2}{2!} + \frac{x^4}{4!} - \frac{x^6}{6!} + \cdots\right)^2$$

$$= 1 - 2 \cdot \frac{x^2}{2} + \left(-\frac{x^2}{2!}\right)^2 + 2\frac{x^4}{4!} - 2\frac{x^6}{6!} - 2\frac{x^2}{2!} \cdot \frac{x^4}{4!} \cdots$$

$$= 1 - x^2 + \left(\frac{1}{4} + \frac{1}{12}\right)x^4 - x^6\left(\frac{1}{3 \cdot 5 \cdot 4!} + \frac{1}{4!}\right) \cdots$$

$$= 1 - x^2 + \frac{x^4}{3} - \frac{2}{45}x^6 \cdots.$$

$$(q(x))^2 = \left(x - \frac{x^3}{3!} + \frac{x^5}{5!} - \cdots\right)^2 = x^2\left(1 - \frac{x^2}{3!} + \frac{x^4}{5!} - \cdots\right)^2$$

$$= x^2\left(1 - 2\frac{x^2}{3!} + \left(-\frac{x^2}{3!}\right)^2 + 2\frac{x^4}{5!} \cdots\right)$$

$$= x^2\left(1 - \frac{x^2}{3} + x^4\left(\frac{1}{(3!)^2} + \frac{1}{5 \cdot 4 \cdot 3}\right) \cdots\right)$$

$$= x^2 \left(1 - \frac{x^2}{3} + \frac{2}{45}x^4 \cdots \right)$$

$$= x^2 - \frac{x^4}{3} + \frac{2}{45}x^6 \cdots.$$

Thus, up to terms in x^6, we have

$$(p(x))^2 + (q(x))^2 = 1.$$

(b) The result of part (a) suggests that $p(x)$ and $q(x)$ could be the sine and cosine. Since $p(x)$ is even and $q(x)$ is odd, we guess that $p(x) = \cos x$ and $q(x) = \sin x$.

Solutions for Chapter 9 Review

Exercises

1. As n increases, the term $4n$ is much larger than 3 and $7n$ is much larger than 5. Thus dividing the numerator and denominator by n and using the fact that $\lim_{n\to\infty} 1/n = 0$, we have

$$\lim_{n\to\infty} \frac{3 + 4n}{5 + 7n} = \lim_{n\to\infty} \frac{(3/n) + 4}{(5/n) + 7} = \frac{4}{7}.$$

Thus, the sequence converges to $4/7$.

5. If $b = 1$, then the sum is 6. If $b \neq 1$, we use the formula for the sum of a finite geometric series. This is a six-term geometric series ($n = 6$) with initial term $a = b^5$ and constant ratio $x = b$:

$$\text{Sum} = \frac{a(1 - x^n)}{1 - x} = \frac{b^5(1 - b^6)}{1 - b}.$$

9. We use the integral test to determine whether this series converges or diverges. To do so we determine whether the corresponding improper integral $\int_1^\infty \frac{3x^2 + 2x}{x^3 + x^2 + 1}\,dx$ converges or diverges. The integral can be calculated using the substitution $w = x^3 + x^2 + 1$, $dw = (3x^2 + 2x)\,dx$.

$$\int_1^\infty \frac{3x^2 + 2x}{x^3 + x^2 + 1}\,dx = \lim_{b\to\infty} \int_1^b \frac{3x^2 + 2x}{x^3 + x^2 + 1}\,dx$$

$$= \lim_{b\to\infty} \ln|x^3 + x^2 + 1| \Big|_1^b$$

$$= \lim_{b\to\infty} \left(\ln|b^3 + b^2 + 1| - \ln 3 \right) = \infty.$$

Since the integral $\int_1^\infty \frac{3x^2 + 2x}{x^3 + x^2 + 1}\,dx$ diverges, we conclude from the integral test that the series $\sum_{n=1}^\infty \frac{3n^2 + 2n}{n^3 + n^2 + 1}$ diverges.

13. Let $a_n = 1/(n\sin^2 n)$. Since $0 < \sin^2 n < 1$, for any positive integer n, we have $n\sin^2 n < n$, so $\frac{1}{n\sin^2 n} > \frac{1}{n}$, thus

$$a_n > \frac{1}{n}.$$

The harmonic series $\sum_{n=1}^\infty \frac{1}{n}$ diverges, so the comparison test tells us that the series $\sum_{n=1}^\infty \frac{1}{n\sin^2 n}$ also diverges.

17. The nth term $a_n = 1/(\sqrt{n^3 - 1})$ behaves like $1/\sqrt{n^3} = 1/n^{3/2}$ for large n, so we take $b_n = 1/n^{3/2}$. We have

$$\lim_{n \to \infty} \frac{a_n}{b_n} = \lim_{n \to \infty} \frac{1/\sqrt{n^3 - 1}}{1/n^{3/2}} = \lim_{n \to \infty} \frac{n^{3/2}}{\sqrt{n^3 - 1}} = \lim_{n \to \infty} \frac{n^{3/2}}{n^{3/2}\sqrt{1 - 1/n^3}} = \lim_{n \to \infty} \frac{1}{\sqrt{1 - 1/n^3}} = \frac{1}{\sqrt{1 - 0}} = 1.$$

The limit comparison test applies with $c = 1$. The p-series $\sum 1/n^{3/2}$ converges because $p = 3/2 > 1$. Therefore $\sum 1/\sqrt{n^3 - 1}$ also converges.

21. Let $a_n = 1/\sqrt{n^2 + 1}$. Then replacing n by $n + 1$ we have $a_{n+1} = 1/\sqrt{(n+1)^2 + 1}$. Since $\sqrt{(n+1)^2 + 1} > \sqrt{n^2 + 1}$, we have

$$\frac{1}{\sqrt{(n+1)^2 + 1}} < \frac{1}{\sqrt{n^2 + 1}},$$

so

$$0 < a_{n+1} < a_n.$$

In addition, $\lim_{n \to \infty} a_n = 0$ so $\displaystyle\sum_{n=0}^{\infty} \frac{(-1)^n}{\sqrt{n^2 + 1}}$ converges by the alternating series test.

25. This is an alternating series. Let $a_n = 1/(\sqrt{n} + 1)$. Then $\lim_{n \to \infty} a_n = 0$. Now replace n by $n + 1$ to give $a_{n+1} = 1/(\sqrt{n+1} + 1)$. Since $\sqrt{n+1} + 1 > \sqrt{n} + 1$, we have $\dfrac{1}{\sqrt{n+1} + 1} < \dfrac{1}{\sqrt{n} + 1}$, so

$$0 < a_{n+1} = \frac{1}{\sqrt{n+1} + 1} < \frac{1}{\sqrt{n} + 1} = a_n.$$

Therefore, the alternating series test tells us that the series $\displaystyle\sum_{n=1}^{\infty} \frac{(-1)^{n-1}}{\sqrt{n} + 1}$ converges.

29. We use the ratio test. Since $a_n = (2n)!/(n!)^2$, replacing n by $n + 1$ gives $a_{n+1} = (2n + 2)!/((n+1)!)^2$. Thus

$$\frac{a_{n+1}}{a_n} = \frac{\dfrac{(2n+2)!}{((n+1)!)^2}}{\dfrac{(2n)!}{(n!)^2}} = \frac{(2n+2)!}{(n+1)!(n+1)!} \cdot \frac{n!n!}{(2n)!}.$$

Since $(2n + 2)! = (2n + 2)(2n + 1)(2n)!$ and $(n + 1)! = (n + 1)n!$, we have

$$\frac{a_{n+1}}{a_n} = \frac{(2n+2)(2n+1)}{(n+1)(n+1)},$$

therefore

$$L = \lim_{n \to \infty} \frac{a_{n+1}}{a_n} = 4.$$

As $L > 1$ the ratio test tells us that the series $\displaystyle\sum_{n=1}^{\infty} \frac{(2n)!}{(n!)^2}$ diverges.

33. Since there is an n in the numerator and a $\sqrt{n}$ in the denominator, the terms in this series are increasing in magnitude. We have

$$\lim_{n \to \infty} \left| \frac{n+1}{\sqrt{n}} (-1)^n \right| = \lim_{n \to \infty} \frac{n+1}{\sqrt{n}} = \infty,$$

so $\lim_{n \to \infty} (-1)^n (n + 1)/\sqrt{n}$ does not approach zero. Therefore, the series diverges by Property 3 of Theorem 9.2.

37. The nth term $a_n = 1/(n^3 - 3)$ behaves like $1/n^3$ for large n, so we take $b_n = 1/n^3$. We have

$$\lim_{n \to \infty} \frac{a_n}{b_n} = \lim_{n \to \infty} \frac{1/(n^3 - 3)}{1/n^3} = \lim_{n \to \infty} \frac{n^3}{n^3 - 3} = 1.$$

The limit comparison test applies with $c = 1$. The p-series $\sum 1/n^3$ converges because $p = 3 > 1$. Therefore $\sum 1/(n^3 - 3)$ also converges.

41. Since $\ln n$ grows much more slowly than n, we suspect that $(\ln n)^2 < n$ for large n. This can be confirmed with L'Hopital's rule.

$$\lim_{n \to \infty} \frac{(\ln n)^2}{n} = \lim_{n \to \infty} \frac{2(\ln n)/n}{1} = \lim_{n \to \infty} \frac{2(\ln n)}{n} = 0.$$

Therefore, for large n, we have $(\ln n)^2/n < 1$, and hence for large n,

$$\frac{1}{n} < \frac{1}{(\ln n)^2}.$$

Thus $\sum_{n=2}^{\infty} 1/(\ln n)^2$ diverges by comparison with the divergent harmonic series $\sum 1/n$.

45. Let $C_n = 1/(n! + 1)$. Then replacing n by $n + 1$ gives $C_{n+1} = 1/((n + 1)! + 1)$. Using the ratio test, we have

$$\frac{|a_{n+1}|}{|a_n|} = |x| \frac{|C_{n+1}|}{|C_n|} = |x| \frac{1/((n + 1)! + 1)}{1/(n! + 1)} = |x| \frac{n! + 1}{(n + 1)! + 1}.$$

Since $n!$ and $(n + 1)!$ dominate the constant term 1 as $n \to \infty$ and $(n + 1)! = (n + 1) \cdot n!$ we have

$$\lim_{n \to \infty} \frac{|a_{n+1}|}{|a_n|} = 0.$$

Thus the radius of convergence is $R = \infty$.

Problems

49. We use the ratio test:

$$\left| \frac{a_{n+1}}{a_n} \right| = \left| \frac{x^{n+1}}{3^{n+1}(n+1)^2} \cdot \frac{3^n n^2}{x^n} \right| = \left(\frac{n}{n+1} \right)^2 \cdot \frac{|x|}{3}.$$

Since $n/(n + 1) \to 1$ as $n \to \infty$, we have

$$\lim_{n \to \infty} \left| \frac{a_{n+1}}{a_n} \right| = \frac{|x|}{3}.$$

We have $|x|/3 < 1$ when $|x| < 3$. The radius of convergence is 3 and the series converges for $-3 < x < 3$.

We check the endpoints. For $x = -3$, we have

$$\sum_{n=1}^{\infty} \frac{x^n}{3^n n^2} = \sum_{n=1}^{\infty} \frac{(-3)^n}{3^n n^2} = \sum_{n=1}^{\infty} \frac{(-1)^n}{n^2}.$$

We know $\sum \frac{1}{n^2}$ is a p-series with $p = 2$ so it converges. Therefore the alternating series $\sum \frac{(-1)^n}{n^2}$ also converges. For $x = 3$, we have

$$\sum_{n=1}^{\infty} \frac{x^n}{3^n n^2} = \sum_{n=1}^{\infty} \frac{3^n}{3^n n^2} = \sum_{n=1}^{\infty} \frac{1}{n^2}.$$

This is a p-series with $p = 2$ and it converges. The series converges at both its endpoints and the interval of convergence is $-3 \leq x \leq 3$.

53. (a) Using an argument similar to Example 5 in Section 9.5, we take

$$a_n = (-1)^n \frac{t^{2n}}{(2n)!},$$

so, replacing n by $n + 1$,

$$a_{n+1} = (-1)^{n+1} \frac{t^{2(n+1)}}{(2(n + 1))!} = (-1)^{n+1} \frac{t^{2n+2}}{(2n + 2)!}.$$

Thus,

$$\frac{|a_{n+1}|}{|a_n|} = \frac{|(-1)^{n+1} t^{2n+2}/(2n + 2)!|}{|(-1)^n t^{2n}/(2n)!|} = \frac{t^2}{(2n + 2)(2n + 1)},$$

so

$$\lim_{n \to \infty} \frac{|a_{n+1}|}{|a_n|} = \lim_{n \to \infty} \frac{t^2}{(2n + 2)(2n + 1)} = 0.$$

The radius of convergence is therefore ∞, so the series converges for all t. Therefore, the domain of h is all real numbers.

(b) Since h involves only even powers,

$$h(t) = 1 - \frac{t^2}{2!} + \frac{t^4}{4!} - \frac{t^6}{6!} + \cdots,$$

h is an even function.

(c) Differentiating term-by-term, we have

$$h'(t) = 0 - 2\frac{t}{2!} + 4\frac{t^3}{4!} - 6\frac{t^6}{6!} + \cdots$$
$$= -t + \frac{t^3}{3!} - \frac{t^5}{5!} + \cdots.$$

$$h''(t) = -1 + 3\frac{t^2}{3!} - 5\frac{t^4}{5!} + \cdots$$
$$= -1 + \frac{t^2}{2!} - \frac{t^4}{4!} + \cdots.$$

So we see $h''(t) = -h(t)$.

57. (a)

$$\text{Present value of first coupon} = \frac{50}{1.05}$$
$$\text{Present value of second coupon} = \frac{50}{(1.05)^2}, \text{etc.}$$

$$\text{Total present value} = \underbrace{\frac{50}{1.05} + \frac{50}{(1.05)^2} + \cdots + \frac{50}{(1.05)^{10}}}_{\text{coupons}} + \underbrace{\frac{1000}{(1.05)^{10}}}_{\text{principal}}$$

$$= \frac{50}{1.05}\left(1 + \frac{1}{1.05} + \cdots + \frac{1}{(1.05)^9}\right) + \frac{1000}{(1.05)^{10}}$$

$$= \frac{50}{1.05}\left(\frac{1 - \left(\frac{1}{1.05}\right)^{10}}{1 - \frac{1}{1.05}}\right) + \frac{1000}{(1.05)^{10}}$$

$$= 386.087 + 613.913$$

$$= \$1000$$

(b) When the interest rate is 5%, the present value equals the principal.

(c) When the interest rate is more than 5%, the present value is smaller than it is when interest is 5% and must therefore be less than the principal. Since the bond will sell for around its present value, it will sell for less than the principal; hence the description *trading at discount*.

(d) When the interest rate is less than 5%, the present value is more than the principal. Hence the bond will be selling for more than the principal, and is described as *trading at a premium*.

61. (a)

$$\text{Total amount of money deposited} = 100 + 92 + 84.64 + \cdots$$
$$= 100 + 100(0.92) + 100(0.92)^2 + \cdots$$
$$= \frac{100}{1 - 0.92} = 1250 \quad \text{dollars}$$

(b) Credit multiplier $= 1250/100 = 12.50$

The 12.50 is the factor by which the bank has increased its deposits, from \$100 to \$1250.

65. If $\sum(a_n + b_n)$ converged, then $\sum(a_n + b_n) - \sum a_n = \sum b_n$ would converge by Theorem 9.2. Since we know that $\sum b_n$ does not converge, we conclude that $\sum(a_n + b_n)$ diverges.

69. We have $a_n + (a_n/2) = (3/2)a_n$, so the series $\sum(a_n + a_n/2)$ converges since it is a constant multiple of the convergent series $\sum a_n$.

73. Using a right-hand sum, we have

$$\frac{1}{2} + \frac{1}{3} + \frac{1}{4} + \cdots + \frac{1}{n} < \int_1^n \frac{dx}{x} = \ln n.$$

If a computer could add a million terms in one second, then it could add

$$60\frac{\text{sec}}{\text{min}} \cdot 60\frac{\text{min}}{\text{hour}} \cdot 24\frac{\text{hour}}{\text{day}} \cdot 365\frac{\text{days}}{\text{year}} \cdot 1 \text{ million } \frac{\text{terms}}{\text{sec}}$$

terms per year. Thus,

$$1 + \frac{1}{2} + \frac{1}{3} \cdots + \frac{1}{n} < 1 + \ln n = 1 + \ln(60 \cdot 60 \cdot 24 \cdot 365 \cdot 10^6) \approx 32.082 < 33.$$

So the sum after one year is about 32.

CHECK YOUR UNDERSTANDING

1. False. The first 1000 terms could be the same for two different sequences and yet one sequence converges and the other diverges. For example, $s_n = 0$ for all n is a convergent sequence, but

$$t_n = \begin{cases} 0 & \text{if } n \leq 1000 \\ n & \text{if } n > 1000 \end{cases}$$

is a divergent sequence.

5. False. The terms s_n tend to the limit of the sequence which may not be zero. For example, $s_n = 1 + 1/n$ is a convergent sequence and s_n tends to 1 as n increases.

9. False. The sequence $-1, 1, -1, 1, \ldots$ given by $s_n = (-1)^n$ alternates in sign but does not converge.

13. True. This power series has an interval of convergence about $x = 0$. If the power series converges for $x = 2$, the radius of convergence is 2 or more. Thus, $x = 1$ is well within the interval of convergence, so the series converges at $x = 1$.

17. True. This is one of the statements of the comparison test.

21. False. For example, if $a_n = 1/n^2$, then

$$\lim_{n \to \infty} \frac{|a_{n+1}|}{|a_n|} = \lim_{n \to \infty} \frac{1/(n+1)^2}{1/n^2} = \lim_{n \to \infty} \frac{n^2}{(n+1)^2} = 1.$$

However, $\sum 1/n^2$ converges.

25. False. The terms in the series do not go to zero:

$$2^{(-1)^1} + 2^{(-1)^2} + 2^{(-1)^3} + 2^{(-1)^4} + 2^{(-1)^5} + \cdots = 2^{-1} + 2^1 + 2^{-1} + 2^1 + 2^{-1} + \cdots$$
$$= 1/2 + 2 + 1/2 + 2 + 1/2 + \cdots.$$

29. True. Let $c_n = (-1)^n|a_n|$. Then $|c_n| = |a_n|$ so $\sum |c_n|$ converges, and therefore $\sum c_n = \sum (-1)^n|a_n|$ converges.

33. True. If the terms do not tend to zero, the partial sums do not tend to a limit. For example, if the terms are all greater than 0.1, the partial sums will grow without bound.

37. True. If $\sum |a_n|$ is convergent, then so is $\sum a_n$.

41. True. The power series $\sum C_n(x - a)^n$ converges at $x = a$.

45. False. The interval of convergence of $\sum C_n x^n$ is centered at the origin.

CHAPTER TEN

Solutions for Section 10.1

Exercises

1. Let $f(x) = \dfrac{1}{1-x} = (1-x)^{-1}$. Then $f(0) = 1$.

$$
\begin{aligned}
f'(x) &= 1!(1-x)^{-2} & f'(0) &= 1!, \\
f''(x) &= 2!(1-x)^{-3} & f''(0) &= 2!, \\
f'''(x) &= 3!(1-x)^{-4} & f'''(0) &= 3!, \\
f^{(4)}(x) &= 4!(1-x)^{-5} & f^{(4)}(0) &= 4!, \\
f^{(5)}(x) &= 5!(1-x)^{-6} & f^{(5)}(0) &= 5!, \\
f^{(6)}(x) &= 6!(1-x)^{-7} & f^{(6)}(0) &= 6!, \\
f^{(7)}(x) &= 7!(1-x)^{-8} & f^{(7)}(0) &= 7!.
\end{aligned}
$$

$$
\begin{aligned}
P_3(x) &= 1 + x + x^2 + x^3, \\
P_5(x) &= 1 + x + x^2 + x^3 + x^4 + x^5, \\
P_7(x) &= 1 + x + x^2 + x^3 + x^4 + x^5 + x^6 + x^7.
\end{aligned}
$$

5. Let $f(x) = \cos x$. Then $f(0) = \cos(0) = 1$, and

$$
\begin{aligned}
f'(x) &= -\sin x & f'(0) &= 0, \\
f''(x) &= -\cos x & f''(0) &= -1, \\
f'''(x) &= \sin x & f'''(0) &= 0, \\
f^{(4)}(x) &= \cos x & f^{(4)}(0) &= 1, \\
f^{(5)}(x) &= -\sin x & f^{(5)}(0) &= 0, \\
f^{(6)}(x) &= -\cos x & f^{(6)}(0) &= -1.
\end{aligned}
$$

Thus,

$$
P_2(x) = 1 - \frac{x^2}{2!},
$$

$$
P_4(x) = 1 - \frac{x^2}{2!} + \frac{x^4}{4!},
$$

$$
P_6(x) = 1 - \frac{x^2}{2!} + \frac{x^4}{4!} - \frac{x^6}{6!}.
$$

9. Let $f(x) = \dfrac{1}{\sqrt{1+x}} = (1+x)^{-1/2}$. Then $f(0) = 1$.

$$
\begin{aligned}
f'(x) &= -\tfrac{1}{2}(1+x)^{-3/2} & f'(0) &= -\tfrac{1}{2}, \\
f''(x) &= \tfrac{3}{2^2}(1+x)^{-5/2} & f''(0) &= \tfrac{3}{2^2}, \\
f'''(x) &= -\tfrac{3\cdot 5}{2^3}(1+x)^{-7/2} & f'''(0) &= -\tfrac{3\cdot 5}{2^3}, \\
f^{(4)}(x) &= \tfrac{3\cdot 5\cdot 7}{2^4}(1+x)^{-9/2} & f^{(4)}(0) &= \tfrac{3\cdot 5\cdot 7}{2^4}
\end{aligned}
$$

Then,

$$
P_2(x) = 1 - \frac{1}{2}x + \frac{1}{2!}\frac{3}{2^2}x^2 = 1 - \frac{1}{2}x + \frac{3}{8}x^2,
$$

$$
P_3(x) = P_2(x) - \frac{1}{3!}\frac{3\cdot 5}{2^3}x^3 = 1 - \frac{1}{2}x + \frac{3}{8}x^2 - \frac{5}{16}x^3,
$$

$$
P_4(x) = P_3(x) + \frac{1}{4!}\frac{3\cdot 5\cdot 7}{2^4}x^4 = 1 - \frac{1}{2}x + \frac{3}{8}x^2 - \frac{5}{16}x^3 + \frac{35}{128}x^4.
$$

13. Let $f(x) = \sin x$. $f(\frac{\pi}{2}) = 1$.

$$f'(x) = \cos x \qquad f'(\tfrac{\pi}{2}) = 0,$$
$$f''(x) = -\sin x \qquad f''(\tfrac{\pi}{2}) = -1,$$
$$f'''(x) = -\cos x \qquad f'''(\tfrac{\pi}{2}) = 0,$$
$$f^{(4)}(x) = \sin x \qquad f^{(4)}(\tfrac{\pi}{2}) = 1.$$

So,

$$P_4(x) = 1 + 0 - \frac{1}{2!}\left(x - \frac{\pi}{2}\right)^2 + 0 + \frac{1}{4!}\left(x - \frac{\pi}{2}\right)^4$$
$$= 1 - \frac{1}{2!}\left(x - \frac{\pi}{2}\right)^2 + \frac{1}{4!}\left(x - \frac{\pi}{2}\right)^4.$$

Problems

17. Since $P_2(x)$ is the second degree Taylor polynomial for $f(x)$ about $x = 0$, $P_2(0) = f(0)$, which says $a = f(0)$. Since

$$\frac{d}{dx}P_2(x)\bigg|_{x=0} = f'(0),$$

$b = f'(0)$; and since

$$\frac{d^2}{dx^2}P_2(x)\bigg|_{x=0} = f''(0),$$

$2c = f''(0)$. In other words, a is the y-intercept of $f(x)$, b is the slope of the tangent line to $f(x)$ at $x = 0$ and c tells us the concavity of $f(x)$ near $x = 0$. So $c < 0$ since f is concave down; $b > 0$ since f is increasing; $a > 0$ since $f(0) > 0$.

21. Using the fact that

$$f(x) \approx P_2(x) = f(0) + f'(0)x + \frac{f''(0)}{2!}x^2$$

and identifying coefficients with those given for $P_2(x)$, we obtain the following:

(a) $f(0)$ = constant term which equals 5,   so $f(0) = 5$.
(b) $f'(0)$ = coefficient of x which equals -7,   so $f'(0) = -7$.
(c) $\frac{f''(0)}{2!}$ = coefficient of x^2 which equals 8,   so $f''(0) = 16$.

25.

$$f(x) = 4x^2 - 7x + 2 \quad f(0) = 2$$
$$f'(x) = 8x - 7 \qquad f'(0) = -7$$
$$f''(x) = 8 \qquad f''(0) = 8,$$

so $P_2(x) = 2 + (-7)x + \frac{8}{2}x^2 = 4x^2 - 7x + 2$. We notice that $f(x) = P_2(x)$ in this case.

29.

$$\lim_{x \to 0} \frac{1 - \cos x}{x^2} = \lim_{x \to 0} \frac{1 - (1 - \frac{x^2}{2!} + \frac{x^4}{4!})}{x^2} = \lim_{x \to 0}\left(\frac{1}{2} - \frac{x^2}{4!}\right) = \frac{1}{2}.$$

33. (a) $f(x) = e^{x^2}$.

$f'(x) = 2xe^{x^2}$, $f''(x) = 2(1 + 2x^2)e^{x^2}$, $f'''(x) = 4(3x + 2x^3)e^{x^2}$,
$f^{(4)}(x) = 4(3 + 6x^2)e^{x^2} + 4(3x + 2x^3)2xe^{x^2}$.
The Taylor polynomial about $x = 0$ is

$$P_4(x) = 1 + \frac{0}{1!}x + \frac{2}{2!}x^2 + \frac{0}{3!}x^3 + \frac{12}{4!}x^4$$
$$= 1 + x^2 + \frac{1}{2}x^4.$$

(b) $f(x) = e^x$. The Taylor polynomial of degree 2 is

$$Q_2(x) = 1 + \frac{x}{1!} + \frac{x^2}{2!} = 1 + x + \frac{1}{2}x^2.$$

If we substitute x^2 for x in the Taylor polynomial for e^x of degree 2, we will get $P_4(x)$, the Taylor polynomial for e^{x^2} of degree 4:

$$Q_2(x^2) = 1 + x^2 + \frac{1}{2}(x^2)^2$$

$$= 1 + x^2 + \frac{1}{2}x^4$$

$$= P_4(x).$$

(c) Let $Q_{10}(x) = 1 + \frac{x}{1!} + \frac{x^2}{2!} + \cdots + \frac{x^{10}}{10!}$ be the Taylor polynomial of degree 10 for e^x about $x = 0$. Then

$$P_{20}(x) = Q_{10}(x^2)$$

$$= 1 + \frac{x^2}{1!} + \frac{(x^2)^2}{2!} + \cdots + \frac{(x^2)^{10}}{10!}$$

$$= 1 + \frac{x^2}{1!} + \frac{x^4}{2!} + \cdots + \frac{x^{20}}{10!}.$$

(d) Let $e^x \approx Q_5(x) = 1 + \frac{x}{1!} + \cdots + \frac{x^5}{5!}$. Then

$$e^{-2x} \approx Q_5(-2x)$$

$$= 1 + \frac{-2x}{1!} + \frac{(-2x)^2}{2!} + \frac{(-2x)^3}{3!} + \frac{(-2x)^4}{4!} + \frac{(-2x)^5}{5!}$$

$$= 1 - 2x + 2x^2 - \frac{4}{3}x^3 + \frac{2}{3}x^4 - \frac{4}{15}x^5.$$

Solutions for Section 10.2

Exercises

1. Differentiating $(1+x)^{3/2}$:

$$\begin{array}{ll} f(x) = (1+x)^{3/2} & f(0) = 1, \\ f'(x) = (3/2)(1+x)^{1/2} & f'(0) = \frac{3}{2}, \\ f''(x) = (1/2)(3/2)(1+x)^{-1/2} = (3/4)(1+x)^{-1/2} & f''(0) = \frac{3}{4}, \\ f'''(x) = (-1/2)(3/4)(1+x)^{-3/2} = (-3/8)(1+x)^{-3/2} & f'''(0) = -\frac{3}{8}. \end{array}$$

$$f(x) = (1+x)^{3/2} = 1 + \frac{3}{2} \cdot x + \frac{(3/4)x^2}{2!} + \frac{(-3/8)x^3}{3!} + \cdots$$

$$= 1 + \frac{3x}{2} + \frac{3x^2}{8} - \frac{x^3}{16} + \cdots$$

5.

$$\begin{array}{ll} f(x) = \frac{1}{1-x} = (1-x)^{-1} & f(0) = 1, \\ f'(x) = -(1-x)^{-2}(-1) = (1-x)^{-2} & f'(0) = 1, \\ f''(x) = -2(1-x)^{-3}(-1) = 2(1-x)^{-3} & f''(0) = 2, \\ f'''(x) = -6(1-x)^{-4}(-1) = 6(1-x)^{-4} & f'''(0) = 6. \end{array}$$

$$f(x) = \frac{1}{1-x} = 1 + 1 \cdot x + \frac{2x^2}{2!} + \frac{6x^3}{3!} + \cdots$$

$$= 1 + x + x^2 + x^3 + \cdots$$

9.

$$
\begin{array}{ll}
f(x) = \sin x & f(\tfrac{\pi}{4}) = \tfrac{\sqrt{2}}{2}, \\
f'(x) = \cos x & f'(\tfrac{\pi}{4}) = \tfrac{\sqrt{2}}{2}, \\
f''(x) = -\sin x & f''(\tfrac{\pi}{4}) = -\tfrac{\sqrt{2}}{2}, \\
f'''(x) = -\cos x & f'''(\tfrac{\pi}{4}) = -\tfrac{\sqrt{2}}{2}.
\end{array}
$$

$$
\begin{aligned}
\sin x &= \frac{\sqrt{2}}{2} + \frac{\sqrt{2}}{2}\left(x - \frac{\pi}{4}\right) - \frac{\sqrt{2}}{2}\frac{(x - \frac{\pi}{4})^2}{2!} - \frac{\sqrt{2}}{2}\frac{(x - \frac{\pi}{4})^3}{3!} - \cdots \\
&= \frac{\sqrt{2}}{2} + \frac{\sqrt{2}}{2}\left(x - \frac{\pi}{4}\right) - \frac{\sqrt{2}}{4}\left(x - \frac{\pi}{4}\right)^2 - \frac{\sqrt{2}}{12}\left(x - \frac{\pi}{4}\right)^3 - \cdots
\end{aligned}
$$

13.

$$
\begin{array}{ll}
f(x) = \frac{1}{x} & f(1) = 1 \\
f'(x) = -\frac{1}{x^2} & f'(1) = -1 \\
f''(x) = \frac{2}{x^3} & f''(1) = 2 \\
f'''(x) = -\frac{6}{x^4} & f'''(1) = -6
\end{array}
$$

$$
\begin{aligned}
\frac{1}{x} &= 1 - (x - 1) + \frac{2(x-1)^2}{2!} - \frac{6(x-1)^3}{3!} + \cdots \\
&= 1 - (x - 1) + (x - 1)^2 - (x - 1)^3 + \cdots .
\end{aligned}
$$

17. The general term can be written as $(-1)^n x^n$ for $n \geq 0$.

21. The general term can be written as $(-1)^k x^{2k+1}/(2k + 1)$ for $k \geq 0$.

Problems

25. (a)

$$
\begin{array}{ll}
f(x) = \ln(1 + 2x) & f(0) = 0 \\
f'(x) = \frac{2}{1+2x} & f'(0) = 2 \\
f''(x) = -\frac{4}{(1+2x)^2} & f''(0) = -4 \\
f'''(x) = \frac{16}{(1+2x)^3} & f'''(0) = 16
\end{array}
$$

$$
\ln(1 + 2x) = 2x - 2x^2 + \frac{8}{3}x^3 + \cdots
$$

(b) To get the expression for $\ln(1 + 2x)$ from the series for $\ln(1 + x)$, substitute $2x$ for x in the series

$$
\ln(1 + x) = x - \frac{x^2}{2} + \frac{x^3}{3} - \frac{x^4}{4} + \cdots
$$

to get

$$
\begin{aligned}
\ln(1 + 2x) &= 2x - \frac{(2x)^2}{2} + \frac{(2x)^3}{3} - \frac{(2x)^4}{4} + \cdots \\
&= 2x - 2x^2 + \frac{8x^3}{3} - 4x^4 + \cdots
\end{aligned}
$$

(c) Since the interval of convergence for $\ln(1 + x)$ is $-1 < x < 1$, substituting $2x$ for x suggests the interval of convergence of $\ln(1 + 2x)$ is $-1 < 2x < 1$, or $-\frac{1}{2} < x < \frac{1}{2}$.

29. The Taylor series for $\ln(1 - x)$ is

$$
\ln(1 - x) = -x - \frac{x^2}{2} - \frac{x^3}{3} - \cdots - \frac{x^n}{n} - \cdots,
$$

so

$$
\lim_{n \to \infty} \frac{|a_{n+1}|}{|a_n|} = |x| \lim_{n \to \infty} \frac{1/(n+1)}{1/n} = |x| \lim_{n \to \infty} \left|\frac{n}{n+1}\right| = |x|.
$$

Thus the series converges for $|x| < 1$, and the radius of convergence is 1. Note: This series can be obtained from the series for $\ln(1 + x)$ by replacing x by $-x$ and has the same radius of convergence as the series for $\ln(1 + x)$.

33. This is the series for $1/(1 - x)$ with x replaced by $1/4$, so the series converges to $1/(1 - (1/4)) = 4/3$.

37. This is the series for e^x with $x = 3$ substituted. Thus

$$1 + 3 + \frac{9}{2!} + \frac{27}{3!} + \frac{81}{4!} + \cdots = 1 + 3 + \frac{3^2}{2!} + \frac{3^3}{3!} + \frac{3^4}{4!} + \cdots = e^3.$$

41. Since $x - \frac{1}{2}x^2 + \frac{1}{3}x^3 + \cdots = \ln(1 + x)$, we solve $\ln(1 + x) = 0.2$, giving $1 + x = e^{0.2}$, so $x = e^{0.2} - 1$.

45. We define $e^{i\theta}$ to be

$$e^{i\theta} = 1 + i\theta + \frac{(i\theta)^2}{2!} + \frac{(i\theta)^3}{3!} + \frac{(i\theta)^4}{4!} + \frac{(i\theta)^5}{5!} + \frac{(i\theta)^6}{6!} + \cdots$$

Suppose we consider the expression $\cos\theta + i\sin\theta$, with $\cos\theta$ and $\sin\theta$ replaced by their Taylor series:

$$\cos\theta + i\sin\theta = \left(1 - \frac{\theta^2}{2!} + \frac{\theta^4}{4!} - \frac{\theta^6}{6!} + \cdots\right) + i\left(\theta - \frac{\theta^3}{3!} + \frac{\theta^5}{5!} - \cdots\right)$$

Reordering terms, we have

$$\cos\theta + i\sin\theta = 1 + i\theta - \frac{\theta^2}{2!} - \frac{i\theta^3}{3!} + \frac{\theta^4}{4!} + \frac{i\theta^5}{5!} - \frac{\theta^6}{6!} - \cdots$$

Using the fact that $i^2 = -1$, $i^3 = -i$, $i^4 = 1$, $i^5 = i, \cdots$, we can rewrite the series as

$$\cos\theta + i\sin\theta = 1 + i\theta + \frac{(i\theta)^2}{2!} + \frac{(i\theta)^3}{3!} + \frac{(i\theta)^4}{4!} + \frac{(i\theta)^5}{5!} + \frac{(i\theta)^6}{6!} + \cdots$$

Amazingly enough, this series is the Taylor series for e^x with $i\theta$ substituted for x. Therefore, we have shown that

$$\cos\theta + i\sin\theta = e^{i\theta}.$$

Solutions for Section 10.3

Exercises

1. Substitute $y = -x$ into $e^y = 1 + y + \frac{y^2}{2!} + \frac{y^3}{3!} + \cdots$. We get

$$e^{-x} = 1 + (-x) + \frac{(-x)^2}{2!} + \frac{(-x)^3}{3!} + \cdots$$

$$= 1 - x + \frac{x^2}{2!} - \frac{x^3}{3!} + \cdots.$$

5. Since $\frac{d}{dx}(\arcsin x) = \frac{1}{\sqrt{1-x^2}} = 1 + \frac{1}{2}x^2 + \frac{3}{8}x^4 + \frac{5}{16}x^6 + \cdots$, integrating gives

$$\arcsin x = c + x + \frac{1}{6}x^3 + \frac{3}{40}x^5 + \frac{5}{112}x^7 + \cdots.$$

Since $\arcsin 0 = 0$, $c = 0$.

9. We substitute $3t$ into the series for $\sin x$ and multiply by t. Since

$$\sin x = x - \frac{x^3}{3!} + \frac{x^5}{5!} - \frac{x^7}{7!} + \cdots,$$

substituting $3t$ gives

$$\sin(3t) = (3t) - \frac{(3t)^3}{3!} + \frac{(3t)^5}{5!} - \frac{(3t)^7}{7!} + \cdots$$
$$= 3t + \frac{-9}{2}t^3 + \frac{81}{40}t^5 + \frac{-243}{560}t^7 + \cdots,$$

so

$$t\sin(3t) = 3t^2 - \frac{9}{2}t^4 + \frac{81}{40}t^6 - \frac{243}{560}t^8 + \cdots.$$

13. Multiplying out gives $(1+x)^3 = 1 + 3x + 3x^2 + x^3$. Since this polynomial equals the original function for all x, it must be the Taylor series. The general term is $0 \cdot x^n$ for $n \geq 4$.

17. Using the binomial expansion for $(1+x)^{-1}$ with $x = -r/a$:

$$\frac{1}{a-r} = \frac{1}{a - a\left(\frac{r}{a}\right)} = \frac{1}{a\left(1 - \frac{r}{a}\right)} = \frac{1}{a}\left(1 + \left(-\frac{r}{a}\right)\right)^{-1}$$
$$= \frac{1}{a}\left(1 + (-1)\left(-\frac{r}{a}\right) + \frac{(-1)(-2)}{2!}\left(-\frac{r}{a}\right)^2 + \frac{(-1)(-2)(-3)}{3!}\left(-\frac{r}{a}\right)^3 + \cdots\right)$$
$$= \frac{1}{a}\left(1 - \left(-\frac{r}{a}\right) + \left(-\frac{r}{a}\right)^2 - \left(-\frac{r}{a}\right)^3 + \cdots\right)$$
$$= \frac{1}{a}\left(1 + \left(\frac{r}{a}\right) + \left(\frac{r}{a}\right)^2 + \left(\frac{r}{a}\right)^3 + \cdots\right).$$

21.

$$\frac{a}{\sqrt{a^2 + x^2}} = \frac{a}{a(1 + \frac{x^2}{a^2})^{\frac{1}{2}}} = \left(1 + \frac{x^2}{a^2}\right)^{-\frac{1}{2}}$$
$$= 1 + \left(-\frac{1}{2}\right)\frac{x^2}{a^2} + \frac{1}{2!}\left(-\frac{1}{2}\right)\left(-\frac{3}{2}\right)\left(\frac{x^2}{a^2}\right)^2$$
$$+ \frac{1}{3!}\left(-\frac{1}{2}\right)\left(-\frac{3}{2}\right)\left(-\frac{5}{2}\right)\left(\frac{x^2}{a^2}\right)^3 + \cdots$$
$$= 1 - \frac{1}{2}\left(\frac{x}{a}\right)^2 + \frac{3}{8}\left(\frac{x}{a}\right)^4 - \frac{5}{16}\left(\frac{x}{a}\right)^6 + \cdots$$

Problems

25. Notice that $\sum px^{p-1}$, is the derivative, term-by-term, of a geometric series:

$$\sum_{p=1}^{\infty} px^{p-1} = 1 \cdot x^0 + 2 \cdot x^1 + 3 \cdot x^2 + \cdots = \frac{d}{dx}\underbrace{(x + x^2 + x^3 + \cdots)}_{\text{Geometric series}}.$$

For $|x| < 1$, the sum of the geometric series with first term x and common ratio x is

$$x + x^2 + x^3 + \cdots = \frac{x}{1-x}.$$

Differentiating gives

$$\sum_{p=1}^{\infty} px^{p-1} = \frac{d}{dx}\left(\frac{x}{1-x}\right) = \frac{1(1-x) - x(-1)}{(1-x)^2} = \frac{1}{(1-x)^2}.$$

29.

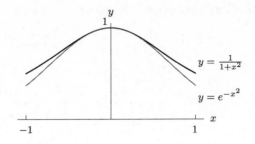

(a)

$$e^{-x^2} = 1 - x^2 + \frac{x^4}{2!} - \frac{x^6}{3!} + \cdots$$

$$\frac{1}{1+x^2} = 1 - x^2 + x^4 - x^6 + \cdots$$

Notice that the first two terms are the same in both series.

(b) $\dfrac{1}{1+x^2}$ is greater.

(c) Even, because the only terms involved are of even degree.

(d) The coefficients for e^{-x^2} become extremely small for higher powers of x, and we can "counteract" the effect of these powers for large values of x. The series for $\frac{1}{1+x^2}$ has no such coefficients.

33.

$$E = kQ \left(\frac{1}{(R-1)^2} - \frac{1}{(R+1)^2} \right)$$

$$= \frac{kQ}{R^2} \left(\frac{1}{(1 - \frac{1}{R})^2} - \frac{1}{(1 + \frac{1}{R})^2} \right)$$

Since $\left| \frac{1}{R} \right| < 1$, we can expand the two terms using the binomial expansion:

$$\frac{1}{(1 - \frac{1}{R})^2} = \left(1 - \frac{1}{R} \right)^{-2}$$

$$= 1 - 2 \left(-\frac{1}{R} \right) + (-2)(-3)\frac{(-\frac{1}{R})^2}{2!} + (-2)(-3)(-4)\frac{(-\frac{1}{R})^3}{3!} + \cdots$$

$$\frac{1}{(1 + \frac{1}{R})^2} = \left(1 + \frac{1}{R} \right)^{-2}$$

$$= 1 - 2 \left(\frac{1}{R} \right) + (-2)(-3)\frac{(\frac{1}{R})^2}{2!} + (-2)(-3)(-4)\frac{(\frac{1}{R})^3}{3!} + \cdots$$

Substituting, we get:

$$E = \frac{kQ}{R^2} \left[1 + \frac{2}{R} + \frac{3}{R^2} + \frac{4}{R^3} + \cdots - \left(1 - \frac{2}{R} + \frac{3}{R^2} - \frac{4}{R^3} + \cdots \right) \right] \approx \frac{kQ}{R^2} \left(\frac{4}{R} + \frac{8}{R^3} \right),$$

using only the first two non-zero terms.

37. (a) Factoring the expression for $t_1 - t_2$, we get

$$\Delta t = t_1 - t_2 = \frac{2l_2}{c(1 - v^2/c^2)} - \frac{2l_1}{c\sqrt{1 - v^2/c^2}} - \frac{2l_2}{c\sqrt{1 - v^2/c^2}} + \frac{2l_1}{c(1 - v^2/c^2)}$$

$$= \frac{2(l_1 + l_2)}{c(1 - v^2/c^2)} - \frac{2(l_1 + l_2)}{c\sqrt{1 - v^2/c^2}}$$

$$= \frac{2(l_1 + l_2)}{c} \left(\frac{1}{1 - v^2/c^2} - \frac{1}{\sqrt{1 - v^2/c^2}} \right).$$

Expanding the two terms within the parentheses in terms of v^2/c^2 gives

$$\left(1 - \frac{v^2}{c^2}\right)^{-1} = 1 + \frac{v^2}{c^2} + \frac{(-1)(-2)}{2!}\left(\frac{-v^2}{c^2}\right)^2 + \frac{(-1)(-2)(-3)}{3!}\left(\frac{-v^2}{c^2}\right)^3 + \cdots$$

$$= 1 + \frac{v^2}{c^2} + \frac{v^4}{c^4} + \frac{v^6}{c^6} + \cdots$$

$$\left(1 - \frac{v^2}{c^2}\right)^{-1/2} = 1 + \frac{1}{2}\frac{v^2}{c^2} + \frac{\left(\frac{-1}{2}\right)\left(\frac{-3}{2}\right)}{2!}\left(\frac{-v^2}{c^2}\right)^2 + \frac{\left(\frac{-1}{2}\right)\left(\frac{-3}{2}\right)\left(\frac{-5}{2}\right)}{3!}\left(\frac{-v^2}{c^2}\right)^3 + \cdots$$

$$= 1 + \frac{1}{2}\frac{v^2}{c^2} + \frac{3}{8}\frac{v^4}{c^4} + \frac{5}{16}\frac{v^6}{c^6} + \cdots$$

Thus, we have

$$\Delta t = \frac{2(l_1 + l_2)}{c}\left(1 + \frac{v^2}{c^2} + \frac{v^4}{c^4} + \frac{v^6}{c^6} + \cdots - 1 - \frac{1}{2}\frac{v^2}{c^2} - \frac{3}{8}\frac{v^4}{c^4} - \frac{5}{16}\frac{v^6}{c^6} - \cdots\right)$$

$$= \frac{2(l_1 + l_2)}{c}\left(\frac{1}{2}\frac{v^2}{c^2} + \frac{5}{8}\frac{v^4}{c^4} + \frac{11}{16}\frac{v^6}{c^6} + \cdots\right)$$

$$\Delta t \approx \frac{(l_1 + l_2)}{c}\left(\frac{v^2}{c^2} + \frac{5}{4}\frac{v^4}{c^4}\right).$$

(b) For small v. we can neglect all but the first nonzero term, so

$$\Delta t \approx \frac{(l_1 + l_2)}{c} \cdot \frac{v^2}{c^2} = \frac{(l_1 + l_2)}{c^3}v^2.$$

Thus, Δt is proportional to v^2 with constant of proportionality $(l_1 + l_2)/c^3$.

41. (a) We take the left-hand Riemann sum with the formula

Left-hand sum $= (1 + 0.9608 + 0.8521 + 0.6977 + 0.5273)(0.2) = 0.8076.$

Similarly,

Right-hand sum $= (0.9608 + 0.8521 + 0.6977 + 0.5273 + 0.3679)(0.2) = 0.6812.$

(b) Since

$$e^x = 1 + x + \frac{x^2}{2!} + \frac{x^3}{3!} + \dots,$$

$$e^{-x^2} \approx 1 + (-x^2) + \frac{(-x^2)^2}{2!} + \frac{(-x^2)^3}{3!}$$

$$= 1 - x^2 + \frac{x^4}{2} - \frac{x^6}{6}.$$

(c)

$$\int_0^1 e^{-x^2}\,dx \approx \int_0^1 \left(1 - x^2 + \frac{x^4}{2} - \frac{x^6}{6}\right)dx$$

$$= \left(x - \frac{x^3}{3} + \frac{x^5}{10} - \frac{x^7}{42}\right)\Bigg|_0^1 = 0.74286.$$

(d) We can improve the left and right sum values by averaging them to get 0.74439 or by increasing the number of subdivisions. We can improve on the estimate using the Taylor approximation by taking more terms.

Solutions for Section 10.4

Exercises

1. Let $f(x) = e^x$, so $f(0.1) = e^{0.1}$. The error bound in the Taylor approximation of degree 3 for $f(0.1) = 3^{0.1}$ about $x = 0$ is:
$$|E_3| = |f(0.1) - P_3(0.1)| \leq \frac{M \cdot |0.1 - 0|^4}{4!} = \frac{M(0.1)^4}{24},$$
where $|f^{(4)}(x)| \leq M$ for $0 \leq x \leq 0.1$. Now, $f^{(4)}(x) = e^x$. By looking at the graph of e^x, we see that $|f^{(4)}(x)|$ is maximized for x between 0 and 0.1 when $x = 0.1$. Thus,
$$|f^{(4)}| \leq e^{0.1},$$
so
$$|E_3| \leq \frac{e^{0.1} \cdot (0.1)^4}{24} \approx 0.00000460.$$

5. Let $f(x) = \ln(1 + x)$. The error bound in the Taylor approximation of degree 3 about $x = 0$ is:
$$|E_4| = |f(0.5) - P_3(0.5)| \leq \frac{M \cdot |0.5 - 0|^4}{4!} = \frac{M(0.5)^4}{24},$$
where $|f^{(4)}(x)| \leq M$ for $0 \leq x \leq 0.5$. Since
$$f^{(4)}(x) = \frac{3!}{(1 + x)^4}$$
and the denominator attains its minimum when $x = 0$, we have $|f^{(4)}(x)| \leq 3!$, so
$$|E_4| \leq \frac{3! \, (0.5)^4}{24} \approx 0.016.$$

Problems

9. (a) The Taylor polynomial of degree 0 about $t = 0$ for $f(t) = e^t$ is simply $P_0(x) = 1$. Since $e^t \geq 1$ on $[0, 0.5]$, the approximation is an underestimate.

(b) Using the zero degree error bound, if $|f'(t)| \leq M$ for $0 \leq t \leq 0.5$, then
$$|E_0| \leq M \cdot |t| \leq M(0.5).$$
Since $|f'(t)| = |e^t| = e^t$ is increasing on $[0, 0.5]$,
$$|f'(t)| \leq e^{0.5} < \sqrt{4} = 2.$$
Therefore
$$|E_0| \leq (2)(0.5) = 1.$$
(Note: By looking at a graph of $f(t)$ and its 0th degree approximation, it is easy to see that the greatest error occurs when $t = 0.5$, and the error is $e^{0.5} - 1 \approx 0.65 < 1$. So our error bound works.)

13. (a) (i) The vertical distance between the graph of $y = \cos x$ and $y = P_{10}(x)$ at $x = 6$ is no more than 4, so
$$|\text{Error in } P_{10}(6)| \leq 4.$$
Since at $x = 6$ the $\cos x$ and $P_{20}(x)$ graphs are indistinguishable in this figure, the error must be less than the smallest division we can see, which is about 0.2 so,
$$|\text{Error in } P_{20}(6)| \leq 0.2.$$

(ii) The maximum error occurs at the ends of the interval, that is, at $x = -9, x = 9$. At $x = 9$, the graphs of $y = \cos x$ and $y = P_{20}(x)$ are no more than 1 apart, so
$$\left| \begin{matrix} \text{Maximum error in } P_{20}(x) \\ \text{for } -9 \leq x \leq 9 \end{matrix} \right| \leq 1.$$

(b) We are looking for the largest x-interval on which the graphs of $y = \cos x$ and $y = P_{10}(x)$ are indistinguishable. This is hard to estimate accurately from the figure, though $-4 \leq x \leq 4$ certainly satisfies this condition.

17. (a) See Figure 10.1. The graph of E_1 looks like a parabola. Since the graph of E_1 is sandwiched between the graph of $y = x^2$ and the x axis, we have

$$|E_1| \leq x^2 \quad \text{for} \quad |x| \leq 0.1.$$

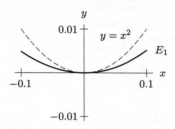

Figure 10.1

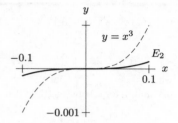

Figure 10.2

(b) See Figure 10.2. The graph of E_2 looks like a cubic, sandwiched between the graph of $y = x^3$ and the x axis, so

$$|E_2| \leq x^3 \quad \text{for} \quad |x| \leq 0.1.$$

(c) Using the Taylor expansion

$$e^x = 1 + x + \frac{x^2}{2!} + \frac{x^3}{3!} + \cdots$$

we see that

$$E_1 = e^x - (1 + x) = \frac{x^2}{2!} + \frac{x^3}{3!} + \frac{x^4}{4!} + \cdots.$$

Thus for small x, the $x^2/2!$ term dominates, so

$$E_1 \approx \frac{x^2}{2!},$$

and so E_1 is approximately a quadratic.

Similarly

$$E_2 = e^x - (1 + x + \frac{x^2}{2}) = \frac{x^3}{3!} + \frac{x^4}{4!} + \cdots.$$

Thus for small x, the $x^3/3!$ term dominates, so

$$E_2 \approx \frac{x^3}{3!}$$

and so E_2 is approximately a cubic.

Solutions for Section 10.5

Exercises

1. No, a Fourier series has terms of the form $\cos nx$, not $\cos^n x$.

5.

$$a_0 = \frac{1}{2\pi} \int_{-\pi}^{\pi} f(x)\, dx = \frac{1}{2\pi} \left[\int_{-\pi}^{0} -1\, dx + \int_{0}^{\pi} 1\, dx \right] = 0$$

$$a_1 = \frac{1}{\pi} \int_{-\pi}^{\pi} f(x) \cos x\, dx = \frac{1}{\pi} \left[\int_{-\pi}^{0} -\cos x\, dx + \int_{0}^{\pi} \cos x\, dx \right]$$

$$= \frac{1}{\pi} \left[-\sin x \Big|_{-\pi}^{0} + \sin x \Big|_{0}^{\pi} \right] = 0.$$

Similarly, a_2 and a_3 are both 0.

(In fact, notice $f(x) \cos nx$ is an odd function, so $\int_{-\pi}^{\pi} f(x) \cos nx = 0$.)

$$b_1 = \frac{1}{\pi} \int_{-\pi}^{\pi} f(x) \sin x \, dx = \frac{1}{\pi} \left[\int_{-\pi}^{0} -\sin x \, dx + \int_{0}^{\pi} \sin x \, dx \right]$$

$$= \frac{1}{\pi} \left[\cos x \Big|_{-\pi}^{0} + (-\cos x) \Big|_{0}^{\pi} \right] = \frac{4}{\pi}$$

$$b_2 = \frac{1}{\pi} \int_{-\pi}^{\pi} f(x) \sin 2x \, dx = \frac{1}{\pi} \left[\int_{-\pi}^{0} -\sin 2x \, dx + \int_{0}^{\pi} \sin 2x \, dx \right]$$

$$= \frac{1}{\pi} \left[\frac{1}{2} \cos 2x \Big|_{-\pi}^{0} + (-\frac{1}{2} \cos 2x) \Big|_{0}^{\pi} \right] = 0.$$

$$b_3 = \frac{1}{\pi} \int_{-\pi}^{\pi} f(x) \sin 3x \, dx = \frac{1}{\pi} \left[\int_{-\pi}^{0} -\sin 3x \, dx + \int_{0}^{\pi} \sin 3x \, dx \right]$$

$$= \frac{1}{\pi} \left[\frac{1}{3} \cos 3x \Big|_{-\pi}^{0} + (-\frac{1}{3} \cos 3x) \Big|_{0}^{\pi} \right] = \frac{4}{3\pi}.$$

Thus, $F_1(x) = F_2(x) = \frac{4}{\pi} \sin x$ and $F_3(x) = \frac{4}{\pi} \sin x + \frac{4}{3\pi} \sin 3x$.

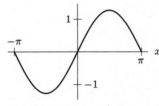

$F_1(x) = F_2(x) = \frac{4}{\pi} \sin x$

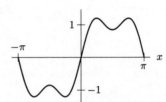

$F_3(x) = \frac{4}{\pi} \sin x + \frac{4}{3\pi} \sin 3x$

9.

$$a_0 = \frac{1}{2\pi} \int_{-\pi}^{\pi} h(x) \, dx = \frac{1}{2\pi} \int_{0}^{\pi} x \, dx = \frac{\pi}{4}$$

As in Problem 10, we use the integral table (III-15 and III-16) to find formulas for a_n and b_n.

$$a_n = \frac{1}{\pi} \int_{-\pi}^{\pi} h(x) \cos(nx) \, dx = \frac{1}{\pi} \int_{0}^{\pi} x \cos nx \, dx = \frac{1}{\pi} \left(\frac{x}{n} \sin(nx) + \frac{1}{n^2} \cos(nx) \right) \Big|_{0}^{\pi}$$

$$= \frac{1}{\pi} \left(\frac{1}{n^2} \cos(n\pi) - \frac{1}{n^2} \right)$$

$$= \frac{1}{n^2 \pi} \left(\cos(n\pi) - 1 \right).$$

Note that since $\cos(n\pi) = (-1)^n$, $a_n = 0$ if n is even and $a_n = -\frac{2}{n^2\pi}$ if n is odd.

$$b_n = \frac{1}{\pi} \int_{-\pi}^{\pi} h(x) \cos(nx) \, dx = \frac{1}{\pi} \int_{0}^{\pi} x \sin x \, dx$$

$$= \frac{1}{\pi} \left(-\frac{x}{n} \cos(nx) + \frac{1}{n^2} \sin(nx) \right) \Big|_{0}^{\pi}$$

$$= \frac{1}{\pi} \left(-\frac{\pi}{n} \cos(n\pi) \right)$$

$$= -\frac{1}{n} \cos(n\pi)$$

$$= \frac{1}{n} (-1)^{n+1} \quad \text{if } n \geq 1$$

We have that the nth Fourier polynomial for h (for $n \geq 1$) is

$$H_n(x) = \frac{\pi}{4} + \sum_{i=1}^{n} \left(\frac{1}{i^2\pi} \left(\cos(i\pi) - 1 \right) \cdot \cos(ix) + \frac{(-1)^{i+1} \sin(ix)}{i} \right).$$

This can also be written as

$$H_n(x) = \frac{\pi}{4} + \sum_{i=1}^{n} \frac{(-1)^{i+1} \sin(ix)}{i} + \sum_{i=1}^{\left[\frac{n}{2}\right]} \frac{-2}{(2i-1)^2\pi} \cos((2i-1)x)$$

where $\left[\frac{n}{2}\right]$ denotes the biggest integer smaller than or equal to $\frac{n}{2}$. In particular, we have the graphs in Figure 10.3.

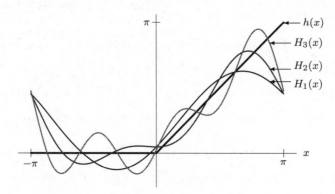

Figure 10.3

Problems

13. Since the period is 2, we make the substitution $t = \pi x - \pi$. Thus, $x = \frac{t+\pi}{\pi}$. We find the Fourier coefficients. Notice that all of the integrals are the same as in Problem 12 except for an extra factor of 2. Thus, $a_0 = 1$, $a_n = 0$, and $b_n = \frac{4}{\pi n}(-1)^{n+1}$, so:

$$G_4(t) = 1 + \frac{4}{\pi} \sin t - \frac{2}{\pi} \sin 2t + \frac{4}{3\pi} \sin 3t - \frac{1}{\pi} \sin 4t.$$

Again, we substitute back in to get a Fourier polynomial in terms of x:

$$F_4(x) = 1 + \frac{4}{\pi} \sin(\pi x - \pi) - \frac{2}{\pi} \sin(2\pi x - 2\pi)$$

$$+ \frac{4}{3\pi} \sin(3\pi x - 3\pi) - \frac{1}{\pi} \sin(4\pi x - 4\pi)$$

$$= 1 - \frac{4}{\pi} \sin(\pi x) - \frac{2}{\pi} \sin(2\pi x) - \frac{4}{3\pi} \sin(3\pi x) - \frac{1}{\pi} \sin(4\pi x).$$

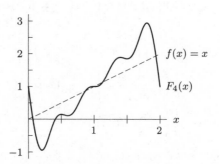

Notice in this case, the terms in our series are $\sin(n\pi x)$, not $\sin(2\pi n x)$, as in Problem 12. In general, the terms will be $\sin(n\frac{2\pi}{b}x)$, where b is the period.

17. Since each square in the graph has area $\left(\frac{\pi}{4}\right) \cdot (0.2)$,

$$a_0 = \frac{1}{2\pi} \int_{-\pi}^{\pi} f(x)\, dx$$

$$= \frac{1}{2\pi} \cdot \left(\frac{\pi}{4}\right) \cdot (0.2) \text{ [Number of squares under graph above } x\text{-axis}$$

$$\qquad\qquad - \text{ Number of squares above graph below } x \text{ axis]}$$

$$\approx \frac{1}{2\pi} \cdot \left(\frac{\pi}{4}\right) \cdot (0.2) \cdot [13 + 11 - 14] = 0.25.$$

Approximate the Fourier coefficients using Riemann sums.

$$a_1 = \frac{1}{\pi} \int_{-\pi}^{\pi} f(x) \cos x\, dx$$

$$\approx \frac{1}{\pi} \left[f(-\pi) \cos(-\pi) + f\left(-\frac{\pi}{2}\right) \cos\left(-\frac{\pi}{2}\right) + f(0) \cos(0) + f\left(\frac{\pi}{2}\right) \cos\left(\frac{\pi}{2}\right) \right] \cdot \frac{\pi}{2}$$

$$= \frac{1}{\pi} [(0.92)(-1) + (1)(0) + (-1.7)(1) + (0.7)(0)] \cdot \frac{\pi}{2}$$

$$= -1.31$$

Similarly for b_1:

$$b_1 = \frac{1}{\pi} \int_{-\pi}^{\pi} f(x) \sin x\, dx$$

$$\approx \frac{1}{\pi} \left[f(-\pi) \sin(-\pi) + f\left(-\frac{\pi}{2}\right) \sin\left(-\frac{\pi}{2}\right) + f(0) \sin(0) + f\left(\frac{\pi}{2}\right) \sin\left(\frac{\pi}{2}\right) \right] \cdot \frac{\pi}{2}$$

$$= \frac{1}{\pi} [(0.92)(0) + (1)(-1) + (-1.7)(0) + (0.7)(1)] \cdot \frac{\pi}{2}$$

$$= -0.15.$$

So our first Fourier approximation is

$$F_1(x) = 0.25 - 1.31 \cos x - 0.15 \sin x.$$

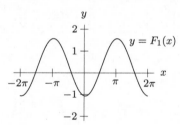

Similarly for a_2:

$$a_2 = \frac{1}{\pi} \int_{-\pi}^{\pi} f(x) \cos 2x\, dx$$

$$\approx \frac{1}{\pi} \left[f(-\pi) \cos(-2\pi) + f\left(-\frac{\pi}{2}\right) \cos(-\pi) + f(0) \cos(0) + f\left(\frac{\pi}{2}\right) \cos(-\pi) \right] \cdot \frac{\pi}{2}$$

$$= \frac{1}{\pi} [(0.92)(1) + (1)(-1) + (-1.7)(1) + (0.7)(-1)] \cdot \frac{\pi}{2}$$

$$= -1.24$$

Similarly for b_2:

$$b_2 = \frac{1}{\pi} \int_{-\pi}^{\pi} f(x) \sin 2x\, dx$$

$$\approx \frac{1}{\pi} \left[f(-\pi) \sin(-2\pi) + f\left(-\frac{\pi}{2}\right) \sin(-\pi) + f(0) \sin(0) + f\left(\frac{\pi}{2}\right) \sin(-\pi) \right] \cdot \frac{\pi}{2}$$

$$= \frac{1}{\pi} [(0.92)(0) + (1)(0) + (-1.7)(0) + (0.7)(0)] \cdot \frac{\pi}{2}$$

$$= 0.$$

So our second Fourier approximation is

$$F_2(x) = 0.25 - 1.31 \cos x - 0.15 \sin x - 1.24 \cos 2x.$$

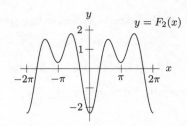

As you can see from comparing our graphs of F_1 and F_2 to the original, our estimates of the Fourier coefficients are not very accurate.

There are other methods of estimating the Fourier coefficients such as taking other Riemann sums, using Simpson's rule, and using the trapezoid rule. With each method, the greater the number of subdivisions, the more accurate the estimates of the Fourier coefficients.

The actual function graphed in the problem was

$$y = \frac{1}{4} - 1.3 \cos x - \frac{\sin\left(\frac{3}{5}\right)}{\pi} \sin x - \frac{2}{\pi} \cos 2x - \frac{\cos 1}{3\pi} \sin 2x$$

$$= 0.25 - 1.3 \cos x - 0.18 \sin x - 0.63 \cos 2x - 0.057 \sin 2x.$$

21. (a)

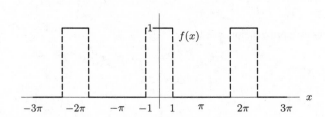

The energy of the pulse train f is

$$E = \frac{1}{\pi} \int_{-\pi}^{\pi} (f(x))^2 \, dx = \frac{1}{\pi} \int_{-1}^{1} 1^2 = \frac{1}{\pi}(1 - (-1)) = \frac{2}{\pi}.$$

Next, find the Fourier coefficients:

$$a_0 = \text{average value of } f \text{ on } [-\pi, \pi] = \frac{1}{2\pi}(\text{Area}) = \frac{1}{2\pi}(2) = \frac{1}{\pi},$$

$$a_k = \frac{1}{\pi} \int_{-\pi}^{\pi} f(x) \cos kx \, dx = \frac{1}{\pi} \int_{-1}^{1} \cos kx \, dx = \frac{1}{k\pi} \sin kx \Big|_{-1}^{1}$$

$$= \frac{1}{k\pi}(\sin k - \sin(-k)) = \frac{1}{k\pi}(2 \sin k),$$

$$b_k = \frac{1}{\pi} \int_{-\pi}^{\pi} f(x) \sin kx \, dx = \frac{1}{\pi} \int_{-1}^{1} \sin kx \, dx = -\frac{1}{k\pi} \cos kx \Big|_{-1}^{1}$$

$$= -\frac{1}{k\pi}(\cos k - \cos(-k)) = \frac{1}{k\pi}(0) = 0.$$

The energy of f contained in the constant term is

$$A_0^2 = 2a_0^2 = 2\left(\frac{1}{\pi}\right)^2 = \frac{2}{\pi^2}$$

which is

$$\frac{A_0^2}{E} = \frac{2/\pi^2}{2/\pi} = \frac{1}{\pi} \approx 0.3183 = 31.83\% \quad \text{of the total.}$$

The fraction of energy contained in the first harmonic is

$$\frac{A_1^2}{E} = \frac{a_1^2}{E} = \frac{\left(\frac{2\sin 1}{\pi}\right)^2}{\frac{2}{\pi}} \approx 0.4508 = 45.08\%.$$

The fraction of energy contained in both the constant term and the first harmonic together is

$$\frac{A_0^2}{E} + \frac{A_1^2}{E} \approx 0.7691 = 76.91\%.$$

(b) The fraction of energy contained in the second harmonic is

$$\frac{A_2^2}{E} = \frac{a_2^2}{E} = \frac{\left(\frac{\sin 2}{\pi}\right)^2}{\frac{2}{\pi}} \approx 0.1316 = 13.16\%$$

so the fraction of energy contained in the constant term and first two harmonics is

$$\frac{A_0^2}{E} + \frac{A_1^2}{E} + \frac{A_2^2}{E} \approx 0.7691 + 0.1316 = 0.9007 = 90.07\%.$$

Therefore, the constant term and the first two harmonics are needed to capture 90% of the energy of f.

(c)

$$F_3(x) = \frac{1}{\pi} + \frac{2\sin 1}{\pi}\cos x + \frac{\sin 2}{\pi}\cos 2x + \frac{2\sin 3}{3\pi}\cos 3x$$

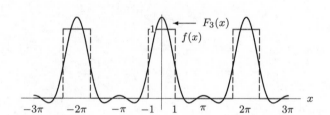

25. The easiest way to do this is to use Problem 24.

$$\int_{-\pi}^{\pi} \sin^2 mx\, dx = \int_{-\pi}^{\pi} (1 - \cos^2 mx)\, dx = \int_{-\pi}^{\pi} dx - \int_{-\pi}^{\pi} \cos^2 mx\, dx$$
$$= 2\pi - \pi \quad \text{using Problem 24}$$
$$= \pi.$$

Solutions for Chapter 10 Review

Exercises

1. $e^x \approx 1 + e(x-1) + \dfrac{e}{2}(x-1)^2$

5. $f'(x) = 3x^2 + 14x - 5$, $f''(x) = 6x + 14$, $f'''(x) = 6$. The Taylor polynomial about $x = 1$ is

$$P_3(x) = 4 + \frac{12}{1!}(x-1) + \frac{20}{2!}(x-1)^2 + \frac{6}{3!}(x-1)^3$$
$$= 4 + 12(x-1) + 10(x-1)^2 + (x-1)^3.$$

Notice that if you multiply out and collect terms in $P_3(x)$, you will get $f(x)$ back.

9. Substituting $y = t^2$ in $\sin y = y - \dfrac{y^3}{3!} + \dfrac{y^5}{5!} - \dfrac{y^7}{7!} + \cdots$ gives

$$\sin t^2 = t^2 - \frac{t^6}{3!} + \frac{t^{10}}{5!} - \frac{t^{14}}{7!} + \cdots$$

13. We use the binomial series to expand $1/\sqrt{1-z^2}$ and multiply by z^2. Since

$$\frac{1}{\sqrt{1+x}} = (1+x)^{-1/2} = 1 - \frac{1}{2}x + \frac{(-1/2)(-3/2)}{2!}x^2 + \frac{(-1/2)(-3/2)(-5/2)}{3!}x^3 + \cdots$$
$$= 1 - \frac{1}{2}x + \frac{3}{8}x^2 - \frac{5}{16}x^3 + \cdots.$$

Substituting $x = -z^2$ gives

$$\frac{z^2}{\sqrt{1-z^2}} = 1 - \frac{1}{2}(-z^2) + \frac{3}{8}(-z^2)^2 - \frac{5}{16}(-z^2)^3 + \cdots$$
$$= 1 + \frac{1}{2}z^2 + \frac{3}{8}z^4 + \frac{15}{16}z^6 + \cdots.$$

Multiplying by z^2, we have

$$\frac{z^2}{\sqrt{1-z^2}} = z^2 + \frac{1}{2}z^4 + \frac{3}{8}z^6 + \frac{15}{16}z^8 + \cdots.$$

17. Using the binomial expansion for $(1+x)^{-3/2}$ with $x = r/a$:

$$\frac{1}{(a+r)^{3/2}} = \frac{1}{\left(a + a\left(\frac{r}{a}\right)\right)^{3/2}} = \frac{1}{\left(a\left(1+\frac{r}{a}\right)\right)^{3/2}} = \frac{1}{a^{3/2}}\left(1 + \left(\frac{r}{a}\right)\right)^{-3/2}$$
$$= \frac{1}{a^{3/2}}\left(1 + (-3/2)\left(\frac{r}{a}\right) + \frac{(-3/2)(-5/2)}{2!}\left(\frac{r}{a}\right)^2 + \frac{(-3/2)(-5/2)(-7/2)}{3!}\left(\frac{r}{a}\right)^3 + \cdots\right)$$
$$= \frac{1}{a^{3/2}}\left(1 - \frac{3}{2}\left(\frac{r}{a}\right) + \frac{15}{8}\left(\frac{r}{a}\right)^2 - \frac{35}{16}\left(\frac{r}{a}\right)^3 + \cdots\right).$$

Problems

21. This is the series for e^x with $x = -2$ substituted. Thus

$$1 - 2 + \frac{4}{2!} - \frac{8}{3!} + \frac{16}{4!} + \cdots = 1 + (-2) + \frac{(-2)^2}{2!} + \frac{(-2)^3}{3!} + \frac{(-2)^4}{4!} + \cdots = e^{-2}.$$

25. Using the fact that

$$f(x) \approx P_3(x) = f(0) + f'(0)x + \frac{f''(0)}{2!}x^2 + \frac{f'''(0)}{3!}x^3$$

and identifying coefficients with those given for $P_3(x)$, we obtain the following:

(a) $f(0) =$ constant term which equals 2,   so $f(0) = 2$.
(b) $f'(0) =$ coefficient of x which equals -1,   so $f'(0) = -1$.
(c) $\frac{f''(0)}{2!} =$ coefficient of x^2 which equals $-1/3$,   so $f''(0) = -2/3$.
(d) $\frac{f'''(0)}{3!} =$ coefficient of x^3 which equals 2,   so $f'''(0) = 12$.

29. The graph in Figure 10.4 suggests that the Taylor polynomials converge to $f(x) = \dfrac{1}{1+x}$ on the interval $(-1, 1)$. The Taylor expansion is

$$f(x) = \frac{1}{1+x} = 1 - x + x^2 - x^3 + x^4 - \cdots,$$

so the ratio test gives

$$\lim_{n \to \infty} \frac{|a_{n+1}|}{|a_n|} = \lim_{n \to \infty} \frac{|(-1)^{n+1}x^{n+1}|}{|(-1)^n x^n|} = |x|.$$

Thus, the series converges if $|x| < 1$; that is $-1 < x < 1$.

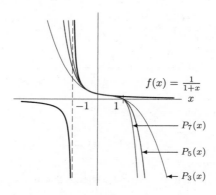

Figure 10.4

33. (a) Since $\sqrt{4 - x^2} = 2\sqrt{1 - x^2/4}$, we use the Binomial expansion

$$\sqrt{4 - x^2} \approx 2\left(1 + \frac{1}{2}\left(-\frac{x^2}{4}\right) + \frac{1}{2!}\left(\frac{1}{2}\right)\left(-\frac{1}{2}\right)\left(-\frac{x^2}{4}\right)^2\right)$$

$$= 2\left(1 - \frac{x^2}{8} - \frac{x^4}{128}\right) = 2 - \frac{x^2}{4} - \frac{x^4}{64}.$$

(b) Substituting the Taylor series in the integral gives

$$\int_0^1 \sqrt{4 - x^2}\, dx \approx \int_0^1 \left(2 - \frac{x^2}{4} - \frac{x^4}{64}\right) dx = 2x - \frac{x^3}{12} - \frac{x^5}{320}\Big|_0^1 = 1.9135.$$

(c) Since $x = 2\sin t$, we have $dx = 2\cos t\, dt$; in addition $t = 0$ when $x = 0$ and $t = \pi/6$ when $x = 1$. Thus

$$\int_0^1 \sqrt{4 - x^2}\, dx = \int_0^{\pi/6} \sqrt{4 - 4\sin^2 t} \cdot 2\cos t\, dt$$

$$= \int_0^{\pi/6} 2 \cdot 2\sqrt{1 - \sin^2 t}\cos t\, dt = 4\int_0^{\pi/6} \cos^2 t\, dt.$$

Using the table of integrals, we find

$$4\int_0^{\pi/6} \cos^2 t\, dt = 4 \cdot \frac{1}{2}(\cos t \sin t + t)\Big|_0^{\pi/6} = 2\left(\cos\frac{\pi}{6}\sin\frac{\pi}{6} + \frac{\pi}{6}\right) = \frac{\sqrt{3}}{2} + \frac{\pi}{3}.$$

(d) Using a calculator, $(\sqrt{3}/3) + (\pi/3) = 1.9132$, so the answers to parts (b) and (c) agree to three decimal places.

37. (a) To find when V takes on its minimum values, set $\frac{dV}{dr} = 0$. So

$$-V_0 \frac{d}{dr}\left(2\left(\frac{r_0}{r}\right)^6 - \left(\frac{r_0}{r}\right)^{12}\right) = 0$$

$$-V_0\left(-12r_0^6 r^{-7} + 12r_0^{12}r^{-13}\right) = 0$$

$$12r_0^6 r^{-7} = 12r_0^{12}r^{-13}$$

$$r_0^6 = r^6$$

$$r = r_0.$$

Rewriting $V'(r)$ as $\dfrac{12r_0^6 V_0}{r^7}\left(1 - \left(\dfrac{r_0}{r}\right)^6\right)$, we see that $V'(r) > 0$ for $r > r_0$ and $V'(r) < 0$ for $r < r_0$. Thus,
$V = -V_0(2(1)^6 - (1)^{12}) = -V_0$ is a minimum.
(Note: We discard the negative root $-r_0$ since the distance r must be positive.)

(b)

$$V(r) = -V_0\left(2\left(\frac{r_0}{r}\right)^6 - \left(\frac{r_0}{r}\right)^{12}\right) \qquad\qquad V(r_0) = -V_0$$

$$V'(r) = -V_0(-12r_0^6 r^{-7} + 12r_0^{12}r^{-13}) \qquad\qquad V'(r_0) = 0$$

$$V''(r) = -V_0(84r_0^6 r^{-8} - 156r_0^{12}r^{-14}) \qquad\qquad V''(r_0) = 72V_0 r_0^{-2}$$

The Taylor series is thus:

$$V(r) = -V_0 + 72V_0 r_0^{-2}\cdot(r - r_0)^2\cdot\frac{1}{2} + \cdots$$

(c) The difference between V and its minimum value $-V_0$ is

$$V - (-V_0) = 36V_0\frac{(r - r_0)^2}{r_0^2} + \cdots$$

which is approximately proportional to $(r - r_0)^2$ since terms containing higher powers of $(r - r_0)$ have relatively small values for r near r_0.

(d) From part (a) we know that $dV/dr = 0$ when $r = r_0$, hence $F = 0$ when $r = r_0$. Since, if we discard powers of $(r - r_0)$ higher than the second,

$$V(r) \approx -V_0\left(1 - 36\frac{(r - r_0)^2}{r_0^2}\right)$$

giving

$$F = -\frac{dV}{dr} \approx 72\cdot\frac{r - r_0}{r_0^2}(-V_0) = -72V_0\frac{r - r_0}{r_0^2}.$$

So F is approximately proportional to $(r - r_0)$.

41. (a) Notice $g'(0) = 0$ because g has a critical point at $x = 0$. So, for $n \geq 2$,

$$g(x) \approx P_n(x) = g(0) + \frac{g''(0)}{2!}x^2 + \frac{g'''(0)}{3!}x^3 + \cdots + \frac{g^{(n)}(0)}{n!}x^n.$$

(b) The Second Derivative test says that if $g''(0) > 0$, then 0 is a local minimum and if $g''(0) < 0$, 0 is a local maximum.

(c) Let $n = 2$. Then $P_2(x) = g(0) + \dfrac{g''(0)}{2!}x^2$. So, for x near 0,

$$g(x) - g(0) \approx \frac{g''(0)}{2!}x^2.$$

If $g''(0) > 0$, then $g(x) - g(0) \geq 0$, as long as x stays near 0. In other words, there exists a small interval around $x = 0$ such that for any x in this interval $g(x) \geq g(0)$. So $g(0)$ is a local minimum.
The case when $g''(0) < 0$ is treated similarly; then $g(0)$ is a local maximum.

45. (a) Expand $f(x)$ into its Fourier series:

$$f(x) = a_0 + a_1 \cos x + a_2 \cos 2x + a_3 \cos 3x + \cdots + a_k \cos kx + \cdots$$
$$+ b_1 \sin x + b_2 \sin 2x + b_3 \sin 3x + \cdots + b_k \sin kx + \cdots$$

Then differentiate term-by-term:

$$f'(x) = -a_1 \sin x - 2a_2 \sin 2x - 3a_3 \sin 3x - \cdots - ka_k \sin kx - \cdots$$
$$+ b_1 \cos x + 2b_2 \cos 2x + 3b_3 \cos 3x + \cdots + kb_k \cos kx + \cdots$$

Regroup terms:

$$f'(x) = +b_1 \cos x + 2b_2 \cos 2x + 3b_3 \cos 3x + \cdots + kb_k \cos kx + \cdots$$
$$- a_1 \sin x - 2a_2 \sin 2x - 3a_3 \sin 3x - \cdots - ka_k \sin kx - \cdots$$

which forms a Fourier series for the derivative $f'(x)$. The Fourier coefficient of $\cos kx$ is kb_k and the Fourier coefficient of $\sin kx$ is $-ka_k$. Note that there is no constant term as you would expect from the formula ka_k with $k = 0$. Note also that if the kth harmonic f is absent, so is that of f'.

(b) If the amplitude of the kth harmonic of f is

$$A_k = \sqrt{a_k^2 + b_k^2}, \quad k \geq 1,$$

then the amplitude of the kth harmonic of f' is

$$\sqrt{(kb_k)^2 + (-ka_k)^2} = \sqrt{k^2(b_k^2 + a_k^2)} = k\sqrt{a_k^2 + b_k^2} = kA_k.$$

(c) The energy of the kth harmonic of f' is k^2 times the energy of the kth harmonic of f.

CAS Challenge Problems

49. (a) The Taylor polynomials of degree 7 are

$$\text{For } \sin x, \qquad P_7(x) = x - \frac{x^3}{6} + \frac{x^5}{120} - \frac{x^7}{5040}$$
$$\text{For } \sin x \cos x, \qquad Q_7(x) = x - \frac{2\,x^3}{3} + \frac{2\,x^5}{15} - \frac{4\,x^7}{315}$$

(b) The coefficient of x^3 in $Q_7(x)$ is $-2/3$, and the coefficient of x^3 in $P_7(x)$ is $-1/6$, so the ratio is

$$\frac{-2/3}{-1/6} = 4.$$

The corresponding ratios for x^5 and x^7 are

$$\frac{2/15}{1/120} = 16 \quad \text{and} \quad \frac{-4/315}{-1/5040} = 64.$$

(c) It appears that the ratio is always a power of 2. For x^3, it is $4 = 2^2$; for x^5, it is $16 = 2^4$; for x^7, it is $64 = 2^6$. This suggests that in general, for the coefficient of x^n, it is 2^{n-1}.

(d) From the identity $\sin(2x) = 2 \sin x \cos x$, we expect that $P_7(2x) = 2Q_7(x)$. So, if a_n is the coefficient of x^n in $P_7(x)$, and if b_n is the coefficient of x^n in $Q_7(x)$, then, since the x^n terms $P_7(2x)$ and $2Q_7(x)$ must be equal, we have

$$a_n(2x)^n = 2b_n x^n.$$

Dividing both sides by x^n and combining the powers of 2, this gives the pattern we observed. For $a_n \neq 0$,

$$\frac{b_n}{a_n} = 2^{n-1}.$$

CHECK YOUR UNDERSTANDING

1. False. For example, both $f(x) = x^2$ and $g(x) = x^2 + x^3$ have $P_2(x) = x^2$.

5. False. The Taylor series for $\sin x$ about $x = \pi$ is calculated by taking derivatives and using the formula

$$f(a) + f'(a)(x - a) + \frac{f''(a)}{2!}(x - a)^2 + \cdots.$$

The series for $\sin x$ about $x = \pi$ turns out to be

$$-(x - \pi) + \frac{(x - \pi)^3}{3!} - \frac{(x - \pi)^5}{5!} + \cdots.$$

9. False. The derivative of $f(x)g(x)$ is not $f'(x)g'(x)$. If this statement were true, the Taylor series for $(\cos x)(\sin x)$ would have all zero terms.

13. True. For large x, the graph of $P_{10}(x)$ looks like the graph of its highest powered term, $x^{10}/10!$. But e^x grows faster than any power, so e^x gets further and further away from $x^{10}/10! \approx P_{10}(x)$.

17. True. Since f is even, $f(x)\sin(mx)$ is odd for any m, so

$$b_m = \frac{1}{\pi} \int_{-\pi}^{\pi} f(x)\sin x(mx)\, dx = 0.$$

21. False. The quadratic approximation to $f_1(x)f_2(x)$ near $x = 0$ is

$$f_1(0)f_2(0) + (f_1'(0)f_2(0) + f_1(0)f_2'(0))x + \frac{f_1''(0)f_2(0) + 2f_1'(0)f_2'(0) + f_1(0)f_2''(0)}{2}x^2.$$

On the other hand, we have
$$L_1(x) = f_1(0) + f_1'(0)x, \quad L_2(x) = f_2(0) + f_2'(0)x,$$

so

$$L_1(x)L_2(x) = (f_1(0) + f_1'(0)x)(f_2(0) + f_2'(0)x) = f_1(0)f_2(0) + (f_1'(0)f_2(0) + f_2'(0)f_1(0))x + f_1'(0)f_2'(0)x^2.$$

The first two terms of the right side agree with the quadratic approximation to $f_1(x)f_2(x)$ near $x = 0$, but the term of degree 2 does not.

For example, the linear approximation to e^x is $1 + x$, but the quadratic approximation to $(e^x)^2 = e^{2x}$ is $1 + 2x + 2x^2$, not $(1 + x)^2 = 1 + 2x + x^2$.

CHAPTER ELEVEN

Solutions for Section 11.1

Exercises

1. (a) (III) An island can only sustain the population up to a certain size. The population will grow until it reaches this limiting value.
   (b) (V) The ingot will get hot and then cool off, so the temperature will increase and then decrease.
   (c) (I) The speed of the car is constant, and then decreases linearly when the breaks are applied uniformly.
   (d) (II) Carbon-14 decays exponentially.
   (e) (IV) Tree pollen is seasonal, and therefore cyclical.

5. We know that at time $t = 0$ the value of y is 8. Since we are told that $dy/dt = 0.5y$, we know that at time $t = 0$ the derivative of y is $.5(8) = 4$. Thus as t goes from 0 to 1, y will increase by 4, so at $t = 1$, $y = 8 + 4 = 12$.
   Likewise, at $t = 1$, we get $dy/dt = 0.5(12) = 6$ so that at $t = 2$, we obtain $y = 12 + 6 = 18$.
   At $t = 2$, we have $dy/dt = 0.5(18) = 9$ so that at $t = 3$, we obtain $y = 18 + 9 = 27$.
   At $t = 3$, we have $dy/dt = 0.5(27) = 13.5$ so that at $t = 4$, we obtain $y = 27 + 13.5 = 40.5$.
   Thus we get the values in the following table

t	0	1	2	3	4
y	8	12	18	27	40.5

9. If $y = \sin 2t$, then $\frac{dy}{dt} = 2\cos 2t$, and $\frac{d^2y}{dt^2} = -4\sin 2t$.
   Thus $\frac{d^2y}{dt^2} + 4y = -4\sin 2t + 4\sin 2t = 0$.

Problems

13. (a) If $y = Cx^n$ is a solution to the given differential equation, then we must have

    $$x\frac{d(Cx^n)}{dx} - 3(Cx^n) = 0$$
    $$x(Cnx^{n-1}) - 3(Cx^n) = 0$$
    $$Cnx^n - 3Cx^n = 0$$
    $$C(n-3)x^n = 0.$$

    Thus, if $C = 0$, we get $y = 0$ is a solution, for every n. If $C \neq 0$, then $n = 3$, and so $y = Cx^3$ is a solution.
    (b) Because $y = 40$ for $x = 2$, we cannot have $C = 0$. Thus, by part (a), we get $n = 3$. The solution to the differential equation is

    $$y = Cx^3.$$

    To determine C if $y = 40$ when $x = 2$, we substitute these values into the equation.

    $$40 = C \cdot 2^3$$
    $$40 = C \cdot 8$$
    $$C = 5.$$

    So, now both C and n are fixed at specific values.

Solutions for Section 11.2

Exercises

1. There are many possible answers. One possibility is shown in Figures 11.1 and 11.2.

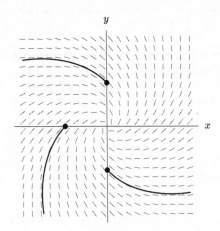

Figure 11.1

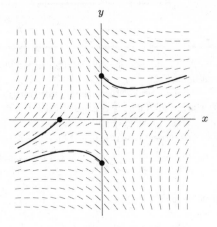

Figure 11.2

Problems

5. (a) See Figure 11.3.

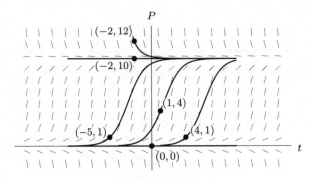

Figure 11.3

(b) If $0 < P < 10$, the solution is increasing; if $P > 10$, it is decreasing. So P tends to 10.

9. (a) II   (b) VI   (c) IV   (d) I   (e) III   (f) V

Solutions for Section 11.3

Exercises

1. (a)

Table 11.1 *Euler's method for*
$y' = x + y$ *with* $y(0) = 1$

x	y	$\Delta y =$(slope)Δx
0	1	$0.1 = (1)(0.1)$
0.1	1.1	$0.12 = (1.2)(0.1)$
0.2	1.22	$0.142 = (1.42)(0.1)$
0.3	1.362	$0.1662 = (1.662)(0.1)$
0.4	1.5282	

So $y(0.4) \approx 1.5282$.

(b)

Table 11.2 *Euler's method for*
$y' = x + y$ *with* $y(-1) = 0$

x	y	$\Delta y =$(slope)Δx
-1	0	$-0.1 = (-1)(0.1)$
-0.9	-0.1	$-0.1 = (-1)(0.1)$
-0.8	-0.2	$-0.1 = (-1)(0.1)$
-0.7	-0.3	
$\vdots$	$\vdots$	Notice that y
0	-1	decreases by 0.1
$\vdots$	$\vdots$	for every step
0.4	-1.4	

So $y(0.4) = -1.4$. (This answer is exact.)

Problems

5. (a) $\Delta x = 0.5$

Table 11.3 *Euler's method for*
$y' = 2x$, *with* $y(0) = 1$

x	y	$\Delta y =$(slope)Δx
0	1	$0 = (2 \cdot 0)(0.5)$
0.5	1	$0.5 = (2 \cdot 0.5)(0.5)$
1	1.5	

$\Delta x = 0.25$

Table 11.4 *Euler's method for*
$y' = 2x$, *with* $y(0) = 1$

x	y	$\Delta y =$(slope)Δx
0	1	$0 = (2 \cdot 0)(0.25)$
0.25	1	$0.125 = (2 \cdot 0.25)(0.25)$
0.50	1.125	$0.25 = (2 \cdot 0.5)(0.25)$
0.75	1.375	$0.375 = (2 \cdot 0.75)(0.25)$
1	1.75	

(b) General solution is $y = x^2 + C$, and $y(0) = 1$ gives $C = 1$. Thus, the solution is $y = x^2 + 1$. So the true value of y when $x = 1$ is $y = 1^2 + 1 = 2$.

(c) When $\Delta x = 0.5$, error $= 0.5$.

When $\Delta x = 0.25$, error $= 0.25$.

Thus, decreasing Δx by a factor of 2 has decreased the error by a factor of 2, as expected.

9. (a) Using one step, $\frac{\Delta B}{\Delta t} = 0.05$, so $\Delta B = \left(\frac{\Delta B}{\Delta t}\right)\Delta t = 50$. Therefore we get an approximation of $B \approx 1050$ after one year.

(b) With two steps, $\Delta t = 0.5$ and we have

Table 11.5

t	B	$\Delta B = (0.05B)\Delta t$
0	1000	25
0.5	1025	25.63
1.0	1050.63	

(c) Keeping track to the nearest hundredth with $\Delta t = 0.25$, we have

Table 11.6

t	B	$\Delta B = (0.05B)\Delta t$
0	1000	12.5
0.25	1012.5	12.66
0.5	1025.16	12.81
0.75	1037.97	12.97
1	1050.94	

(d) In part (a), we get our approximation by making a single increment, ΔB, where ΔB is just $0.05B$. If we think in terms of interest, ΔB is just like getting one end of the year interest payment. Since ΔB is 0.05 times the balance B, it is like getting 5% interest at the end of the year.

(e) Part (b) is equivalent to computing the final amount in an account that begins with $1000 and earns 5% interest compounded twice annually. Each step is like computing the interest after 6 months. When $t = 0.5$, for example, the interest is $\Delta B = (0.05B) \cdot \frac{1}{2}$, and we add this to $1000 to get the new balance.

Similarly, part (c) is equivalent to the final amount in an account that has an initial balance of $1000 and earns 5% interest compounded quarterly.

Solutions for Section 11.4

Exercises

1. Separating variables gives

$$\int \frac{1}{P}dP = -\int 2dt,$$

so

$$\ln|P| = -2t + C.$$

Therefore

$$P = \pm e^{-2t+C} = Ae^{-2t}.$$

The initial value $P(0) = 1$ gives $1 = A$, so

$$P = e^{-2t}.$$

5. Separating variables gives

$$\int \frac{dy}{y} = -\int \frac{1}{3}\, dx$$

$$\ln|y| = -\frac{1}{3}x + C.$$

Solving for y, we have

$$y = Ae^{-\frac{1}{3}x}, \text{ where } A = \pm e^C.$$

Since $y(0) = A = 10$, we have

$$y = 10e^{-\frac{1}{3}x}.$$

9. Separating variables gives

$$\int \frac{dz}{z} = \int 5\, dt$$

$$\ln|z| = 5t + C.$$

Solving for z, we have

$$z = Ae^{5t}, \text{ where } A = \pm e^C.$$

Using the fact that $z(1) = 5$, we have $z(1) = Ae^5 = 5$, so $A = 5/e^5$. Therefore,

$$z = \frac{5}{e^5}e^{5t} = 5e^{5t-5}.$$

13. Separating variables gives

$$\int \frac{dy}{y - 200} = \int 0.5 dt$$

$$\ln|y - 200| = 0.5t + C$$

$$y = 200 + Ae^{0.5t}, \quad \text{where } A = \pm e^C.$$

The initial condition, $y(0) = 50$, gives

$$50 = 200 + A, \quad \text{so} \quad A = -150.$$

Thus,

$$y = 200 - 150e^{0.5t}.$$

17. Factoring out the 0.1 gives

$$\frac{dm}{dt} = 0.1m + 200 = 0.1(m + 2000)$$

$$\int \frac{dm}{m + 2000} = \int 0.1\, dt,$$

so

$$\ln|m + 2000| = 0.1t + C,$$

and

$$m = Ae^{0.1t} - 2000, \text{ where } A = \pm e^C.$$

Using the initial condition, $m(0) = Ae^{(0.1)\cdot 0} - 2000 = 1000$, gives $A = 3000$. Thus

$$m = 3000e^{0.1t} - 2000.$$

21. Separating variables gives

$$\frac{dz}{dt} = te^z$$

$$e^{-z} dz = t dt$$

$$\int e^{-z} dz = \int t \, dt,$$

so

$$-e^{-z} = \frac{t^2}{2} + C.$$

Since the solution passes through the origin, $z = 0$ when $t = 0$, we must have

$$-e^{-0} = \frac{0}{2} + C, \text{ so } C = -1.$$

Thus

$$-e^{-z} = \frac{t^2}{2} - 1,$$

or

$$z = -\ln\left(1 - \frac{t^2}{2}\right).$$

25. Separating variables gives

$$\frac{dw}{d\theta} = \theta w^2 \sin\theta^2$$

$$\int \frac{dw}{w^2} = \int \theta \sin\theta^2 \, d\theta,$$

so

$$-\frac{1}{w} = -\frac{1}{2}\cos\theta^2 + C.$$

According to the initial conditions, $w(0) = 1$, so $-1 = -\frac{1}{2} + C$ and $C = -\frac{1}{2}$. Thus,

$$-\frac{1}{w} = -\frac{1}{2}\cos\theta^2 - \frac{1}{2}$$

$$\frac{1}{w} = \frac{\cos\theta^2 + 1}{2}$$

$$w = \frac{2}{\cos\theta^2 + 1}.$$

Problems

29. Separating variables gives

$$\int \frac{dR}{R} = \int k \, dt.$$

Integrating gives

$$\ln|R| = kt + C,$$

so

$$|R| = e^{kt+C} = e^{kt}e^C$$

$$R = Ae^{kt}, \quad \text{where } A = \pm e^C \quad \text{or} \quad A = 0.$$

33. Separating variables gives

$$\int \frac{dP}{P-a} = \int k\,dt.$$

Integrating yields

$$\ln|P-a| = kt + C,$$

so

$$P = a + Ae^{kt} \quad \text{where } A = \pm e^C \quad \text{or } A = 0.$$

37. Separating variables and integrating gives

$$\int \frac{1}{R^2 + 1}\,dR = \int a\,dx$$

or

$$\arctan R = ax + C$$

so that

$$R = \tan(ax + C).$$

41. Separating variables gives

$$\frac{dx}{dt} = \frac{x\ln x}{t},$$

so

$$\int \frac{dx}{x\ln x} = \int \frac{dt}{t},$$

and thus

$$\ln|\ln x| = \ln t + C,$$

so

$$|\ln x| = e^C e^{\ln t} = e^C t.$$

Therefore

$$\ln x = At, \quad \text{where } A = \pm e^C \quad \text{or} \quad A = 0, \quad \text{so} \quad x = e^{At}.$$

45. (a), (b)

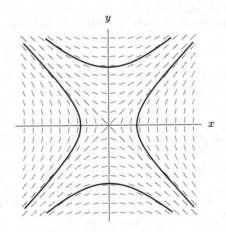

(c) Since $dy/dx = x/y$, we have

$$\int y\,dy = \int x\,dx,$$

and thus

$$\frac{y^2}{2} = \frac{x^2}{2} + C,$$

or

$$y^2 - x^2 = 2C.$$

This is the equation of the hyperbolas in part (b).

Solutions for Section 11.5

Exercises

1. (a) (I)
    (b) (IV)
    (c) (II) and (IV)
    (d) (II) and (III)

5. (a) The equilibrium solutions occur where the slope $y' = 0$, which occurs on the slope field where the lines are horizontal, or (looking at the equation) at $y = 2$ and $y = -1$. Looking at the slope field, we can see that $y = 2$ is stable, since the slopes at nearby values of y point toward it, whereas $y = -1$ is unstable.
    (b) Draw solution curves passing through the given points by starting at these points and following the flow of the slopes, as shown in Figure 11.4.

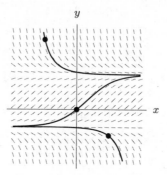

Figure 11.4

Problems

9. (a) Since we are told that the rate at which the quantity of the drug decreases is proportional to the amount of the drug left in the body, we know the differential equation modeling this situation is

$$\frac{dQ}{dt} = kQ.$$

Since we are told that the quantity of the drug is decreasing, we know that $k < 0$.
    (b) We know that the general solution to the differential equation

$$\frac{dQ}{dt} = kQ$$

is

$$Q = Ce^{kt}.$$

    (c) We are told that the half life of the drug is 3.8 hours. This means that at $t = 3.8$, the amount of the drug in the body is half the amount that was in the body at $t = 0$, or, in other words,

$$0.5Q(0) = Q(3.8).$$

Solving this equation gives

$$0.5Q(0) = Q(3.8)$$
$$0.5Ce^{k(0)} = Ce^{k(3.8)}$$
$$0.5C = Ce^{k(3.8)}$$
$$0.5 = e^{k(3.8)}$$
$$\ln(0.5) = k(3.8)$$
$$\frac{\ln(0.5)}{3.8} = k$$
$$k \approx -0.182.$$

(d) From part (c) we know that the formula for Q is

$$Q = Ce^{-0.182t}.$$

We are told that initially there are 10 mg of the drug in the body. Thus at $t = 0$, we get

$$10 = Ce^{-0.182(0)}$$

so

$$C = 10.$$

Thus our equation becomes

$$Q(t) = 10e^{-0.182t}.$$

Substituting $t = 12$, we get

$$Q(t) = 10e^{-0.182t}$$
$$Q(12) = 10e^{-0.182(12)}$$
$$= 10e^{-2.184}$$
$$Q(12) \approx 1.126 \text{ mg.}$$

13. According to Newton's Law of Cooling, the temperature, T, of the roast as a function of time, t, satisfies

$$T'(t) = k(350 - T)$$
$$T(0) = 40.$$

Solving this differential equation, we get that $T = 350 - 310e^{-kt}$ for some $k > 0$. To find k, we note that at $t = 1$ we have $T = 90$, so

$$90 = 350 - 310e^{-k(1)}$$
$$\frac{260}{310} = e^{-k}$$
$$k = -\ln\left(\frac{260}{310}\right)$$
$$\approx 0.17589.$$

Thus, $T = 350 - 310e^{-0.17589t}$. Solving for t when $T = 140$, we have

$$140 = 350 - 310e^{-0.17589t}$$
$$\frac{210}{310} = e^{-0.17589t}$$
$$t = \frac{\ln(210/310)}{-0.17589}$$
$$t \approx 2.21 \text{ hours.}$$

17. Michigan:

$$\frac{dQ}{dt} = -\frac{r}{V}Q = -\frac{158}{4.9 \times 10^3}Q \approx -0.032Q$$

so

$$Q = Q_0 e^{-0.032t}.$$

We want to find t such that

$$0.1Q_0 = Q_0 e^{-0.032t}$$

so

$$t = \frac{-\ln(0.1)}{0.032} \approx 72 \text{ years.}$$

Ontario:

$$\frac{dQ}{dt} = -\frac{r}{V}Q = \frac{-209}{1.6 \times 10^3}Q = -0.131Q$$

so

$$Q = Q_0 e^{-0.131t}.$$

We want to find t such that

$$0.1Q_0 = Q_0 e^{-0.131t}$$

so

$$t = \frac{-\ln(0.1)}{0.131} \approx 18 \text{ years.}$$

Lake Michigan will take longer because it is larger (4900 km^3 compared to 1600 km^3) and water is flowing through it at a slower rate (158 km^3/year compared to 209 km^3/year).

21. (a) $\dfrac{dT}{dt} = -k(T - A)$, where $A = 68°$F is the temperature of the room, and t is time since 9 am.

(b)

$$\int \frac{dT}{T - A} = -\int k\,dt$$
$$\ln|T - A| = -kt + C$$
$$T = A + Be^{-kt}.$$

Using $A = 68$, and $T(0) = 90.3$, we get $B = 22.3$. Thus

$$T = 68 + 22.3e^{-kt}.$$

At $t = 1$, we have

$$89.0 = 68 + 22.3e^{-k}$$
$$21 = 22.3e^{-k}$$
$$k = -\ln\frac{21}{22.3} \approx 0.06.$$

Thus $T = 68 + 22.3e^{-0.06t}$.

We want to know when T was equal to 98.6°F, the temperature of a live body, so

$$98.6 = 68 + 22.3e^{-0.06t}$$
$$\ln\frac{30.6}{22.3} = -0.06t$$
$$t = \left(-\frac{1}{0.06}\right)\ln\frac{30.6}{22.3}$$
$$t \approx -5.27.$$

The victim was killed approximately $5\frac{1}{4}$ hours prior to 9 am, at 3:45 am.

Solutions for Section 11.6

Exercises

1. (a) If $B = f(t)$, where t is in years,

$$\frac{dB}{dt} = \text{Rate of money earned from interest} + \text{Rate of money deposited}$$
$$\frac{dB}{dt} = 0.10B + 1000.$$

(b) We use separation of variables to solve the differential equation

$$\frac{dB}{dt} = 0.1B + 1000.$$

$$\int \frac{1}{0.1B + 1000} dB = \int dt$$

$$\frac{1}{0.1} \ln |0.1B + 1000| = t + C_1$$

$$0.1B + 1000 = C_2 e^{0.1t}$$

$$B = Ce^{0.1t} - 10,000$$

For $t = 0$, $B = 0$, hence $C = 10,000$. Therefore, $B = 10,000e^{0.1t} - 10,000$.

Problems

5. (a) Since the rate of change is proportional to the amount present, $dy/dt = ky$ for some constant k.

(b) Solving the differential equation, we have $y = Ae^{kt}$, where A is the initial amount. Since 100 grams become 54.9 grams in one hour, $54.9 = 100e^k$, so $k = \ln(54.9/100) \approx -0.5997$. Thus, after 10 hours, there remains $100e^{(-0.5997)10} \approx 0.2486$ grams.

9. Let the depth of the water at time t be y. Then $\dfrac{dy}{dt} = -k\sqrt{y}$, where k is a positive constant. Separating variables,

$$\int \frac{dy}{\sqrt{y}} = -\int k \, dt,$$

so

$$2\sqrt{y} = -kt + C.$$

When $t = 0$, $y = 36$; $2\sqrt{36} = -k \cdot 0 + C$, so $C = 12$.

When $t = 1$, $y = 35$; $2\sqrt{35} = -k + 12$, so $k \approx 0.17$.

Thus, $2\sqrt{y} \approx -0.17t + 12$. We are looking for t such that $y = 0$; this happens when $t \approx \frac{12}{0.17} \approx 71$ hours, or about 3 days.

13. Let $V(t)$ be the volume of water in the tank at time t, then

$$\frac{dV}{dt} = k\sqrt{V}$$

This is a separable equation which has the solution

$$V(t) = (\frac{kt}{2} + C)^2$$

Since $V(0) = 200$ this gives $200 = C^2$ so

$$V(t) = (\frac{kt}{2} + \sqrt{200})^2.$$

However, $V(1) = 180$ therefore

$$180 = (\frac{k}{2} + \sqrt{200})^2,$$

so that $k = 2\left(\sqrt{180} - \sqrt{200}\right) = -1.45146$. Therefore,

$$V(t) = (-0.726t + \sqrt{200})^2.$$

The tank will be half-empty when $V(t) = 100$, so we solve

$$100 = (-0.726t + \sqrt{200})^2$$

to obtain $t = 5.7$ days. The tank will be half empty in 5.7 days.

The volume after 4 days is $V(4)$ which is approximately 126.32 liters.

17. (a) The balance in the account at the beginning of the month is given by the following sum

$$\left(\begin{array}{c} \text{balance in} \\ \text{account} \end{array} \right) = \left(\begin{array}{c} \text{previous month's} \\ \text{balance} \end{array} \right) + \left(\begin{array}{c} \text{interest on} \\ \text{previous month's balance} \end{array} \right) + \left(\begin{array}{c} \text{monthly deposit} \\ \text{of \$100} \end{array} \right)$$

Denote month i's balance by B_i. Assuming the interest is compounded continuously, we have

$$\left(\begin{array}{c}\text{previous month's}\\\text{balance}\end{array}\right) + \left(\begin{array}{c}\text{interest on previous}\\\text{month's balance}\end{array}\right) = B_{i-1}e^{0.1/12}.$$

Since the interest rate is $10\% = 0.1$ per year, interest is $\frac{0.1}{12}$ per month. So at month i, the balance is

$$B_i = B_{i-1}e^{\frac{0.1}{12}} + 100$$

Explicitly, we have for the five years (60 months) the equations:

$$B_0 = 0$$
$$B_1 = B_0 e^{\frac{0.1}{12}} + 100$$
$$B_2 = B_1 e^{\frac{0.1}{12}} + 100$$
$$B_3 = B_2 e^{\frac{0.1}{12}} + 100$$
$$\vdots \quad \vdots$$
$$B_{60} = B_{59} e^{\frac{0.1}{12}} + 100$$

In other words,

$$B_1 = 100$$
$$B_2 = 100 e^{\frac{0.1}{12}} + 100$$
$$B_3 = (100 e^{\frac{0.1}{12}} + 100) e^{\frac{0.1}{12}} + 100$$
$$\quad = 100 e^{\frac{(0.1)2}{12}} + 100 e^{\frac{0.1}{12}} + 100$$
$$B_4 = 100 e^{\frac{(0.1)3}{12}} + 100 e^{\frac{(0.1)2}{12}} + 100 e^{\frac{(0.1)}{12}} + 100$$
$$\vdots \quad \vdots$$
$$B_{60} = 100 e^{\frac{(0.1)59}{12}} + 100 e^{\frac{(0.1)58}{12}} + \cdots + 100 e^{\frac{(0.1)1}{12}} + 100$$
$$B_{60} = \sum_{k=0}^{59} 100 e^{\frac{(0.1)k}{12}}$$

(b) The sum $B_{60} = \displaystyle\sum_{k=0}^{59} 100 e^{\frac{(0.1)k}{12}}$ can be written as $B_{60} = \displaystyle\sum_{k=0}^{59} 1200 e^{\frac{(0.1)k}{12}} (\frac{1}{12})$ which is the left Riemann sum for

$\displaystyle\int_0^5 1200 e^{0.1t} dt$, with $\Delta t = \dfrac{1}{12}$ and $N = 60$. Evaluating the sum on a calculator gives $B_{60} = 7752.26$.

(c) The situation described by this problem is almost the same as that in Problem 16, except that here the money is being deposited once a month rather than continuously; however the nominal yearly rates are the same. Thus we would expect the balance after 5 years to be approximately the same in each case. This means that the answer to part (b) of this problem should be approximately the same as the answer to part (c) to Problem 16. Since the deposits in this problem start at the end of the first month, as opposed to right away, we would expect the balance after 5 years to be slightly smaller than in Problem 16, as is the case.

Alternatively, we can use the Fundamental Theorem of Calculus to show that the integral can be computed exactly

$$\int_0^5 1200 e^{0.1t} dt = 12000(e^{(0.1)5} - 1) = 7784.66$$

Thus $\int_0^5 1200 e^{0.1t} dt$ represents the exact solution to Problem 16. Since $1200 e^{0.1t}$ is an increasing function, the left hand sum we calculated in part (b) of this problem underestimates the integral. Thus the answer to part (b) of this problem should be less than the answer to part (c) of Problem 16.

21. (a)

$$\frac{dQ}{dt} = r - \alpha Q = -\alpha(Q - \frac{r}{\alpha})$$

$$\int \frac{dQ}{Q - r/\alpha} = = -\alpha \int dt$$

$$\ln\left|Q - \frac{r}{\alpha}\right| = -\alpha t + C$$

$$Q - \frac{r}{\alpha} = Ae^{-\alpha t}$$

When $t = 0$, $Q = 0$, so $A = -\frac{r}{\alpha}$ and

$$Q = \frac{r}{\alpha}(1 - e^{-\alpha t})$$

So,

$$Q_\infty = \lim_{t \to \infty} Q = \frac{r}{\alpha}.$$

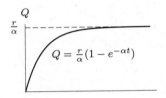

(b) Doubling r doubles Q_∞. Since $Q_\infty = r/\alpha$, the time to reach $\frac{1}{2}Q_\infty$ is obtained by solving

$$\frac{r}{2\alpha} = \frac{r}{\alpha}(1 - e^{-\alpha t})$$

$$\frac{1}{2} = 1 - e^{-\alpha t}$$

$$e^{-\alpha t} = \frac{1}{2}$$

$$t = -\frac{\ln(1/2)}{\alpha} = \frac{\ln 2}{\alpha}.$$

So altering r doesn't alter the time it takes to reach $\frac{1}{2}Q_\infty$. See Figure 11.5.

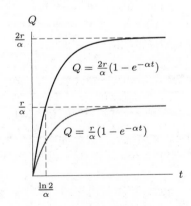

Figure 11.5

(c) Q_∞ is halved by doubling α, and so is the time, $t = \frac{\ln 2}{\alpha}$, to reach $\frac{1}{2}Q_\infty$.

25. (a) Newton's Law of Motion says that

$$\text{Force} = (\text{mass}) \times (\text{acceleration}).$$

Since acceleration, dv/dt, is measured upward and the force due to gravity acts downward,

$$-\frac{mgR^2}{(R+h)^2} = m\frac{dv}{dt}$$

so

$$\frac{dv}{dt} = -\frac{gR^2}{(R+h)^2}.$$

(b) Since $v = \frac{dh}{dt}$, the chain rule gives

$$\frac{dv}{dt} = \frac{dv}{dh} \cdot \frac{dh}{dt} = \frac{dv}{dh} \cdot v.$$

Substituting into the differential equation in part (a) gives

$$v\frac{dv}{dh} = -\frac{gR^2}{(R+h)^2}.$$

(c) Separating variables gives

$$\int v\, dv = -\int \frac{gR^2}{(R+h)^2}\, dh$$

$$\frac{v^2}{2} = \frac{gR^2}{(R+h)} + C$$

Since $v = v_0$ when $h = 0$,

$$\frac{v_0^2}{2} = \frac{gR^2}{(R+0)} + C \quad \text{gives} \quad C = \frac{v_0^2}{2} - gR,$$

so the solution is

$$\frac{v^2}{2} = \frac{gR^2}{(R+h)} + \frac{v_0^2}{2} - gR$$

$$v^2 = v_0^2 + \frac{2gR^2}{(R+h)} - 2gR$$

(d) The escape velocity v_0 ensures that $v^2 \geq 0$ for all $h \geq 0$. Since the positive quantity $\dfrac{2gR^2}{(R+h)} \to 0$ as $h \to \infty$, to ensure that $v^2 \geq 0$ for all h, we must have

$$v_0^2 \geq 2gR.$$

When $v_0^2 = 2gR$ so $v_0 = \sqrt{2gR}$, we say that v_0 is the escape velocity.

Solutions for Section 11.7

Exercises

1. A continuous growth rate of 0.2% means that

$$\frac{1}{P}\frac{dP}{dt} = 0.2\% = 0.002.$$

Separating variables and integrating gives

$$\int \frac{dP}{P} = \int 0.002\, dt$$

$$P = P_0 e^{0.002t} = (6.6 \times 10^6)e^{0.002t}.$$

Problems

5. (a) The logistic model is a reasonable one because at first very few houses have a VCR. As movie rentals become popular and as VCRs get cheaper, more people will buy VCRs. However, we know that the rate of VCR buying will start slowing down at some point as it is impossible for more than 100% of houses to have VCRs.

(b) To find the point of inflection, we must find the year at which the rate of VCR buying changes from increasing to decreasing. The following table shows the rate of change in the years from 1978 to 1990.

Year	1978	1979	1980	1981	1982	1983	1984
% Change per year	0.2	0.6	0.7	1.3	2.4	5.1	10.2
Year	1985	1986	1987	1988	1989	1990	1991
% Change per year	15.2	12.7	9.3	6.6	7.3	0	

Looking at the table, we see that the rate of percent change per year changes from increasing to decreasing in the year 1986. At this time 36% of households own VCRs giving $P = (1986, 36)$. Since at the inflection point we expect the vertical coordinate to be $L/2$, we get

$$L/2 = 36$$
$$L = 72\%.$$

Thus we expect the limiting value to be 72%. This fits in well with the data that we have for 1990 and 1991.

(c) Since the general form of a logistic equation is

$$P = \frac{L}{1 + Ce^{-kt}}$$

where L is the limiting value, we have that in our case $L = 75$ and the limiting value is 75%.

9. (a) We know that a logistic curve can be modeled by the function

$$P = \frac{L}{1 + Ce^{-kt}}$$

where $C = (L - P_0)/(P_0)$ and P is the number of people infected by the virus at a particular time t. We know that L is the limiting value, or the maximal number of people infected with the virus, so in our case

$$L = 5000.$$

We are also told that initially there are only ten people infected with the virus so that we get

$$P_0 = 10.$$

Thus we have

$$C = \frac{L - P_0}{P_0}$$
$$= \frac{5000 - 10}{10}$$
$$= 499.$$

We are also told that in the early stages of the virus, infection grows exponentially with $k = 1.78$. Thus we get that the logistic function for people infected is

$$P = \frac{5000}{1 + 499e^{-1.78t}}.$$

(b)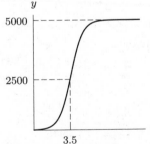

(c) Looking at the graph we see that the the point at which the rate changes from increasing to decreasing, the inflection point, occurs at roughly $t = 3.5$ giving a value of $P = 2500$. Thus after roughly 2500 people have been infected, the rate of infection starts dropping. See above.

13.

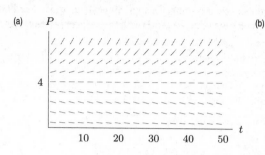

(a)

(b)

(c) There are two equilibrium values, $P = 0$, and $P = 4$. The first, representing extinction, is stable. The equilibrium value $P = 4$ is unstable because the populations increase if greater than 4, and decrease if less than 4. Notice that the equilibrium values can be obtained by setting $dP/dt = 0$:

$$\frac{dP}{dt} = 0.02P^2 - 0.08P = 0.02P(P - 4) = 0$$

so

$$P = 0 \text{ or } P = 4.$$

Solutions for Section 11.8

Exercises

1. Since

$$\frac{dS}{dt} = -aSI,$$

$$\frac{dI}{dt} = aSI - bI,$$

$$\frac{dR}{dt} = bI$$

we have

$$\frac{dS}{dt} + \frac{dI}{dt} + \frac{dR}{dt} = -aSI + aSI - bI + bI = 0.$$

Thus $\frac{d}{dt}(S + I + R) = 0$, so $S + I + R = \text{constant}$.

5. If $w = 2$ and $r = 2$, then $\frac{dw}{dt} = -2$ and $\frac{dr}{dt} = 2$, so initially the number of worms decreases and the number of robins increases. In the long run, however, the populations will oscillate; they will even go back to $w = 2$ and $r = 2$.

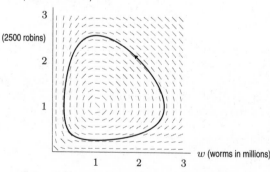

9. The numbers of robins begins to increase while the number of worms remains approximately constant. See Figure 11.6.

The numbers of robins and worms oscillate periodically between 0.2 and 3, with the robin population lagging behind the worm population.

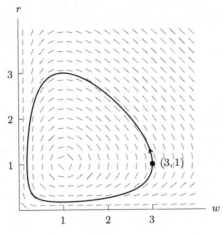

Figure 11.6

13. x decreases quickly while y increases more slowly.

Problems

17. (a) Predator-prey, because x decreases while alone, but is helped by y, whereas y increases logistically when alone, and is harmed by x. Thus x is predator, y is prey.

(b)

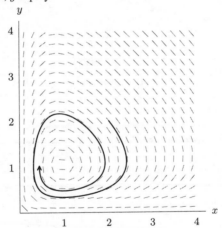

Provided neither initial population is zero, both populations tend to about 1. If x is initially zero, but y is not, then $y \rightarrow \infty$. If y is initially zero, but x is not, then $x \rightarrow 0$.

21. (a) Since the guerrillas are hard to find, the rate at which they are put out of action is proportional to the number of chance encounters between a guerrilla and a conventional soldier, which is in turn proportional to the number of guerrillas and to the number of conventional soldiers. Thus the rate at which guerrillas are put out of action is proportional to the product of the strengths of the two armies.

(b)
$$\frac{dx}{dt} = -xy$$
$$\frac{dy}{dt} = -x$$

(c) Thinking of y as a function of x and x a function of of t, then by the chain rule: $\dfrac{dy}{dt} = \dfrac{dy}{dx}\dfrac{dx}{dt}$ so:

$$\frac{dy}{dx} = \frac{dy/dt}{dx/dt} = \frac{-x}{-xy} = \frac{1}{y}$$

Separating variables:

$$\int y\,dy = \int dx$$

$$\frac{y^2}{2} = x + C$$

The value of C is determined by the initial strengths of the two armies.

(d) The sign of C determines which side wins the battle. Looking at the general solution $\dfrac{y^2}{2} = x + C$, we see that if $C > 0$ the y-intercept is at $\sqrt{2C}$, so y wins the battle by virtue of the fact that it still has troops when $x = 0$. If $C < 0$ then the curve intersects the axes at $x = -C$, so x wins the battle because it has troops when $y = 0$. If $C = 0$, then the solution goes to the point $(0, 0)$, which represents the case of mutual annihilation.

(e) We assume that an army wins if the opposing force goes to 0 first. Figure 11.7 shows that the conventional force wins if $C > 0$ and the guerrillas win if $C < 0$. Neither side wins if $C = 0$ (all soldiers on both sides are killed in this case).

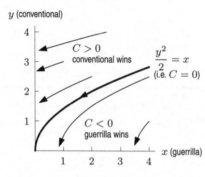

Figure 11.7

Solutions for Section 11.9

Exercises

1. (a) To find the equilibrium points we set

$$20x - 10xy = 0$$
$$25y - 5xy = 0.$$

So, $x = 0$, $y = 0$ is an equilibrium point. Another one is given by

$$10y = 20$$
$$5x = 25.$$

Therefore, $x = 5$, $y = 2$ is the other equilibrium point.

(b) At $x = 2$, $y = 4$,

$$\frac{dx}{dt} = 20x - 10xy = 40 - 80 = -40$$

$$\frac{dy}{dt} = 25y - 5xy = 100 - 40 = 60.$$

Since these are not both zero, this point is not an equilibrium point.

Problems

5. We first find the nullclines. Vertical nullclines occur where $\frac{dx}{dt} = 0$, which happens when $x = 0$ or $y = \frac{1}{3}(2 - x)$. Horizontal nullclines occur where $\frac{dy}{dt} = y(1 - 2x) = 0$, which happens when $y = 0$ or $x = \frac{1}{2}$. These nullclines are shown in Figure 11.8.

Equilibrium points (also shown in Figure 11.8) occur at the intersections of vertical and horizontal nullclines. There are three such points for this system of equations; $(0, 0)$, $(\frac{1}{2}, \frac{1}{2})$ and $(2, 0)$.

The nullclines divide the positive quadrant into four regions as shown in Figure 11.8. Trajectory directions for these regions are shown in Figure 11.9.

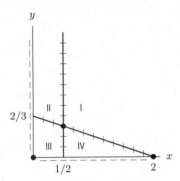

Figure 11.8: Nullclines and equilibrium points (dots)

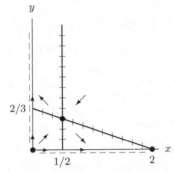

Figure 11.9: General directions of trajectories and equilibrium points (dots)

9. We assume that x, $y \geq 0$ and then find the nullclines. $\frac{dx}{dt} = x(1 - \frac{x}{2} - y) = 0$ when $x = 0$ or $y + \frac{x}{2} = 1$. $\frac{dy}{dt} = y(1 - \frac{y}{3} - x) = 0$ when $y = 0$ or $x + \frac{y}{3} = 1$.
We find the equilibrium points. They are $(2, 0)$, $(0, 3)$, $(0, 0)$, and $(\frac{4}{5}, \frac{3}{5})$. The nullclines and equilibrium points are shown in Figure 11.10.

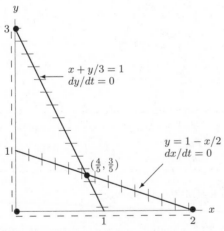

Figure 11.10: Nullclines and equilibrium points (dots)

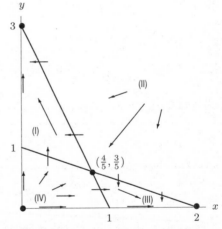

Figure 11.11: General directions of trajectories and equilibrium points (dots)

Figure 11.11 shows that if the initial point is in sector (I), the trajectory heads toward the equilibrium point $(0, 3)$. Similarly, if the trajectory begins in sector (III), then it heads toward the equilibrium $(2, 0)$ over time. If the trajectory begins in sector (II) or (IV), it can go to any of the three equilibrium points $(2, 0)$, $(0, 3)$, or $(\frac{4}{5}, \frac{3}{5})$.

13. (a)

$$\frac{dx}{dt} = 0 \text{ when } x = \frac{10.5}{0.45} = 23.3$$

$$\frac{dy}{dt} = 0 \text{ when } 8.2x - 0.8y - 142 = 0$$

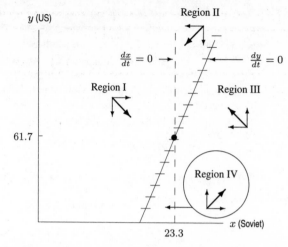

Figure 11.12: Nullclines and equilibrium point (dot) for US-Soviet arms race

There is an equilibrium point where the trajectories cross at $x = 23.3$, $y = 61.7$

In region I, $\dfrac{dx}{dt} > 0$, $\dfrac{dy}{dt} < 0$.

In region II, $\dfrac{dx}{dt} < 0$, $\dfrac{dy}{dt} < 0$.

In region III, $\dfrac{dx}{dt} < 0$, $\dfrac{dy}{dt} > 0$.

In region IV, $\dfrac{dx}{dt} > 0$, $\dfrac{dy}{dt} > 0$.

(b)

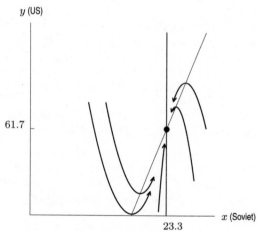

Figure 11.13: Trajectories for US-Soviet arms race.

(c) All the trajectories tend toward the equilibrium point $x = 23.3$, $y = 61.7$. Thus the model predicts that in the long run the arms race will level off with the Soviet Union spending 23.3 billion dollars a year on arms and the US 61.7 billion dollars.

(d) As the model predicts, yearly arms expenditure did tend toward 23 billion for the Soviet Union and 62 billion for the US.

Solutions for Section 11.10

Exercises

1. If $y = 2\cos t + 3\sin t$, then $y' = -2\sin t + 3\cos t$ and $y'' = -2\cos t - 3\sin t$. Thus, $y'' + y = 0$.

5. If $y(t) = A\sin(\omega t) + B\cos(\omega t)$ then

$$y' = \omega A\cos(\omega t) - \omega B\sin(\omega t)$$
$$y'' = -\omega^2 A\sin(\omega t) - \omega^2 B\cos(\omega t)$$

therefore

$$y'' + \omega^2 y = -\omega^2 A\sin(\omega t) - \omega^2 B\cos(2t) + \omega^2(A\sin(\omega t) + B\cos(\omega t)) = 0$$

for all values of A and B, so the given function is a solution.

9. The amplitude is $\sqrt{3^2 + 7^2} = \sqrt{58}$.

Problems

13. First, we note that the solutions of:
(a) $x'' + x = 0$ are $x = A\cos t + B\sin t$;
(b) $x'' + 4x = 0$ are $x = A\cos 2t + B\sin 2t$;
(c) $x'' + 16x = 0$ are $x = A\cos 4t + B\sin 4t$.
This follows from what we know about the general solution to $x'' + \omega^2 x = 0$.
The period of the solutions to (a) is 2π, the period of the solutions to (b) is π, and the period of the solutions of (c) is $\frac{\pi}{2}$. Since the t-scales are the same on all of the graphs, we see that graphs (I) and (IV) have the same period, which is twice the period of graph (III). Graph (II) has twice the period of graphs (I) and (IV). Since each graph represents a solution, we have the following:

- equation (a) goes with graph (II)
  equation (b) goes with graphs (I) and (IV)
  equation (c) goes with graph (III)
- The graph of (I) passes through $(0, 0)$, so $0 = A\cos 0 + B\sin 0 = A$. Thus, the equation is $x = B\sin 2t$. Since the amplitude is 2, we see that $x = 2\sin 2t$ is the equation of the graph. Similarly, the equation for (IV) is $x = -3\sin 2t$. The graph of (II) also passes through $(0, 0)$, so, similarly, the equation must be $x = B\sin t$. In this case, we see that $B = -1$, so $x = -\sin t$.
  Finally, the graph of (III) passes through $(0, 1)$, and 1 is the maximum value. Thus, $1 = A\cos 0 + B\sin 0$, so $A = 1$. Since it reaches a local maximum at $(0, 1)$, $x'(0) = 0 = -4A\sin 0 + 4B\cos 0$, so $B = 0$. Thus, the solution is $x = \cos 4t$.

17. At $t = 0$, we find that $y = -1$, which is clearly the lowest point on the path. Since $y' = 3\sin 3t$, we see that $y' = 0$ when $t = 0$. Thus, at $t = 0$ the object is at rest, although it will move up after $t = 0$.

21. (a) Since a mass of 3 kg stretches the spring by 2 cm, the spring constant k is given by

$$3g = 2k \quad \text{so} \quad k = \frac{3g}{2}.$$

See Figure 11.14.

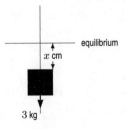

Figure 11.14

Suppose we measure the displacement x from the equilibrium; then, using

$$\text{Mass} \cdot \text{Acceleration} = \text{Force}$$

gives

$$3x'' = -kx = -\frac{3gx}{2}$$

$$x'' + \frac{g}{2}x = 0$$

Since at time $t = 0$, the brick is 5 cm below the equilibrium and not moving, the initial conditions are $x(0) = 5$ and $x'(0) = 0$.

(b) The solution to the differential equation is

$$x = A\cos\left(\sqrt{\frac{g}{2}}t\right) + B\sin\left(\sqrt{\frac{g}{2}}t\right).$$

Since $x(0) = 5$, we have

$$x = A\cos(0) + B\sin(0) = 5 \quad \text{so} \quad A = 5.$$

In addition,

$$x'(t) = -5\sqrt{\frac{g}{2}}\sin\left(\sqrt{\frac{g}{2}}t\right) + B\sqrt{\frac{g}{2}}\cos\left(\sqrt{\frac{g}{2}}t\right)$$

so

$$x'(0) = -5\sqrt{\frac{g}{2}}\sin(0) + B\sqrt{\frac{g}{2}}\cos(0) = 0 \quad \text{so} \quad B = 0.$$

Thus,

$$x = 5\cos\sqrt{\frac{g}{2}}t.$$

25. (a) $36\dfrac{d^2Q}{dt^2} + \dfrac{Q}{9} = 0$ so $\dfrac{d^2Q}{dt^2} = -\dfrac{Q}{324}.$
Thus,

$$Q = C_1\cos\frac{1}{18}t + C_2\sin\frac{1}{18}t.$$
$$Q(0) = 0 = C_1\cos 0 + C_2\sin 0 = C_1,$$
$$\text{so} \quad C_1 = 0.$$

So, $Q = C_2\sin\dfrac{1}{18}t$, and

$$Q' = I = \frac{1}{18}C_2\cos\frac{1}{18}t.$$
$$Q'(0) = I(0) = 2 = \frac{1}{18}C_2\cos\left(\frac{1}{18}\cdot 0\right) = \frac{1}{18}C_2,$$
$$\text{so} \quad C_2 = 36.$$

Therefore, $Q = 36\sin\dfrac{1}{18}t.$

(b) As in part (a), $Q = C_1\cos\dfrac{1}{18}t + C_2\sin\dfrac{1}{18}t.$
According to the initial conditions:

$$Q(0) = 6 = C_1\cos 0 + C_2\sin 0 = C_1,$$
$$\text{so} \quad C_1 = 6.$$

So $Q = 6\cos\dfrac{1}{18}t + C_2\sin\dfrac{1}{18}t.$
Thus,

$$Q' = I = -\frac{1}{3}\sin\frac{1}{18}t + \frac{1}{18}C_2\cos\frac{1}{18}t.$$
$$Q'(0) = I(0) = 0 = -\frac{1}{3}\sin\left(\frac{1}{18}\cdot 0\right) + \frac{1}{18}C_2\cos\left(\frac{1}{18}\cdot 0\right) = \frac{1}{18}C_2,$$
$$\text{so} \quad C_2 = 0.$$

Therefore, $Q = 6\cos\dfrac{1}{18}t.$

Solutions for Section 11.11

Exercises

1. The characteristic equation is $r^2 + 4r + 3 = 0$, so $r = -1$ or -3.
Therefore $y(t) = C_1 e^{-t} + C_2 e^{-3t}$.

5. The characteristic equation is $r^2 + 7 = 0$, so $r = \pm\sqrt{7}i$.
Therefore $s(t) = C_1 \cos\sqrt{7}t + C_2 \sin\sqrt{7}t$.

9. The characteristic equation is $r^2 + r + 1 = 0$, so $r = -\frac{1}{2} \pm \frac{\sqrt{3}}{2}i$.
Therefore $p(t) = C_1 e^{-t/2} \cos\frac{\sqrt{3}}{2}t + C_2 e^{-t/2} \sin\frac{\sqrt{3}}{2}t$.

13. The characteristic equation is

$$r^2 + 5r + 6 = 0$$

which has the solutions $r = -2$ and $r = -3$ so that

$$y(t) = Ae^{-3t} + Be^{-2t}$$

The initial condition $y(0) = 1$ gives

$$A + B = 1$$

and $y'(0) = 0$ gives

$$-3A - 2B = 0$$

so that $A = -2$ and $B = 3$ and

$$y(t) = -2e^{-3t} + 3e^{-2t}$$

17. The characteristic equation is $r^2 + 6r + 5 = 0$, so $r = -1$ or -5.
Therefore $y(t) = C_1 e^{-t} + C_2 e^{-5t}$.
$y'(t) = -C_1 e^{-t} - 5C_2 e^{-5t}$
$y'(0) = 0 = -C_1 - 5C_2$
$y(0) = 1 = C_1 + C_2$
    Therefore $C_2 = -1/4$, $C_1 = 5/4$ and $y(t) = \frac{5}{4}e^{-t} - \frac{1}{4}e^{-5t}$.

21. The characteristic equation is

$$r^2 + 5r + 6 = 0$$

which has the solutions $r = -2$ and $r = -3$ so that

$$y(t) = Ae^{-2t} + Be^{-3t}$$

The initial condition $y(0) = 1$ gives

$$A + B = 1$$

and $y(1) = 0$ gives

$$Ae^{-2} + Be^{-3} = 0$$

so that $A = \dfrac{1}{1-e}$ and $B = -\dfrac{e}{1-e}$ and

$$y(t) = \frac{1}{1-e}e^{-2t} + \frac{-e}{1-e}e^{-3t}$$

Problems

25. (a) $x'' + 4x = 0$ represents an undamped oscillator, and so goes with (IV).

(b) $x'' - 4x = 0$ has characteristic equation $r^2 - 4 = 0$ and so $r = \pm 2$. The solution is $C_1 e^{-2t} + C_2 e^{2t}$. This represents non-oscillating motion, so it goes with (II).

(c) $x'' - 0.2x' + 1.01x = 0$ has characteristic equation $r^2 - 0.2 + 1.01 = 0$ so $b^2 - 4ac = 0.04 - 4.04 = -4$, and $r = 0.1 \pm i$. So the solution is

$$C_1 e^{(0.1+i)t} + C_2 e^{(0.1-i)t} = e^{0.1t}(A\sin t + B\cos t).$$

The negative coefficient in the x' term represents an amplifying force. This is reflected in the solution by $e^{0.1t}$, which increases as t increases, so this goes with (I).

(d) $x'' + 0.2x' + 1.01x$ has characteristic equation $r^2 + 0.2r + 1.01 = 0$ so $b^2 - 4ac = -4$. This represents a damped oscillator. We have $r = -0.1 \pm i$ and so the solution is $x = e^{-0.1t}(A\sin t + B\cos t)$, which goes with (III).

29. Recall that $s'' + bs' + cs = 0$ is overdamped if the discriminant $b^2 - 4c > 0$, critically damped if $b^2 - 4c = 0$, and underdamped if $b^2 - 4c < 0$. Since $b^2 - 4c = b^2 - 20$, the solution is overdamped if $b > 2\sqrt{5}$ or $b < -2\sqrt{5}$, critically damped if $b = \pm 2\sqrt{5}$, and underdamped if $-2\sqrt{5} < b < 2\sqrt{5}$.

33. The frictional force is $F_{\text{drag}} = -c\frac{ds}{dt}$. Thus spring (iv) has the smallest frictional force.

37. (a) If $r_1 = \frac{-b - \sqrt{b^2 - 4c}}{2}$ then $r_1 < 0$ since both b and $\sqrt{b^2 - 4c}$ are positive.

If $r_2 = \frac{-b + \sqrt{b^2 - 4c}}{2}$, then $r_2 < 0$ because

$$b = \sqrt{b^2} > \sqrt{b^2 - 4c}.$$

(b) The general solution to the differential equation is of the form

$$y = C_1 e^{r_1 t} + C_2 e^{r_2 t}$$

and since r_1 and r_2 are both negative, y must go to 0 as $t \to \infty$.

41. In this case, the differential equation describing the charge is $Q'' + Q' + \frac{1}{4}Q = 0$, so the characteristic equation is $r^2 + r + \frac{1}{4} = 0$. This equation has one root, $r = -\frac{1}{2}$, so the equation for charge is

$$Q(t) = (C_1 + C_2 t)e^{-\frac{1}{2}t},$$
$$Q'(t) = -\frac{1}{2}(C_1 + C_2 t)e^{-\frac{1}{2}t} + C_2 e^{-\frac{1}{2}t}$$
$$= \left(C_2 - \frac{C_1}{2} - \frac{C_2 t}{2} \right) e^{-\frac{1}{2}t}.$$

(a) We have

$$Q(0) = C_1 = 0,$$
$$Q'(0) = C_2 - \frac{C_1}{2} = 2.$$

Thus, $C_1 = 0$, $C_2 = 2$, and

$$Q(t) = 2te^{-\frac{1}{2}t}.$$

(b) We have

$$Q(0) = C_1 = 2,$$
$$Q'(0) = C_2 - \frac{C_1}{2} = 0.$$

Thus, $C_1 = 2$, $C_2 = 1$, and

$$Q(t) = (2 + t)e^{-\frac{1}{2}t}.$$

(c) The resistance was decreased by exactly the amount to switch the circuit from the overdamped case to the critically damped case. Comparing the solutions of parts (a) and (b) in Problems 40, we find that in the critically damped case the net charge goes to 0 much faster as $t \to \infty$.

Solutions for Chapter 11 Review

Exercises

1. This equation is separable, so we integrate, giving

$$\int dP = \int t \, dt$$

so

$$P(t) = \frac{t^2}{2} + C.$$

5. This equation is separable, so we integrate, using the table of integrals or partial fractions, to get

$$\int \frac{1}{R - 3R^2} \, dR = 2 \int dt$$

$$\int \frac{1}{R} \, dR + \int \frac{3}{1 - 3R} \, dR = 2 \int dt$$

so

$$\ln |R| - \ln |1 - 3R| = 2t + C$$

$$\ln \left| \frac{R}{1 - 3R} \right| = 2t + C$$

$$\frac{R}{1 - 3R} = Ae^{2t}$$

$$R = \frac{Ae^{2t}}{1 + 3Ae^{2t}}.$$

9. $1 + y^2 - \frac{dy}{dx} = 0$ gives $\frac{dy}{dx} = y^2 + 1$, so $\int \frac{dy}{1+y^2} = \int dx$ and $\arctan y = x + C$. Since $y(0) = 0$ we have $C = 0$, giving $y = \tan x$.

13. $\frac{dy}{dx} = \frac{0.2y(18+0.1x)}{x(100+0.5y)}$ giving $\int \frac{(100+0.5y)}{0.2y} \, dy = \int \frac{18+0.1x}{x} \, dx$, so

$$\int \left(\frac{500}{y} + \frac{5}{2} \right) dy = \int \left(\frac{18}{x} + \frac{1}{10} \right) dx.$$

Therefore, $500 \ln |y| + \frac{5}{2} y = 18 \ln |x| + \frac{1}{10} x + C$. Since the curve passes through (10,10), $500 \ln 10 + 25 = 18 \ln 10 + 1 + C$, so $C = 482 \ln 10 + 24$. Thus, the solution is

$$500 \ln |y| + \frac{5}{2} y = 18 \ln |x| + \frac{1}{10} x + 482 \ln 10 + 24.$$

We cannot solve for y in terms of x, so we leave the answer in this form.

17. $\frac{df}{dx} = \sqrt{x f(x)}$ gives $\int \frac{df}{\sqrt{f(x)}} = \int \sqrt{x} \, dx$, so $2\sqrt{f(x)} = \frac{2}{3} x^{\frac{3}{2}} + C$. Since $f(1) = 1$, we have $2 = \frac{2}{3} + C$ so $C = \frac{4}{3}$. Thus, $2\sqrt{f(x)} = \frac{2}{3} x^{\frac{3}{2}} + \frac{4}{3}$, so $f(x) = (\frac{1}{3} x^{\frac{3}{2}} + \frac{2}{3})^2$. (Note: this is only defined for $x \geq 0$.)

21. $(1+t^2)y\frac{dy}{dt} = 1 - y$ implies that $\int \frac{y \, dy}{1-y} = \int \frac{dt}{1+t^2}$ implies that $\int \left(-1 + \frac{1}{1-y} \right) dy = \int \frac{dt}{1+t^2}$. Therefore $-y - \ln |1-y| = \arctan t + C$. $y(1) = 0$, so $0 = \arctan 1 + C$, and $C = -\frac{\pi}{4}$, so $-y - \ln |1 - y| = \arctan t - \frac{\pi}{4}$. We cannot solve for y in terms of t.

25. The characteristic equation of $9z'' + z = 0$ is

$$9r^2 + 1 = 0$$

If we write this in the form $r^2 + br + c = 0$, we have that $r^2 + 1/9 = 0$ and

$$b^2 - 4c = 0 - (4)(1/9) = -4/9 < 0$$

This indicates underdamped motion and since the roots of the characteristic equation are $r = \pm \frac{1}{3} i$, the general equation is

$$y(t) = C_1 \cos \left(\frac{1}{3} t \right) + C_2 \sin \left(\frac{1}{3} t \right)$$

Problems

29. Figure (I) shows a line segment at $(4, 0)$ with positive slope. The only possible differential equation is (b), where $y'(4, 0) = \cos 0 = 1$. Note that (a) is not possible as $y'(4, 0) = e^{-16} = 0.0000001$, a much smaller positive slope than that shown.

Figure (II) shows a line segment at $(0, 4)$ with zero slope. The possible differential equations are (d), where $y'(0, 4) = 4(4 - 4) = 0$, and (f), where $y'(0, 4) = 0(3 - 0) = 0$.

Figure (III) shows a line segment at $(4, 0)$ with negative slope of large magnitude. The only possible differential equation is (f), where $y'(4, 0) = 4(3 - 4) = -4$. Note that (c) is not possible as $y'(4, 0) = \cos(4 - 0) = -0.65$, a negative slope of smaller magnitude than that shown.

Figure (IV) shows a line segment at $(4, 0)$ with a negative slope of small magnitude. The only possible differential equation is (c), where $y'(4, 0) = \cos(4 - 0) = -0.65$. Note that (f) is not possible as $y'(4, 0) = 4(3 - 4) = -4$, a negative slope of larger magnitude than that shown.

Figure (V) shows a line segment at $(0, 4)$ with positive slope. Possible differential equations are (a), where $y'(0, 4) = e^{0^2} = 1$, and (c), where $y'(0, 4) = \cos(4 - 4) = 1$.

Figure (VI) shows a line segment at $(0, 4)$ with a negative slope of large magnitude. The only possible differential equation is (e), where $y'(0, 4) = 4(3 - 4) = -4$. Note that (b) is not possible as $y'(0, 4) = \cos 4 = -0.65$, a negative slope of smaller magnitude than that shown.

33. (a) $\frac{dB}{dt} = \frac{r}{100} B$. The constant of proportionality is $\frac{r}{100}$.

(b) Solving, we have

$$\frac{dB}{B} = \frac{r\, dt}{100}$$

$$\int \frac{dB}{B} = \int \frac{r}{100}\, dt$$

$$\ln |B| = \frac{r}{100} t + C$$

$$B = e^{(r/100)t + C} = Ae^{(r/100)t}, \qquad A = e^C.$$

A is the initial amount in the account, since A is the amount at time $t = 0$.

(c)

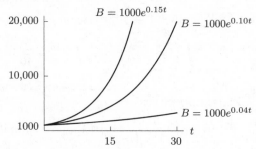

37. Quantity of A left at time $t = $ Quantity of B left at time t equals $(a - x)$.
Thus

$$\text{Rate of formation of } C = k(\text{Quantity of } A)(\text{Quantity of } B)$$

gives

$$\frac{dx}{dt} = k(a - x)(a - x) = k(a - x)^2.$$

Separating gives

$$\int \frac{dx}{(x - a)^2} = \int k\, dt$$

Integrating gives, for some constant K,

$$-(x - a)^{-1} = kt + K.$$

When $t = 0$, $x = 0$ so $K = a^{-1}$. Solving for x:

$$-(x - a)^{-1} = kt + a^{-1}$$

$$x - a = -\frac{1}{kt + a^{-1}}$$

$$x = a - \frac{a}{akt + 1} = \frac{a^2 kt}{akt + 1}$$

41. (a) $\frac{dp}{dt} = kp(B - p)$, where $k > 0$.

  (b) To find when $\frac{dp}{dt}$ is largest, we notice that $\frac{dp}{dt} = kp(B - p)$, as a function of p, is a parabola opening downward with the maximum at $p = \frac{B}{2}$, i.e. when $\frac{1}{2}$ the tin has turned to powder. This is the time when the tin is crumbling fastest.

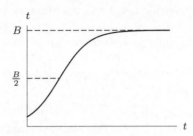

  (c) If $p = 0$ initially, then $\frac{dp}{dt} = 0$, so we would expect p to remain 0 forever. However, since many organ pipes get tin pest, we must reconcile the model with reality. There are two possible ideas which solve this problem. First, we could assume that p is never 0. In other words, we assume that all tin pipes, no matter how new, must contain some small amount of tin pest. Assuming this means that all organ pipes must deteriorate due to tin pest eventually. Another explanation is that the powder forms at a slow rate even if there was none present to begin with. Since not all organ pipes suffer, it is possible that the conversion is catalyzed by some other impurities not present in all pipes.

CAS Challenge Problems

45. (a) We find the equilibrium solutions by setting $dP/dt = 0$, that is, $P(P - 1)(2 - P) = 0$, which gives three solutions, $P = 0$, $P = 1$, and $P = 2$.

  (b) To get your computer algebra system to check that P_1 and P_2 are solutions, substitute one of them into the equation and form an expression consisting of the difference between the right and left hand sides, then ask the CAS to simplify that expression. Do the same for the other function. In order to avoid too much typing, define P_1 and P_2 as functions in your system.

  (c) Substituting $t = 0$ gives

$$P_1(0) = 1 - \frac{1}{\sqrt{4}} = 1/2$$

$$P_2(0) = 1 + \frac{1}{\sqrt{4}} = 3/2.$$

  We can find the limits using a computer algebra system. Alternatively, setting $u = e^t$, we can use the limit laws to calculate

$$\lim_{t \to \infty} \frac{e^t}{\sqrt{3 + e^{2t}}} = \lim_{u \to \infty} \frac{u}{\sqrt{3 + u^2}} = \lim_{u \to \infty} \sqrt{\frac{u^2}{3 + u^2}}$$

$$= \sqrt{\lim_{u \to \infty} \frac{u^2}{3 + u^2}} = \sqrt{\lim_{u \to \infty} \frac{1}{\frac{3}{u^2} + 1}}$$

$$= \sqrt{\frac{1}{\lim_{u \to \infty} \frac{3}{u^2} + 1}} = \sqrt{\frac{1}{0 + 1}} = 1.$$

  Therefore, we have

$$\lim_{t \to \infty} P_1(t) = 1 - 1 = 0$$

$$\lim_{t \to \infty} P_2(t) = 1 + 1 = 2.$$

  To predict these limits without having a formula for P, looking at the original differential equation. We see if $0 < P < 1$, then $P(P - 1)(2 - P) < 0$, so $P' < 0$. Thus, if $0 < P(0) < 1$, then $P'(0) < 0$, so P is initially decreasing, and tends toward the equilibrium solution $P = 0$. On the other hand, if $1 < P < 2$, then $P(P - 1)(2 - P) > 0$, so $P' > 0$. So, if $1 < P(0) < 2$, then $P'(0) > 0$, so P is initially increasing and tends toward the equilibrium solution $P = 2$.

CHECK YOUR UNDERSTANDING

1. True. The general solution to $y' = -ky$ is $y = Ce^{-kt}$.

5. True. No matter what initial value you pick, the solution curve has the x-axis as an asymptote.

9. True. Rewrite the equation as $dy/dx = xy + x = x(y + 1)$. Since the equation now has the form $dy/dx = f(x)g(y)$, it can be solved by separation of variables.

13. False. This is a logistic equation with equilibrium values $P = 0$ and $P = 2$. Solution curves do not cross the line $P = 2$ and do not go from $(0, 1)$ to $(1, 3)$.

17. True. Since $f'(x) = g(x)$, we have $f''(x) = g'(x)$. Since $g(x)$ is increasing, $g'(x) > 0$ for all x, so $f''(x) > 0$ for all x. Thus the graph of f is concave up for all x.

21. False. Let $g(x) = 0$ for all x and let $f(x) = 17$. Then $f'(x) = g(x)$ and $\lim_{x\to\infty} g(x) = 0$, but $\lim_{x\to\infty} f(x) = 17$.

25. True. The slope of the graph of f is $dy/dx = 2x - y$. Thus when $x = a$ and $y = b$, the slope is $2a - b$.

29. False. Since $f'(1) = 2(1) - 5 = -3$, the point $(1, 5)$ could not be a critical point of f.

33. True. We will use the hint. Let $w = g(x) - f(x)$. Then:

$$\frac{dw}{dx} = g'(x) - f'(x) = (2x - g(x)) - (2x - f(x)) = f(x) - g(x) = -w.$$

Thus $dw/dx = -w$. This equation is the equation for exponential decay and has the general solution $w = Ce^{-x}$. Thus,

$$\lim_{x\to\infty} (g(x) - f(x)) = \lim_{x\to\infty} Ce^{-x} = 0.$$

37. If we differentiate implicitly the equation for the family, we get $2x - 2y\,dy/dx = 0$. When we solve, we get the differential equation we want $dy/dx = x/y$.

APPENDIX

Solutions for Section A

1. The graph is

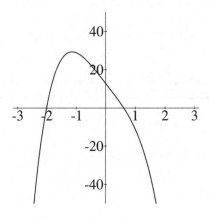

   (a) The range appears to be $y \leq 30$.
   (b) The function has two zeros.

5. The largest root is at about 2.5.

9. Using a graphing calculator, we see that when x is around 0.45, the graphs intersect.

13. (a) Only one real zero, at about $x = -1.15$.
    (b) Three real zeros: at $x = 1$, and at about $x = 1.41$ and $x = -1.41$.

17. (a) Since f is continuous, there must be one zero between $\theta = 1.4$ and $\theta = 1.6$, and another between $\theta = 1.6$ and $\theta = 1.8$. These are the only clear cases. We might also want to investigate the interval $0.6 \leq \theta \leq 0.8$ since $f(\theta)$ takes on values close to zero on at least part of this interval. Now, $\theta = 0.7$ is in this interval, and $f(0.7) = -0.01 < 0$, so f changes sign twice between $\theta = 0.6$ and $\theta = 0.8$ and hence has two zeros on this interval (assuming f is not *really* wiggly here, which it's not). There are a total of 4 zeros.
    (b) As an example, we find the zero of f between $\theta = 0.6$ and $\theta = 0.7$. $f(0.65)$ is positive; $f(0.66)$ is negative. So this zero is contained in $[0.65, 0.66]$. The other zeros are contained in the intervals $[0.72, 0.73]$, $[1.43, 1.44]$, and $[1.7, 1.71]$.
    (c) You've found all the zeros. A picture will confirm this; see Figure A.1.

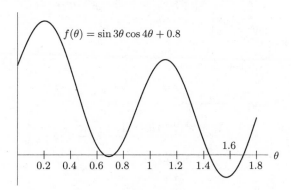

$f(\theta) = \sin 3\theta \cos 4\theta + 0.8$

Figure A.1

21.

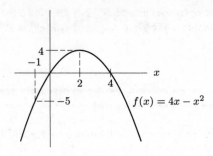

Bounded and $-5 \le f(x) \le 4$.

Solutions for Section B

1. $2e^{i\pi/2}$

5. $0e^{i\theta}$, for any θ.

9. $-3 - 4i$

13. $\frac{1}{4} - \frac{9i}{8}$

17. $5^3(\cos \frac{3\pi}{2} + i \sin \frac{3\pi}{2}) = -125i$

21. One value of $\sqrt[3]{i}$ is $\sqrt[3]{e^{i\frac{\pi}{2}}} = (e^{i\frac{\pi}{2}})^{\frac{1}{3}} = e^{i\frac{\pi}{6}} = \cos \frac{\pi}{6} + i \sin \frac{\pi}{6} = \frac{\sqrt{3}}{2} + \frac{i}{2}$

25. One value of $(-4+4i)^{2/3}$ is $[\sqrt{32}e^{(i3\pi/4)}]^{(2/3)} = (\sqrt{32})^{2/3}e^{(i\pi/2)} = 2^{5/3}\cos\frac{\pi}{2} + i2^{5/3}\sin\frac{\pi}{2} = 2i\sqrt[3]{4}$

29. We have

$$i^{-1} = \frac{1}{i} = \frac{1}{i} \cdot \frac{i}{i} = -i,$$

$$i^{-2} = \frac{1}{i^2} = -1,$$

$$i^{-3} = \frac{1}{i^3} = \frac{1}{-i} \cdot \frac{i}{i} = i,$$

$$i^{-4} = \frac{1}{i^4} = 1.$$

The pattern is

$$i^n = \begin{cases} -i & n = -1, -5, -9, \cdots \\ -1 & n = -2, -6, -10, \cdots \\ i & n = -3, -7, -11, \cdots \\ 1 & n = -4, -8, -12, \cdots . \end{cases}$$

Since 36 is a multiple of 4, we know $i^{-36} = 1$.
Since $41 = 4 \cdot 10 + 1$, we know $i^{-41} = -i$.

33. To confirm that $z = \dfrac{a + bi}{c + di}$, we calculate the product

$$z(c + di) = \left(\frac{ac + bd}{c^2 + d^2} = \frac{bc - ad}{c^2 + d^2}i \right)(c + di)$$

$$= \frac{ac^2 + bcd - bcd + ad^2 + (bc^2 - acd + acd + bd^2)i}{c^2 + d^2}$$

$$= \frac{a(c^2 + d^2) + b(c^2 + d^2)i}{c^2 + d^2} = a + bi.$$

37. True, since $\sqrt{a}$ is real for all $a \ge 0$.

41. True. We can write any nonzero complex number z as $re^{i\beta}$, where r and β are real numbers with $r > 0$. Since $r > 0$, we can write $r = e^c$ for some real number c. Therefore, $z = re^{i\beta} = e^c e^{i\beta} = e^{c+i\beta} = e^w$ where $w = c + i\beta$ is a complex number.

45. Using Euler's formula, we have:

$$e^{i(2\theta)} = \cos 2\theta + i \sin 2\theta$$

On the other hand,

$$e^{i(2\theta)} = \left(e^{i\theta}\right)^2 = (\cos\theta + i\sin\theta)^2 = (\cos^2\theta - \sin^2\theta) + i(2\cos\theta\sin\theta)$$

Equating real parts, we find

$$\cos 2\theta = \cos^2\theta - \sin^2\theta.$$

49. Replacing θ by $(x + y)$ in the formula for $\sin\theta$:

$$\sin(x+y) = \frac{1}{2i}\left(e^{i(x+y)} - e^{-i(x+y)}\right) = \frac{1}{2i}\left(e^{ix}e^{iy} - e^{-ix}e^{-iy}\right)$$
$$= \frac{1}{2i}\left((\cos x + i\sin x)(\cos y + i\sin y) - (\cos(-x) + i\sin(-x))(\cos(-y) + i\sin(-y))\right)$$
$$= \frac{1}{2i}\left((\cos x + i\sin x)(\cos y + i\sin y) - (\cos x - i\sin x)(\cos y - i\sin y)\right)$$
$$= \sin x \cos y + \cos x \sin y.$$

Solutions for Section C

1. (a) $f'(x) = 3x^2 + 6x + 3 = 3(x + 1)^2$. Thus $f'(x) > 0$ everywhere except at $x = -1$, so it is increasing everywhere except perhaps at $x = -1$. The function is in fact increasing at $x = -1$ since $f(x) > f(-1)$ for $x > -1$, and $f(x) < f(-1)$ for $x < -1$.

(b) The original equation can have at most one root, since it can only pass through the x-axis once if it never decreases. It must have one root, since $f(0) = -6$ and $f(1) = 1$.

(c) The root is in the interval $[0, 1]$, since $f(0) < 0 < f(1)$.

(d) Let $x_0 = 1$.

$$x_0 = 1$$
$$x_1 = 1 - \frac{f(1)}{f'(1)} = 1 - \frac{1}{12} = \frac{11}{12} \approx 0.917$$
$$x_2 = \frac{11}{12} - \frac{f\left(\frac{11}{12}\right)}{f'\left(\frac{11}{12}\right)} \approx 0.913$$
$$x_3 = 0.913 - \frac{f(0.913)}{f'(0.913)} \approx 0.913.$$

Since the digits repeat, they should be accurate. Thus $x \approx 0.913$.

5. Let $f(x) = \sin x - 1 + x$; we want to find all zeros of f, because $f(x) = 0$ implies $\sin x = 1 - x$. Graphing $\sin x$ and $1 - x$ in Figure C.2, we see that $f(x)$ has one solution at $x \approx \frac{1}{2}$.

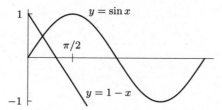

Figure C.2

Letting $x_0 = 0.5$, and using Newton's method, we have $f'(x) = \cos x + 1$, so that

$$x_1 = 0.5 - \frac{\sin(0.5) - 1 + 0.5}{\cos(0.5) + 1} \approx 0.511,$$

$$x_2 = 0.511 - \frac{\sin(0.511) - 1 + 0.511}{\cos(0.511) + 1} \approx 0.511.$$

Thus $\sin x = 1 - x$ has one solution at $x \approx 0.511$.

9. Let $f(x) = \ln x - \frac{1}{x}$, so $f'(x) = \frac{1}{x} + \frac{1}{x^2}$.
Now use Newton's method with an initial guess of $x_0 = 2$.

$$x_1 = 2 - \frac{\ln 2 - \frac{1}{2}}{\frac{1}{2} + \frac{1}{4}} \approx 1.7425,$$

$$x_2 \approx 1.763,$$

$$x_3 \approx 1.763.$$

Thus $x \approx 1.763$ is a solution. Since $f'(x) > 0$ for positive x, f is increasing: it must be the only solution.